GEOMETRY OF CLASSICAL FIELDS

GEOMETRY OF CLASSICAL FIELDS

Ernst Binz
Fakultät für Mathematik und Informatik
Universität Mannheim
Mannheim, F.R.G.

Jędrzej Śniatycki
Department of Mathematics and Statistics
University of Calgary
Calgary, Canada

Hans Fischer
Department of Mathematics
University of Massachusetts
Amherst, U.S.A.

DOVER PUBLICATIONS, INC.
Mineola, New York

Bibliographical Note

This Dover edition, first published in 2006, is an unabridged republication of the first edition published by Elsevier Science Publishers B.V., New York, 1988.

Library of Congress Cataloging-in-Publication Data

Binz, Ernst, 1939–
Geometry of classical fields / Ernst Binz, Jedrzej Sniatycki, Hans Fischer.
p. cm.
Originally published: Amsterdam ; New York : North-Holland, 1988, in series: North-Holland mathematics studies ; 154 [and] Notas de matemática (Rio de Janeiro, Brazil) ; no. 123.
Includes bibliographical references and index.
ISBN 0-486-45053-8 (pbk.)
1. Geometry, Differential. 2. Field theory (Physics). I. Sniatycki, Jedrzej. II. Fischer, Hans, 1939– III. Title.

QA641.B585 2006
516.3'6—dc22

2006041124

Manufactured in the United States of America
Dover Publications, Inc., 31 East 2nd Street, Mineola, N.Y. 11501

TABLE OF CONTENTS

INTRODUCTION

Classical field theory is a discipline which for centuries supplied mathematicians with motivations to develop new analytical techniques, and provided physicistss with explanations of many fascinating phenomena in nature. It shares with mechanics its basic approach to the fundamental laws of physics. However, its developements make twists and turnes which may affect mechanics very deeply. For example, Maxwell's discovery of the laws of electromagnetism led Einstein to special theory of relativity, which drastically changed our understanding of mechanics. At present, classical field theory covers a vast area encompassing mechanics of continua as well as non–mechanical phenomena like electromagnetism and gravitation.

The successes of quantum electrodynamics in obtaining amazingly accurate theoretical predictions led to attempts to explain the phenomena of strong and weak interactions in the framework of classical field theory. In the early stages of quantum field theory, one started with a classical field theory and formulated a corresponding quantum theory by a process of quantization. In later developements, the underlying classical theory became less prominent. Recent theories unifying strong, weak and electromagnetic interactions again emphasize the underlying classical theory of Yang–Mills fields.

There are two approaches to the problem of quantization of classical fields: Feynman integral and canonical quantization. The first approach is based on the principle of stationary action in classical field theory. The second one starts with a Hamiltonian formulation of dynamics of classical fields. Thus, classical field theory is in some sense close to quantum field theory along a path joining the principle of stationary action and a Hamiltonian formalism. For this reason we have concentrated our attention on the problems in classical field theory one encounters following this path. We have left out of our considerations other important but highly specialized geometric aspects of field theory, like monopoles, symmetry breaking, anomalies and ghosts. However, we expect to discuss them in the next volume. We hope that the approach to field theory presented here will form the foundations for a successful analysis of geometry of classical fields.

Classical fields are systems with an infinite number of degrees of freedom. This leads to problems of functional analytic nature in both classical and quantum theories. On a classical level these problems can be dealt with by the theory of manifolds of maps. On a quantum level one has additional difficulties, related to renormalizability of the theory, which have to dealt with by specialized techniques. The existence of infinite dimensional symmetry groups leads to another class of problems in field theory. On a classical level they appear as constraints on the admissible Cauchy data. This means that the physical system we want to describe has a smaller number of degrees of freedom than that which is needed in order to formulate the theory in terms of the principle of stationary action. This difficulty is carried over to quantum field theory where one has to eliminate the spurious degrees of freedom. On the other hand, the existence of an infinite dimensional group of gauge symmetries of Yang–Mills theory enables one to prove renormalizability of the corresponding quantum theory. We hope that a deeper understanding of this geometric aspect of classical field theory may help in dealing with the corresponding problems on a quantum level.

We address this book to the community of mathematicians who are interested in theoretical physics and to theoretical physicists who use differential geometric methods in their modelling. The book splits into two parts. The first part serves mainly as an introduction to the elementary notions of differential geometry, the theory of Lie groups, and manifolds of maps. We do this in order to prepare a language nowadays used to describe globally defined fields as is done in part two. We mostly use a coordinate free language in the introductory chapters of the first part, whereas the calculations in the later chapters are done in a coordinate description, a method used somewhat more in applications to physics. There are several other differences between the presentation of the material in the first and the second part. They are due not only to different approaches adopted by mathematicians and physicists, but also to different styles of the authors and their geographic separation.

Chapter one has very much an introductory character. In contains most of the proofs to the results presented. Other chapters lack this completeness, because otherwise a series of volumes would have to be written.

The idealized bodies, more generally the idealized spaces on which fields (which usually depend on further parameters) act upon are manifolds or manifolds with boundaries. Symmetries of bodies and fields are groups of linear maps or more generally Lie groups and even infinite dimensional differentiable groups.

We, therefore, begin with a chapter on manifolds and Lie groups, pass then to vector bundles and to a chapter on elementary, mostly extrinsic differential geometry. The differential geometry reappears in the following chapter again. However the scope is the approach based on the idea of symmetry, hence based on principal bundles. The first part then closes by an introduction to the elementary notions and basic results in an infinite dimensional manifold, namely of function spaces and especially of manifolds of maps.

Following the basic notions such as manifolds, manifolds with boundaries, flows of vector fields, and the theorem of Frobenius the first chapter presents an introduction to Lie groups. Based on the foundations including the relation of Lie groups with Lie algebras via the exponential map, Cartan's theorem stating that any closed subgroup of a Lie group is a Lie group again, and its infinitesimal version on immersed Lie groups, we present a series of examples. It includes in particular GL(E), SL(E) for a given finite dimensional vector space E, O(p,n–p) the Lorentz group, SO(p,n–p), Pin(p,n–p), Spin(p,n–p), U(n), SU(n), Sp(n). This treatment uses in an elementary fashion technique involving maximal tori, maximal compact subgroups and polar decomposition.

The chapter ends with determining the Lie algebra of the Lie group of all automorphisms of a Lie group and with the semidirect product of two Lie groups.

Since the classical fields defined on a manifold assume their values in a vector bundle, we present in the second chapter some of the standard operations on vector bundles. In particular we show that the pull backs under homotopic maps of isomorphic vector bundles yield isomorphic bundles again. The bundles of k–jets of vector bundles and the canonical one–form on such bundles are treated to the extend that these objects together with the bundle formalism become familiar enough to follow the calculus of variations in chapter nine. Of preparatory character for the following two chapters are also the sections on Riemannian structures and connections on vector bundles.

The above mentioned fields depend heavily on the geometry of the domain of their evolution, hence on the geometry of the manifold on which these fields evolve. For this reason we emphasize the more extrinsic viewpoint of the geometry of manifolds in chapter three. We begin with the lemma of Poincaré and show that in case of an Euclidean ambient space the notion of torsion and symmetry of the second fundamental tensor are closely related with the idea of integrability. We then step to a more general situation and introduce connections, geodesic sprays, geodesics and the exponential in a metric free setting, and turn afterwards to the intrinsic geometry given by a Riemannian metric by introducing the Riemannian spray via the canonical symplectic structure on the cotangent bundle. Different notions of curvature are studied, and following the spirit of extrinsic geometry we derive the equations of Gauss and Codazzi. Finally we describe the Riemannian sprays on a Lie group determined by right–, left–, and biinvariant metrics. The reader is asked to investigate the other concepts and notions introduced in this chapter in the special environment of a Lie group.

The following chapter four takes up the subject of differential geometry again but in a quite different fashion however. The basis of the viewpoint is the theory of principal bundles which formalizes the notion of global symmetries. Vector bundles are hence considered as being associated with principal bundles. This viewpoint is adapted because of its constant use in the theory of symmetry in Lagrangian mechanics and field theory, in particular in gauge theories. Following a general section on principal bundles where fundamental terms such as vertical bundles, fundamental vector fields, equivariant forms etc., are introduced a series of examples is presented. This list includes in particular the principal bundle over a homogeneous space where the total space is a Lie group, as well as the universal bundle. In turn special homogeneous spaces such as the Stiefel and Grassmann manifolds are treated. The general construction of associated bundles precedes the study of the so called homogeneous bundles. Connections, covariant exterior derivations, curvatures, products of connections etc., are intensively studied. A special section is devoted to invariant connections. In the last two sections on linear connections some notions introduced in the preceding chapters reappear and get elucidated by the principal bundle approach.

The next chapter on function spaces and manifolds of maps collects the basic entities to provide a framework in which fields depending on parameters varying in infinite dimensional manifolds can be studied. Such parameters might be, for example, Cauchy data, connections or configurations of moving fluids. These function spaces and in particular manifolds of maps are as fundamental in field theory as the notion of manifold in classical mechanics. We mostly collect results without detailed proof, but indicate often an argument of plausibility. First we consider classical spaces of locally defined functions and distributions, since these sort of spaces are fundamental entities in the theory of partial differential equations. Then we pass to spaces of maps and sections being globally defined and study them equipped with two types of topologies namely the C^k–topology for any $k=1,\ldots,\infty$ as well as some Sobolev type of topologies. An appendix on currents closes the first part of the chapter showing how to generalize the notion of a distribution to global setting. An often used differential calculus is introduced and applied to differential mappings whose domains and ranges are spaces of ksections.

Eells's construction of a C^∞–atlas on the spaces of all smooth maps from one manifold to another is the basic tool for the rest of the chapter. By using the manifold structure of spaces of maps we study the derivative and tangent mappings of well known maps such as the pull back of metrics, of volumes, of one–forms etc. and introduce in this spirit the Lie derivative. We further present the principal bundle of embeddings of a compat manifold into some ambient manifold. The base space is the moduli space of submanifolds of a given diffeomorphism type of the ambient manifold. These objects form a framework in which motions of moving bodies can be studied, and yield an environment in which a theory of deformations of manifolds can be developed. Hence they link with elasticity and general relativity. The chapter continues with the C^∞ and the Sobolev structure on diffeomorphism groups. Then it turns to notions fundamental in gauge theories such as current and gauge groups as well as the automorphism group of a principal bundle and their infinitesimal versions. It closes with a formalism used to describe deformations of differentials of immersions with Euclidean spaces as their ranges. In particular it is shown that all Euclidean immersions isometric to a given one form a Fréchet manifold.

Part II is concerned with a systematic development of a covariant Hamiltonian formulation of field theory starting from the principle of stationary action.

In Chapter 6 we present dynamics of systems with a finite number of degrees of freedom in a way which could serve as a model for a subsequent formulation of dynamics of classical fields. The principle of stationary action is a starting point of the presentation adopted here. For regular time independent Lagrangians it leads in the usual way to the Hamiltonian formalism. Techniques of symplectic geometry and Poisson algebras are basic tools of the Hamiltonian formalism, and they are briefly reviewed. Since the Lagrangians appearing in important examples in field theory are not regular, degenerate Lagrangians and the corresponding constraints are studied in the spirit of the Dirac theory of constraints.

The main emphasis of the presentation adopted here is on symmetries of the theory. The standard Hamiltonian formalism describes the motion of the system under consideration relative to an arbitrarily chosen frame of reference. The choice of a frame of reference introduces an extrinsic object to the theory which obscures the symmetry group. Since our aim is to have an explicitely covariant, intrinsic, Hamiltonian formulation of dynamics we return to the principle of stationary action. An equivalent variational principle, in which the Lagrangian is replaced by the corresponding Cartan form, leads to equations of motion given by the kernel of the exterior differential of the Cartan form. For degenerate Lagrangians the Cartan form pushes forward to the Hamilton form on the primary constraint manifold. We determine conditions under which, for each direction of evolution (corresponding to a choice of the reference frame) the equations of motion split into constraint equations and Hamiltonian equations of evolution. This splitting is covariant, and the corresponding Hamiltonian is given by the evaluation of the Hamilton form on the vector field describing the direction of evolution. For the sake of simplicity we consider here only the case when the direction of evolution is given by an infinitesimal symmetry.

The constraint equations determine the constraint set of the theory which need not have a manifold structure. Using the Second Noether Theorem we relate the constraint set to the action of the group $\mathscr{H}$ of localizable symmetries. We identify the space of $\mathscr{H}$ orbits in the constraint set with the space of physical states of the system under consideration, and use the reduction of Poisson algebras to discuss its structure.

In Chapter 7 we follow the developments of the preceeding chapter to obtain a covariant Hamiltonian formulation of field theory.

The usual transition from the Lagrangian to the Hamiltonian formalism requires an a priori splitting of space–time into space and time. This introduces an external element and obscures the symmetries of a relativistic theory. In order to avoid it, we pass to an equivalent action principle in which the Lagrangian is replaced by the corresponding DeDonder form. In this way one obtains the field equations expresses in terms of the exterior differential of the DeDonder form. This form of field equations enables one to make a covariant passage to equations of motion for Cauchy data, provided we consider a space of Cauchy surfaces stable under the action of the symmetry group. Integration of the DeDonder form over Cauchy data gives rise to a 1–form in the space of Cauchy data which is an analogue of the Cartan form in non–relativistic dynamics. The transision from the obtained Cartan formalism to a covariant Hamiltonian formalism follows the steps described in the preceding chapter.

Since dynamics of classical fields is governed by partial differential equations, one has to consider boundary conditions satisfied by the Cauchy data. One finds that the evolution equation for Cauchy data can be cast into a Hamiltonian form only under appropriate boundary conditions. Here we consider a class of boundary conditions which includes the Dirichlet conditions and the Neumann conditions. The choice of boundary conditions determines the symmetry group of the theory.

Another essential difference between field theory and systems with a finite number of degrees of freedom is that in field theory we have to deal with manifolds of maps. Most existence results in field theory are obtained in appropriate Sobolev spaces. However, in manifolds of maps modelled on Sobolev spaces many interesting geometric objects of the theory are only densely defined. For this reason we restrict our considerations to the category of smooth maps. In this category we have no implicit function theorem. Hence, the results in the theory of finite dimensional manifolds which depend on the implicit function theorem, need not extend. Moreover, all symplectic forms appearing in applications are weak. This implies that not all smooth functions admit Hamiltonian vector fields.

Throughout Chapter 6 we illustrate the constructions discussed here by the Klein–Gordon dynamics of a scalar field. More complicated and interesting examples are discussed in subsequent chapters. Chapter 8 is devoted to Yang–Mills theory, and Chapter 9 to general relativity.

Yang–Mills fields are connections in a principal fibre bundle over a space–time manifold. Matter fields are sections of an associated bundle. Dynamics of interacting Yang–Mills and matter fields is given by a Lagrangian invariant under the group of automorphisms of the principal bundle which preserve the boundary conditions and cover isometries of the space time (conformal isometries in the absence of matter fields). Localizable symmetries are gauge transformations, that is automorphisms of the principal bundle which reduce to the identity on the boundary and cover the identity transformation of space–time. The presence of gauge transformations gives rise to constraints which can be studied in terms of the formalism developed in Chapter 7.

We show that Yang–Mills theory admits a covariant Hamiltonian description of dynamics developed in the preceding section. The conformal invariance of the Yang–Mills Lagrangian in absence of matter enables us to relate a Yang–Mills theory on a manifold with boundary to a Yang–Mills theory on Minkowski space.

Electrodynamics can be interpreted as a Yang–Mills theory for the structure group U(1). In fact, Yang–Mills theory is a generalization of this formulation of electrodynamics to non–abelian structure groups. Since lot of our understanding of the Yang–Mills theory is based on the analogy with electrodynamics we give a detailed discussion of some aspects of electrodynamics. In particular we study the notions of electric and magnetic charges in electrodynamics and their generalization to Yang–Mills theory. We relate the charges to symmetries of boundary conditions for Yang–Mills fields; matter fields are assumed to vanish on the boundary.

In general relativity the dynamical variables are Lorentzian metrics on the space–time manifold. They satisfy Einstein equations which are derivable from a variational principle with a Lagrangian depending on the second derivatives of the metric. Following Palatini, we consider metrics and torsion free connections as independent variables, and obtain a variational principle with a Lagrangian depending on first derivatives. The corresponding DeDonder form pushes forward to the fibre product of the bundles of Lorentzian metrics and of torsion free connections, which enables us to apply the general theory developed for the first order Lagrangians.

The boundary conditions for Einstein equations considered here consist of specifying a Lorentzian metric along the boundary, such that the induced metric on the boundary is Lorentzian. With these boundary conditions general relativity admits a covariant Hamiltonian formulation of dynamics, which is a modification of the ADM formalism.

The symmetry group of general relativity contains diffeomorphisms of the space–time manifold which preserve the boundary conditions. Localizable infinitesimal symmetries are given by vector fields which vanish on the boundary to the first order. The corresponding constraints are given by the vanishing of the superhamiltonian and the supermomentum. The conserved momenta on the set of Cauchy data satisfying the constraint conditions correspond to the Killing vectors of the boundary conditions.

A conformal compactification of the space–time manifold and attaching the hyperboloid of space–like directions at spatial infinity give an embedding of asymptotically flat space–time into manifolds with boundary. This enables us to extend the general formalism to asymptotically flat space–times, and to discuss conservation laws for isolated gravitating systems.

We are very much indebted to Tae Nosal for typing and retyping the manuscript. We are also very grateful to Rudolph Maly, and Wolfgang Menden for helpful discussions, suggestions and for the painstaking job of proof–reading.

This work was partially supported by the Natural Science and Engineering Research Council of Canada Operating Grant and a Killam Residence Fellowship held by the second author (J.Ś.).

PART I

DIFFERENTIAL GEOMETRIC PRELIMINARIES

Chapter 1

MANIFOLDS AND LIE GROUPS

1.1 Manifolds, tangent manifolds

The geometric basis to formulate physical concepts is formed by all finite dimensional topological vector spaces. To describe phenomena of global nature topological spaces which look locally like finite dimensional topological vector spaces, so called manifolds, are favoured. This is due to the idea of passing from local situations to global ones. However, in order to do this a certain rigidity, namely the existence of partitions of unity, has to be required.

Thus let us specify our notion of a manifold among the class $\mathcal{T}$ of all those topological spaces which satisfy the second axiom of countability and the Hausdorff separation axiom.

The concept of a manifold is based on local charts: An open neighbourhood U of a point p of a topological space $M \in \mathcal{T}$ together with a map φ from U into a finite dimensional real vector space E is called a chart if $\varphi(U) \subset E$ is open and

$$\varphi : U \longrightarrow \varphi(U)$$

is a homeomorphism. We call U to be modelled over E. Denote a chart by (U,φ) or if no misunderstandings arise just by U.

Call a topological space in $\mathcal{T}$ to be locally Euclidean if each point admits a local chart. To pull back the concepts of analysis associated with a finite dimensional topological vector space, such as smoothness e.g., we need to have charts fitting together. This is achieved by using the notion of an atlas.

A smooth atlas $\mathcal{A}$ of a locally Euclidean space M is a collection of charts satisfying

1) $\bigcup_{U \in \mathcal{A}} U = M$

and 2) for any pair (U,φ), (V,Ψ) of charts in $\mathcal{A}$

$\Psi \circ \varphi^{-1}: \varphi(U \cap V) \longrightarrow \Psi(U \cap V)$

is smooth.

A smooth atlas is called maximal if $\mathcal{A}$ is saturated with respect to all charts satisfying (2). Obviously every smooth atlas can be extended to a maximal smooth atlas in exactly one way. By a smooth manifold we mean a locally Euclidean space $M \in \mathcal{T}$ equipped with a maximal smooth atlas.

Clearly the dimension of the topological vector space containing the images of those charts of a manifold lying inside a connected component is a constant, called the dimension of the connected component. If all the connected components of a manifold M have the same dimensions, say n, then n is the dimension of M.

Let us illustrate these notions on a few examples. Evidently, every open subset U of a n–dimensional vector space is a smooth manifold of dimension n with respect to the atlas generated by the chart (U,id_U). Moreover, any open subset of a manifold and the cartesian product $M \times N$ of any two manifolds M and N are manifolds again. The atlas of $M \times N$ is generated by charts of the form $(U \times V, \varphi \times \Psi)$, they are modelled over $E \times F$ if U and V are modelled over E and F respectively. This manifold is called the cartesian product of the manifolds M and N.

i) Thus $GL(n,\mathbb{R})$ and $GL(n,\mathbb{C})$, the groups of all invertible members of $End(\mathbb{R}^n)$ and $End(\mathbb{C}^n)$, the vector spaces of all $\mathbb{R}$– and $\mathbb{C}$–linear endomorphisms of $\mathbb{R}^n$ and $\mathbb{C}^n$ respectively, are manifolds. The dimensions are n^2 respectively $2n^2$.

ii) Next consider $f : U \longrightarrow F$, a smooth function defined on an open subset U of a finite dimensional topological vector space E into an other one F. Clearly $pr_E: E \times F \longrightarrow E$, the natural projection onto E, restricted to the graph of f is a chart of the graph which is regarded as a subspace of $E \times F$. Thus the graph equipped with the maximal atlas $\mathcal{A}$ of all those charts which are compatible with (graph f, pr_E) in the sense of (2) from above is a manifold. The dimension of E is the dimension of graph f.

iii) The unit sphere S^{n-1} of an n–dimensional Euclidean space $\mathbb{R}^n$ with respect to a fixed scalar product $<,>$ on $\mathbb{R}^n$ is a manifold with the following atlas: The point set $S^{n-1} \subset \mathbb{R}^n$ is given by the equation

$$\Sigma\, x_i^2 = 1$$

with respect to an orthonormal basis $(e_1,\dots,e_n)$ in $\mathbb{R}^n$. Consider S^{n-1} as a topological subspace of $\mathbb{R}^n$. As an open neighbourhood V of e_1 in S^{n-1} choose the graph of the function assigning the value $x_1 = \sqrt{1 - \sum_{i=2}^{n} x_i^2}$ to any tuple $(x_2,\dots,x_n)$ in an open neighbourhood W of zero in $e_1^{\perp}$, the orthogonal complement of e_1 in $\mathbb{R}^n$. If $\varphi : V \longrightarrow W$ denotes the projection given by $\varphi(x) = (x_2,\dots,x_n)$, then (V,φ) is a chart for S^{n-1}. On the other hand the group $0(n)$ of all orthogonal linear isomorphisms of $\mathbb{R}^n$ acts transitively on S^{n-1}, i.e. given any two points $a,b \in S^{n-1}$ there is a map $\sigma \in 0(n)$, with $\sigma(a) = b$. Thus given $a \in S^{n-1}$ and $\sigma \in 0(n)$ with $\sigma(e_1) = a$

$$(\sigma(V),\varphi \circ \sigma^{-1})$$

is a chart of a. The collection of all such charts forms an atlas $\mathcal{A}$. Saturating $\mathcal{A}$ with all those charts compatible in the sense of (2) with all members of $\mathcal{A}$ yields a maximal atlas $\mathcal{A}'$. Thus S^{n-1} together with $\mathcal{A}'$ is a manifold of dimension $(n-1)$.

iv) Since $S^1 \subset \mathbb{R}^2$ is a manifold, T^n, the n–fold cartesian product of S^1, is a manifold too. Its dimension is n.

Next consider a topological subspace $N \subset M$ which is a manifold by itself. N is called a <u>submanifold</u> if each point $p \in N$ admits open charts (U,φ) in M and $(V,\varphi|V)$ in N modelled on the finite dimensional vector spaces E and F respectively such that

$$1) \ E = F \oplus K$$

for some subspace K of E and

$$2) \ \varphi(U) = \varphi(V) \oplus \varphi(W)$$

where $W \subset M$ such that $\varphi(W) \subset K$ is open.

We continue this section by pulling back the concept of differentiability on finite dimensional vector spaces to manifolds.

Consider two manifolds M and N. A continuous map $f : M \longrightarrow N$ is of class C^k at $p \in M$ provided for any two charts (U,φ) and (V,Ψ) of $p \in M$ and $f(p) \in N$ respectively with

$$f(U) \subset V$$

the map $\Psi \circ f \circ \varphi^{-1}: \varphi(U) \longrightarrow \Psi(V)$ is k–times continuously differentiable. Call f a <u>C^k–map</u> if it is of class C^k everywhere. A bijective map $f : M \longrightarrow N$ is called a <u>C^k–diffeomorphism,</u> if f and f^{-1} are C^k–maps. Denote the collection of all C^k–maps from M into N by $C^k(M,N)$.

One of our main tools is, as indicated at the beginning of this section, the <u>partition of unity.</u>

<u>Lemma 1.1.1</u> Given a local finite open cover of a smooth manifold M there is a family $\varphi_\iota \in C^\infty(M,\mathbb{R})$ where ι varies in an index set $\mathcal{A}$ subordinated to the cover such that

$$\Sigma_\iota \varphi_\iota = 1.$$

For the terminology and the proof which crucially uses the paracompactness of the manifold (a consequence of the axiom of second countability) consult [Hi].

To decide as to whether $f^{-1}(q) \subset M$ is a manifold or not for $q \in N$ we need to introduce the tangent space of a point in M and in turn the tangent manifold of M.

A tangent vector to $p \in M$ has intuitively to be obtained by differentiating appropriate curves in each chart. Therefore we define a <u>tangent vector</u> v at $p \in M$ to be the equivalence class $\overline{(U,\varphi,w,p)}$ of all quadruples of the form

$$(U,\varphi,w,p),$$

where $w \in E$ and (U,φ) is a chart of p subjected to the following equivalence relation:

$$(U,\varphi,w_1,p) \sim (V,\psi,w_2,p) \text{ iff } D(\Psi \circ \varphi^{-1})(\varphi(p)(w_1)) = w_2 .$$

Clearly T_pM, the collection of all tangent vectors at p, is an $\mathbb{R}$–vector space. For each chart (U,φ) the mapping $E \longrightarrow T_pM$ sending $w \in E$ into $\overline{(U,\varphi,w,p)} \in T_pM$ is a linear isomorphism for all $p \in U$. Thus $\dim T_pM = \dim M$. In particular if M is an open submanifold of E there is a natural isomorphism $T_pM \cong E$ for all $p \in M$ defined by the chart (M,id_M). Moreover, for a product manifold $M \times N$ we have a canonical isomorphism

$$T_{(p,q)}(M \times N) \cong T_pM \times T_qN$$

for all $(p,q) \in M \times N$.

Each C^1–map f from a manifold M into another one N induces a linear map

$$Tf(p) : T_pM \longrightarrow T_{f(p)}N$$

assuming on each tangent vector $v = \overline{(U,\varphi,w,p)}$ the value

$$Tf(p)(\overline{(U,\varphi,w,p)}) = \overline{(V,\Psi,D(\Psi \circ f \circ \varphi^{-1})(\varphi(p))(w),f(p))} .$$

Here (V,Ψ) is a chart of $f(p)$ containing $f(U)$.

Given any tangent vector $v \in T_pM$, there is a smooth curve σ into M, defined on a neighbourhood I of zero in $\mathbb{R}$ such that

$$\sigma(0) = p$$

and

$$T\sigma(0)(1) = v.$$

Henceforth $\dot{\sigma}(0)$ abbreviates $T\sigma(0)(1)$. More generally define $\dot{\sigma}(t)$ and $\frac{d}{dt}\sigma(t)$ to be synonymous notions for $T\sigma(t)(1)$.

The collection

$$TM := \bigcup_{p \in M} T_pM$$

of all tangent vectors to M is a manifold with respect to the following atlas:
Let (U,φ) be a chart of $p \in M$ with $\varphi(U) \subset E$. Clearly $\varphi(U) \times E \subset E \times E$ is open. If $\mathcal{A}$ denotes the maximal atlas of M then

$$\{(\bigcup_{p \in U} T_pM, T\varphi) \mid (U,\varphi) \in \mathcal{A}\}$$

is an atlas of TM satisfying the conditions analogous to (1) and (2) from above. Here $T\varphi$ means the mapping which sends $v = \overline{(U,\varphi,w,p)} \in T_pM$ into $(\varphi(p),w) \in \varphi(U) \times E$.

Next let us go back to smooth maps from a manifold M into another one N and answer the question posed earlier. If $f:M \longrightarrow N$ is of class C^1, then the induced map

$$Tf : TM \longrightarrow TN$$

given by $Tf(v) = Tf(p)(v)$ for all $v \in T_pM$ is continuous. The following theorem on regular values is the main tool in the topology of finite dimensional manifolds. Let $f : M \longrightarrow N$ be smooth. $q \in N$ is called <u>regular</u> provided that $Tf(p)$ has maximal rank for all $p \in f^{-1}(q)$.

<u>Theorem 1.1.2</u> For every regular value $q \in N$ the set $f^{-1}(q)$ is a submanifold of M.

<u>Proof</u>: Let (U,φ) and (V,Ψ) be charts of $p \in f^{-1}(q)$ and q respectively where $\varphi(U) \subset E$ and $\Psi(V) \subset F$ for two finite dimensional veector spaces E and F. Without loss of generality we assume $\varphi(p) = 0$, $\Psi(q) = 0$, $f(U) \subset V$ and moreover that $Tf(p)$ is surjective. Then $D(\Psi \circ f \circ \varphi^{-1})(0)$ is surjective as well. Therefore E splits into a direct sum $E_1 \oplus E_2$ such that the restriction of $D(\Psi \circ f \circ \varphi^{-1})(0)$ to E_2 is an isomorphism onto F. By the implicit function theorem there are open neighbourhoods $W_1 \subset E_1$ and $W_2 \subset E_2$ of zero satisfying $W_1 \times W_2 \subset \varphi(U)$ and a smooth map $u: W_1 \longrightarrow W_2$ such that $u(0) = 0$ and $\varphi^{-1}(\text{graph } u)$ is an open neighbourhood of p in $f^{-1}(q)$. Then

$$(\varphi^{-1}(\text{graph } u),\ pr_1 \circ \varphi | \varphi^{-1}(\text{graph } u))$$

is a chart of p, where $pr_1: W_1 \times W_2 \longrightarrow W_1$ denotes the projection onto the first factor. The rest is easily verified.

The next part of this section will be devoted to a short study of the collection of all smooth vector fields. Let us begin it by the remark, that

$$\pi_M : TM \longrightarrow M,$$

the map assigning to each $v_p \in T_pM$ the point p for all $p \in M$ is smooth. A smooth <u>vector field</u> X on M, called simply a vector field on M, is a smooth map $X : M \longrightarrow TM$ satisfying $\pi_M \circ X = id_M$. Locally, i.e. in a chart U modelled over E, a vector field X on M is represented by a map of the form

$$U \xrightarrow{(id, X_U)} U \times E.$$

Call X_U the <u>principal part</u> of X. Clearly ΓTM, the collection of all vector fields on M, is a module over $C^\infty(M,\mathbb{R})$ (cf. [L]).

Each vector field $X \in \Gamma TM$ yields an operator on $C^\infty(M,\mathbb{R})$, called X again, given by

$$X(f) = df \circ X$$

for all $f \in C^\infty(M,\mathbb{R})$ where df determines Tf by $Tf = (f,df)$ since $T\mathbb{R} = \mathbb{R} \times \mathbb{R}$. This operator is $\mathbb{R}$–linear, however,

$$X(f \cdot g) = f \cdot X(g) + g \cdot X(f)$$

for all $f,g \in C^\infty(M,\mathbb{R})$. Thus X is a derivation on $C^\infty(M,\mathbb{R})$. Vice versa any derivation on $C^\infty(M,\mathbb{R})$ is of this form (cf. [W]). Given $p \in M$ we also write $X(p)(f)$ for $X(f)(p)$. For any two vector fields $X,Y \in \Gamma TM$ we have $X(p)(f) = Y(p)(f)$ for all $f \in C^\infty(M,\mathbb{R})$ iff $X(p) = Y(p)$. Thus any tangent vector $v_p \in T_pM$ can be considered as a map

$$v_p : C^\infty(M,\mathbb{R}) \longrightarrow \mathbb{R}.$$

Clearly for any two vector fields X, Y regarded as derivations on $C^\infty(M,\mathbb{R})$ the product is in general not a derivation anymore. However, the commutator $X \circ Y - Y \circ X$ is one. This commutator, denoted by $[X,Y]$ is locally, i.e. represented in a chart U, given by

$$[X,Y]_U(p) = D\,Y_U(p)(X_U(p)) - D\,X_U(p)(Y_U(p))$$

for all $p \in U$. Here $[X,Y]_U$, Y_U and X_U are the principal parts. The Lie bracket [,] thus is bilinear over $\mathbb{R}$ and satisfies

3) $[X,Y] = -[Y,X]$

and

4) $[X,[Y,Z]] + [Z,[X,Y]] + [Y,[Z,X]] = 0$

for all $X,Y,Z \in \Gamma TM$. The module ΓTM together with [ ] as a product is thus a real Lie algebra (cf.[L]), called the Lie algebra of M. Moreover

5) $[X,fY] = f[X,Y] + X(f)Y$

holds for all $X,Y \in \Gamma TM$ and $f \in C^\infty(M,\mathbb{R})$.

Let M,N be two smooth manifolds and $\varphi : M \longrightarrow N$ be smooth. Observe that

$$(T\varphi \circ X)(h) = X(h \circ \varphi)$$

for all $X \in \Gamma TM$ and $h \in C^\infty(N,\mathbb{R})$, which means

$$(T\varphi(p)X(p))(h) = X(p)(h\circ\varphi)$$

for all $p \in M$. $X \in \Gamma TM$ is called <u>φ–related</u> to $Y \in \Gamma TN$ if $T\varphi\circ X = Y\circ\varphi$. Clearly this is the case iff

$$X(h\circ\varphi) = Y(h)\circ\varphi$$

for all $h \in C^\infty(N,\mathbb{R})$.

<u>Lemma 1.1.3</u> Let $X_1,X_2 \in \Gamma TM$ and $Y_1,Y_2 \in \Gamma TN$. If X_i is φ–related to Y_i (i=1,2) then $[X_1,X_2]$ is φ–related to $[Y_1,Y_2]$.

<u>Proof</u>: For all $h \in C^\infty(N,\mathbb{R})$ we have

$$\begin{aligned}&[X_1,X_2](h\circ\varphi)\\ &= X_1(X_2(h\circ\varphi)) - X_2(X_1(h\circ\varphi))\\ &= X_1(Y_2(h)\circ\varphi) - X_2(Y_1(h)\circ\varphi)\\ &= Y_1(Y_2(h))\circ\varphi - Y_2(Y_1(h))\circ\varphi\\ &= [Y_1,Y_2](h)\circ\varphi.\end{aligned}$$

If M is a submanifold of a smooth manifold N, then the Lie bracket of two vector fields $X_1,X_2 \in \Gamma TM$ is given by

$$[X_1,X_2]_M = [\tilde{X}_1,\tilde{X}_2]_N,$$

where $\tilde{X}_i \in \Gamma TN$ is any extension of X_i to N (i=1,2). Here $[\ ]_M$ and $[\ ]_N$ denote the Lie brackets in M and N respectively. This follows immediately from the above lemma, since X_i is related to $\tilde{X}_i$ with respect to the inclusion map $M \longrightarrow N$.

In general we can not find a finite number of nowhere vanishing vector fields $X_1,\dots,X_n \in \Gamma TM$ say, such that any $X \in \Gamma TM$ can be represented by

$$X = \sum_{i=1}^{n} f^i \cdot X_i ,$$

or expressed pointwise

$$X(p) = \Sigma f^i(p) \cdot X_i(p)$$

for all $p \in M$. Here $f^i \in C^\infty(M,\mathbb{R})$. In case such a finite number of vector fields generating the modul ΓTM exist, M is called parallelizable. Any open set of a finite dimensional vector space is parallelizable.

We close the section by introducing the concept of a manifold with boundary. To this end let M again be a second countable topological space which satisfies the separation axiom of Hausdorff.

In order to define charts for such objects we need the notion of a half space. Such a space is conveniently introduced by taking a (continuous) surjective linear map $1 : E \longrightarrow \mathbb{R}$, where E is a finite dimensional vector space. Then define $E_1^+ := \{x \in E \mid 1(x) \geq 0\}$ and $E_1^- := \{x \in E \mid 1(x) \leq 0\}$. Clearly E_1^+ and E_1^- are homeomorphic. Call ker 1 by ∂E_1^+ and refer to it as the boundary of E_1^+. Let $L(E,\mathbb{R})$ be the vector space of all $\mathbb{R}$–valued linear maps of E.

M is called a manifold with boundary, if every point $p \in M$ admits an open set $U \subset M$ and a continuous map $\varphi : U \longrightarrow E_1^+$ for some $1 \in L(E,\mathbb{R})\setminus\{0\}$ for which $\varphi(U) \subset E_1^+$ is open and $\varphi : U \longrightarrow \varphi(U)$ is a homeomorphism. Such maps are called chart maps. Observe, that if $\varphi(p) \in \partial E_1^+$ then $\Psi(p) \in \partial E_k^+$ for any other chart map $\Psi : U \longrightarrow E_k^+$. Therefore let ∂M be the collection of all those points in M whose images under chart maps are in the boundary of half spaces. ∂M is called the boundary of M.

To define the notion of a smooth manifold with boundary consider open sets $U \subset E_1^+$ and $V \subset E_k^+$ and a map

$$\varphi : U \longrightarrow V.$$

Such a map is called smooth if there is a smooth extension $\overline{\varphi} : U_1 \longrightarrow V_k$ of φ where U_1 and $V_k \subset E$ are open and for which the intersections with E_1^+ and E_k^+ are U and V respectively.

An <u>atlas</u> of a manifold with boundary M is a collection $\{(U_\iota, \varphi_\iota)\}_{\iota \in I}$ of chart maps with

$$\bigcup_{\iota \in I} U_\iota = M.$$

Call such an atlas to be <u>smooth</u> if for any choice of $\alpha, \beta \in I$

$$\varphi_\alpha \circ \varphi_\beta^{-1}$$

is smooth at all points for which $\varphi_\alpha \circ \varphi_\beta^{-1}$ is defined. Then M is a <u>smooth manifold with boundary</u> provided it carries a maximal smooth atlas. We easily generalize the geometric concepts introduced for manifolds so far to manifolds with boundary. Observe that $T_pM \cong E$ for all $p \in \partial M$.

Clearly ∂M is empty or a manifold as well, in fact it can be regarded as a submanifold of M. In particular if $X, Y \in \Gamma TM$ satisfy $X|\partial M$, $Y|\partial M \in \Gamma T(\partial M)$ then $[X,Y]|\partial M$ maps into $T(\partial M)$ as well.

1.2 <u>Flows, the theorem of Frobenius</u>

First we will introduce the flow of a given vector field $X \in \Gamma TM$ and discuss some of its properties for general as well as for special types of manifolds.

The flow of X is essentially the totality of all integral curves of X. To make this intuitive statement more precise we represent X in a local chart (U, φ) of a point p_0

modelled on a finite dimensional vector space E. We simplify our study by identifying U with $\varphi(U) \subset E$ and T_pM with E for all $p \in U$. Accordingly we denote the restriction of a vector field X to U by the same symbol as its principal part X_U.

The classical theory for ordinary differential equations as described in [L] ensures us of the existence of an open set $0 \subset U \times \mathbb{R}$ containing $(p_0,0)$ and a smooth map

$$\alpha_U : 0 \longrightarrow U$$

such that

$$\alpha_U(p,0) = p$$

for all $p \in U$ satisfying $(p,0) \in 0$ and

$$\frac{d}{dt}\, \alpha_U(p,t) = X(\alpha_U(p,t))$$

for all $(p,t) \in 0$. α_U is called a flow of X_U on 0 with initial condition $\alpha_U(p,0) = p$. It is unique up to the domain of definition. This implies

$$\alpha_U(p,s+t) = \alpha_U(\alpha_U(p,s),t)$$

whenever both sides are defined.

The relation of the flow α_U with integral curves through p_0 is the following. Call a smooth curve $\beta : I \longrightarrow M$ (defined on an open interval $I \subset \mathbb{R}$) an integral curve of X if

$$\dot{\beta}(t) = X(\beta(t))$$

for all $t \in I$. If $(p_0,0) \in \{p_0\} \times I \subset 0$, then

$$\beta_{p_0} : I \longrightarrow U$$

given by

$$\beta_{p_0}(t) = \alpha_U(p_0,t)$$

is obviously an integral curve of X with the so called initial condition

$$\beta_{p_0}(0) = p_0 \,.$$

Any two integral curves of X with the same initial condition agree on the intersection of their domains. This allows us to extend β_{p_0} to a maximal connected domain $I(p_0)$ in an obvious way. Clearly, if s,t, s+t $\in I(p_0)$ then $t \in I(\beta_{p_0}(s))$ and

$$\beta_{p_0}(s+t) = \beta_{\beta_{p_0}(s)}(t)$$

where $\beta_{\beta_{p_0}(s)}$ denotes the maximal integral curve with initial condition

$$\beta_{\beta_{p_0}(s)}(0) = \beta_{p_0}(s) .$$

Next we describe the global flow of the vector field X. Let

$$0_X = \bigcup_{p \in M} (\{p\} \times I(p)) \subset M \times \mathbb{R} .$$

Call X <u>complete</u> if $0_X = M \times \mathbb{R}$. Any vector field on a compact manifold is complete as shown in [L], where criterions for completeness can be found. If X is a vector field on M with $X(p) \neq 0$ for some $p \in M$, then $M \setminus \{p\}$ is a manifold on which X is incomplete. Whether X is complete or not $0_X \supset M \times \{0\}$ and $0_X \subset M \times \mathbb{R}$ is open, as is seen immediately by the existence of local flows near any point in a chart. Moreover there is a unique smooth map $\alpha_X : 0_X \longrightarrow M$ satisfying

$$\alpha_X(p,0) = p$$

for all $p \in M$ and

$$\frac{d}{dt} \alpha_X(p,t) = X(\alpha_X(p,t))$$

for all $(p,t) \in 0_X$, namely given by

$$\alpha_X(p,t) = \beta_p(t)$$

for all $(p,t) \in 0_X$. This map satisfies the following condition: If $(p,t) \in 0_X$ then

$$(\alpha_X(p,t),s) \in 0_X \text{ iff } (p,t+s) \in 0_X$$

and in addition

$$\alpha_X(p,t+s) = \alpha_X(\alpha_X(p,t),s) .$$

α_X is called the <u>global flow</u> of X (with initial condition $\alpha_X(p,0) = p$).

If X is complete then the map A(X)(t) sending p into $\alpha_X(p,t)$ for some fixed $t \in \mathbb{R}$ is a smooth diffeomorphism of M. Denote by Diff M the group of all smooth diffeomorphisms of M. Clearly

$$A(X) : \mathbb{R} \longrightarrow \text{Diff } M$$

sending any t into A(X)(t) is a group homomorphism. We will see in a later chapter that if M is compact, A(X) is smooth in a well defined sense and will find that A(X) determines α_X uniquely.

Next we prove Frobenius's theorem:

Instead of assigning to each point $p \in M$ a tangent vector we assign to each $p \in M$ a linear subspace $D_p \subset T_pM$. Let $\mathcal{D} = \bigcup_{p\in M} D_p$. Call such an assignment a smooth n–dimensional involutive distribution of tangent subspaces of TM if the following three conditions hold:

1) $\dim D_p = n$ for all $p \in M$, where n is a given natural number.
2) For each $p \in M$ there is a chart (U,φ) and smooth vector fields $Y_1,\dots,Y_n$ on U which span $\mathcal{D}$ at each point $g \in U$.
3) If $X,Y \in \Gamma TM$ map into $\mathcal{D}$, which means that $X(p),Y(p) \in D_p$ for all $p \in M$, then $[X,Y]$ also maps into $\mathcal{D}$.

Example: Consider a manifold M of dimension m which is the disjoint union $\bigcup_{\iota\in I} N_\iota$ of smooth submanifolds N_ι all of the same dimension say n. Here ι varies in an index set. Any $p \in M$ belongs to exactly one N_ι. Let $\mathcal{D}$ be given by $D_p = T_pN_\iota \subset TM$ for all $p \in M$. Then $\mathcal{D}$ satisfies the first of the above three conditions. Now assume condition 2 also holds (this is the case e.g. if $M = U \times V$ for some smooth manifolds U and V each of a fixed dimension). Then also condition 3 holds which is seen as follows. Consider N_ι

passing through p and assume that X,Y map both into $\mathcal{D}$. Hence $X|N_\iota$ and $Y|N_\iota$ are both in ΓTN_ι. Thus there are vector fields $\hat{X},\hat{Y}$ on N_ι such that $\hat{X},\hat{Y}$ and X,Y are respectively related by the inclusion map i from N_ι to M. As we know from lemma 1.1.3 $[\hat{X},\hat{Y}]$ on N_ι and [X,Y] on M are i–related.

Given a smooth involutive distribution $\mathcal{D}$ of tangent subspaces of TM call $N \subset M$ an <u>integral manifold</u> of $\mathcal{D}$ through p if N is a manifold such that the inclusion $i : N \longrightarrow M$ is smooth, $p \in N$ and moreover $TN = \bigcup_{g \in N} D_g$. To simplify notation write $\mathcal{D}|S$ instead of $\bigcup_{g \in S} D_g$ for a subset $S \subset M$.

Let $\mathcal{D}$ be a smooth involutive distribution of tangent subspaces of TM. We will show next that each point $p \in M$ admits an integral manifold. The method to do so is to work locally and phrase the problem in the language of total differential equations.

To this end we choose a chart (U,φ) of $p \in M$, modelled on the finite dimensional topological vector space E. Consider $T\varphi(p)\, D_p \subset E$ and denote it for the sake of simplicity by E_1. Then choose a complement $E_2 \subset E$. Since $E_1 \oplus E_2 = E$ we can assume without loss of generality that $U = V_1 \times V_2$, where $\varphi(V_i) \subset E_i$ is open for $i = 1,2$. Again we identify U with $\varphi(U)$ via φ and hence V_i with $\varphi(V_i)$ for $i = 1,2$. Denote furthermore the finite dimensional vector space of all linear maps from E_1 to E_2 by $L(E_1,E_2)$. The differential equation in mind arises then as follows: $T\varphi{:}TU \longrightarrow U \times E$ is a diffeomorphism which restricted to each tangent space is linear. Choose U small enough to ensure that the projection $pr_1 : E \longrightarrow E_1$ restricted to $T\varphi(g)D_g$ is an isomorphism onto E_1 for all $g \in U$. Refer to this isomorphism by $\pi(g)$. Then $pr_2 \circ \pi(g)^{-1} : E_1 \longrightarrow E_2$ is clearly a linear map, depending smoothly on $g \in U$. Here pr_2 denotes the canonical projection onto E_2. Form next

$$1 : V_1 \times V_2 \longrightarrow L(E_1,E_2),$$

defined as

$$l(x,y) = pr_2 \circ \pi((x,y))^{-1}$$

for all $(x,y) \in V_1 \times V_2$. The total differential equation is then

$$y' = l(x,y).$$

This equation is called completely integrable iff it satisfies the Frobenius condition (cf. [D,1]). We will show now that condition (3) on $\mathcal{D}$ from above implies indeed complete integrability. A solution of this total differential equation is a smooth map $u : V_1' \longrightarrow V_2$, where V_1' is an open neighbourhood of $p_1 = pr_1(p)$, such that

$$u(p_1) = p_2 = pr_2(p)$$

and

$$Du(x) = l(x,u(x))$$

hold for all $x \in V_1'$. The symmetry of the second derivative of u yields a necessary condition on l, the so called Frobenius condition

$$D_1\, l(x,y)(h,k) + D_2\, l(x,y)\, (l(x,y)(h),k) =$$
$$= D_1\, l(x,y)(k,h) + D_2\, l(x,y)\, (l(x,y)(k),h).$$

Here $L(E_1,L(E_1,E_2))$ and $L(E_2,L(E_1,E_2))$ are identified with $L(E_1 \times E_1,E_2)$ and $L(E_2 \times E_1,E_2)$. Accordingly $D_1\, l(x,y)(\cdots)(\cdots) = D_1\, l(x,y)(\cdots,\cdots)$ and $D_2\, l(x,y)(\cdots)(\cdots) = D_2\, l(x,y)(\cdots,\cdots)$ for all $(x,y) \in V_1 \times V_2$. The elements h,k vary in E_1.

The Frobenius condition on l is also sufficient for the existence of a unique, smooth solution u in a neighbourhood of $p_1 \in V_1$, satisfying the initial condition $u(p_1) = p_2$. For details see [D,1].

Since we assume $\mathcal{D}$ to be involutive, the Frobenius condition on l is a direct consequence of this assumption. To see this choose $X,Y, \in \Gamma TM$ which both map into $\mathcal{D}$. Call the principal parts of the local representations of X and Y on U by X and Y again.

Then X and Y are functions of $(x,y) \in V_1 \times V_2$ and both split according to the splitting of $E = E_1 \oplus E_2$ into

$$X = (X_1, X_2) \quad \text{and} \quad Y = (Y_1, Y_2).$$

The fact that both map into $\mathcal{D}$ means that

$$l(x,y)(X_1(x,y)) = X_2(x,y) \text{ and } l(x,y)(Y_1(x,y)) = Y_2(x,y)$$

for all $(x,y) \in V_1 \times V_2$. A simple calculation shows that $[X,Y]$ (splitting into $([X,Y]_1,[X,Y]_2)$ satisfies

$$l(x,y)([X,Y]_1(x,y)) = [X,Y]_2(x,y)$$

iff the above stated Frobenius condition on l holds. The variables h and k are replaced by X_1 and Y_1 respectively. The graph of the solution u with $u(p_1) = p_2$ in U is a smooth manifold N′ for which $TN' = \mathcal{D}|N'$. Since the solution is unique we may set up an extension process similar as in the first part of this section to obtain a maximal connected integral manifold N, which contains $p \in M$ and satisfies

$$TN = \mathcal{D}|N .$$

Due to the uniqueness of the solution of the local total differential equation the integral manifold N is unique.

Theorem 1.2.1 To any smooth n–dimensional involutive distribution $\mathcal{D}$ of tangent subspaces of TM and a point $p \in M$ there is a unique maximal connected integral manifold N of $\mathcal{D}$ passing through p.

Remark: An integral manifold need not to be a submanifold. Compare 16.9.9.3. p.56 in [D,III].

1.3 Lie groups

Definition 1.3.1 Let G be a smooth manifold together with a group operation $\mu : G \times G \longrightarrow G$. Regard $G \times G$ as the product manifold. G is called a Lie group if μ is smooth.

Observe, that for any Lie group G the map sending each element $g \in G$ into g^{-1} is, due to the implicit function theorem, smooth as well.

Furthermore, any connected Lie group G is generated by any open neighbourhood U of the identity. This is easily seen as follows: Let

$$U^{-1} = \{g^{-1} \mid g \in U\}$$

and

$$U^n = \{g_1 \cdots g_n \mid g_i \in U\}$$

where n is a natural number. Both U^{-1} and U^n are open. Thus V being $U \cap U^{-1}$ and V^n are open. Since $H := \bigcup_{n \in \mathbb{N}} V^n$ is an open subgroup of G any left coset $g \cdot H$ with $g \in G$ is open, hence the complement of H is open. Thus H is open and closed meaning $H = G$.

To study the tangent manifold of G we pose a slightly more general situation. We assume that G is a Lie group and M a manifold. Let $G \times M$ be the product manifold. Suppose we have a smooth map

$$\phi : G \times M \longrightarrow M$$

satisfying

i) $\phi(g_1 \cdot g_2, p) = \phi(g_1, \phi(g_2, p))$

and ii) $\phi(e, p) = p$

for any choice of $g_1, g_2 \in G$, any $p \in M$ and the identity $e \in G$. Such a map ϕ is called a (smooth) left <u>action</u> of G on M. (In an analogous way we talk of right actions.) Then $\phi_g: M \longrightarrow M$ given by $\phi_g(p) = \phi(g,p)$ is a smooth diffeomorphism. If Diff M denotes the group of all smooth diffeomorphisms of M then ϕ is uniquely determined by the group homomorphism

$$\hat{\phi} : G \longrightarrow \text{Diff } M$$

assigning to any $g \in G$ the diffeomorphisms ϕ_g. Call ϕ to be <u>effective</u> if $\hat{\phi}$ is injective. Given a point $p \in M$ the subset

$$\{\phi(g,p) \mid g \in G\}$$

of M is called the <u>orbit</u> of G through $p \in M$. If we introduce for all $g \in G$ and $p \in M$ the short hand $g\cdot p$ for $\phi(g,p)$ then the orbit has the form $G\cdot p$. Clearly $G\cdot p$ is a manifold. If G is compact, then $G\cdot p$ is even a submanifold of M. Fix next $p' \in G\cdot p$ and denote by $\phi_{p'} : G \longrightarrow G\cdot p$ the map sending $g \in G$ into $g\cdot p'$. Any $\xi \in T_eG$ determines a so called left invariant vector field $X(\xi)$ on $G\cdot p$ given by

$$X(\xi)(p') = T\,\phi_{p'}(e)(\xi).$$

Let ϕ_p be injective and let $\xi_1,\dots,\xi_n$ be a basis of T_eG. Since $\phi_{p'}$ is a diffeomorphism for all $p' \in G\cdot p$, $T\phi_{p'}(e)(\xi_1),\dots,T\phi_{p'}(e)(\xi_n)$ is a basis for $T_{p'}G\cdot p$. Thus any vector field X on $G\cdot p$ can be represented at $g\cdot p$ by

$$X(g\cdot p) = \Sigma\, f^i(g\cdot p)X(\xi_i)(g\cdot p)\,.$$

As an exercise the reader is left to verify the smoothness of $f^i : G\cdot p \longrightarrow \mathbb{R}$ mapping $g\cdot p$ into $f^i(g\cdot p)$ for any $i = 1,\dots,n$. To summarize we have:

<u>Proposition 1.3.2</u> Let G be a Lie group acting effectively on the left on M. Given any $p \in M$, the tangent manifold $T(G\cdot p)$ of the orbit $G\cdot p$ is parallelizable. As a basis over $C^\infty(G\cdot p,\mathbb{R})$ of $\Gamma T(G\cdot p)$ serve the left invariant vector fields $X(\xi_1),\dots,X(\xi_n)$, given by a basis $\xi_1,\dots,\xi_n$ of T_eG.

Since μ is a smooth effective left action of G on itself $g \in G$ determines by multiplying from the left the <u>left translation</u> L_g. Clearly $G \cdot e = G$, and thus any basis $\xi_1,\dots,\xi_n$ of T_eG determines a basis

$$X(\xi_1),\dots,X(\xi_n)$$

over $C^\infty(G,\mathbb{R})$ for the vector fields on G. Obviously $X(\xi_i)$ is given by $X(\xi_i)(g) = TL_g(e)(\xi_i)$ for all $g \in G$. Analogously we have a right translation, denoted by R_g. The tangent manifold is thus determined by T_eG. Even the Lie algebra of G is determined by T_eG:

Call a vector field $Y \in \Gamma TG$ to be <u>left invariant</u> if for any $g \in G$ it is L_g–related to itself. Clearly in this case $Y = X(Y(e))$. By lemma 1.1.3 the Lie product of two left invariant vector fields is left invariant as well. Therefore

$$TL_g(e)([X(\xi_1),X(\xi_2)](e)) = [X(\xi_1),X(\xi_2)](g)$$

for all $g \in G$. Hence T_eG has a product given by

$$[\xi_1,\xi_2] := [X(\xi_1),X(\xi_2)](e)$$

for all $\xi_1,\xi_2 \in T_eG$. Thus T_eG is a Lie algebra over $\mathbb{R}$.

To simplify notation write $\mathcal{G}$ instead of T_eG. What we just verified is, that the Lie algebra $\mathcal{G}$ over $\mathbb{R}$ determines the Lie algebra ΓTG by using the left action induced by the group multiplication.

Given any two Lie groups G and H, by a <u>homomorphism</u> $h : G \longrightarrow H$ we mean a smooth group homomorphism.

<u>Proposition 1.3.3</u> For any homomorphism $h : G \longrightarrow H$ the map $Th(e)$ is a Lie algebra homomorphism from the Lie algebra $\mathcal{G}$ into the Lie algebra $\mathcal{F} = T_eH$.

Proof: Let $e \in G$ be the neutral element of G and let $\xi_1, \xi_2 \in \mathcal{G}$. Then $X(\xi_i) \in \Gamma TG$ is h–related to $X(Th(e)\xi_i) \in \Gamma TH$ for $i = 1,2$. Thus by lemma 1.1.3

$$\begin{aligned} Th(e)([\xi_1,\xi_2]) &= Th(e)([X(\xi_1),X(\xi_2)](e)) = \\ &= [Th(e)(\xi_1),Th(e)(\xi_2)]. \end{aligned}$$

A converse setting of the above proposition, namely theorem 1.4.3, will be shown later.

We now present the heart of the Lie theory by relating a neighbourhood of the identity of a Lie group G with a zero neighbourhood of $\mathcal{G} = T_eG$ via a smooth map called the exponential map. To do so consider $\xi \in \mathcal{G}$, its left invariant vector field $X(\xi)$ and its maximal flow with initial condition $\alpha(g,0) = g$ introduced in section 1.2. The integral curve $\alpha(\xi)(e)$ of $X(\xi)$ with initial condition $\alpha(\xi)(e)(0) = e$ satisfies the identity

$$\alpha(\alpha(e,s),t) = \alpha(e,s+t)$$

as shown in 1.2. On the other hand we have

$$\alpha(g,t) = g\cdot\alpha(e,t)$$

since α_g, mapping t into $g\cdot\alpha(e,t)$, is an integral curve of $X(\xi)$ with initial condition $\alpha_g(0) = g$. To see this verify

$$\begin{aligned} \frac{d}{dt}(g\cdot\alpha(e,t)) &= TL_g \frac{d}{dt}\alpha(e,t) = (TL_g \circ X(\xi))(\alpha(e,t)) \\ &= (X(\xi)\circ L_g)(\alpha(e,t)) = X(\xi)(g\cdot\alpha(e,t)). \end{aligned}$$

Moreover $\alpha(g,0) = g\cdot\alpha(e,0)$. Since integral curves depend uniquely on the initial conditions we have

$$\alpha(\alpha(e,s),t) = \alpha(e,s)\cdot\alpha(e,t).$$

Thus the maximal integral curve $\alpha(\xi)(e)$ is defined on all of $\mathbb{R}$ and consequently is a group homomorphism.

We therefore call $\alpha(\xi)(e)$ the one parameter group given by ξ and denote it by α_ξ. Since $\dot{\alpha}_\xi(0) = \xi$ we have

Proposition 1.3.4 For each $\xi \in \mathscr{G}$ there is a unique one parameter group

$$\alpha_\xi : \mathbb{R} \longrightarrow G$$

with $\dot{\alpha}_\xi(0) = \xi$.

The above proposition allows us to define the exponential map

$$\exp : \mathscr{G} \longrightarrow G$$

by $\exp \xi = \alpha_\xi(1)$.

Proposition 1.3.5 The exponential map is a diffeomorphism of an open zero neighbourhood U in $\mathscr{G}$ onto exp U, an open neighbourhood of the identity of G.

Proof: For each $\xi \in \mathscr{G}$ we have

$$T\exp(0)\xi = \dot{\alpha}_\xi(0) = \xi.$$

Hence T exp(0) is invertible and thus by the inverse function theorem exp is a diffeomorphism of an open zero neighbourhood $U \subset \mathscr{G}$ onto the open set $\exp(U) \subset G$.

The tangent map of exp at an arbitrarily given element in $\mathscr{G}$ will be given below. To prepare some calculus of the exponential mapping let us introduce next the adjoint representation.

Each element $g \in G$ defines an inner automorphism

$$\tau_g : G \longrightarrow G$$

mapping g' into $\tau_g(g') = g \cdot g' \cdot g^{-1}$. Since τ_g is smooth the tangent map $T\tau_g(g')$: $T_{g'}G \longrightarrow T_{\tau_g(g')}G$ exists and is a linear isomorphism. In particular $T\tau_g(e)$ is a Lie algebra isomorphism. Denote the group of linear automorphisms of $\mathfrak{g}$ by $GL(\mathfrak{g})$. The map

$$\mathrm{Ad} : G \longrightarrow GL(\mathfrak{g})$$

sending each $g \in G$ into $T\tau_g(e)$ is a representation (cf. [W]), called the <u>adjoint representation</u> of G. Let us pause here to illustrate some notions just introduced on an easy example which will be constantly used in the sequel.

Give a finite dimensional vector space V either over $\mathbb{R}$ or $\mathbb{C}$. Then GL(V) is a Lie group with Lie algebra $T_{id}GL(V) = \mathrm{End}\ V$ where the Lie bracket is given by $[\xi_1,\xi_2] = \xi_1 \circ \xi_2 - \xi_2 \circ \xi_1$ for all $\xi_1,\xi_2 \in \mathrm{End}\ V$ (cf. 1.5 Example 2). To demonstrate the nature of exp in this case let e^1 be the limit of the series

$$id + 1 + \frac{1^2}{2!} + \cdots + \frac{1^n}{n!} + \cdots$$

converging uniformly for all $1 \in \mathrm{End}\ V$. Since for each $1 \in \mathrm{End}\ V$ the left invariant vector field X(1) is given by

$$X(1)(g) = g \circ 1$$

for all $g \in GL(V)$ (cf. 1.5 Example 2) its integral curve through $id \in GL(V)$ is of the form

$$t \longmapsto e^{t \cdot 1}$$

for all $t \in \mathbb{R}$. Hence

$$\exp 1 = e^1 .$$

Denoting by det and tr the determinant and the trace respectively we immediately verify for any $1 \in \mathrm{End}\ V$

$$\det \exp 1 = e^{\mathrm{tr}\ 1} .$$

Moreover, for any Lie group homomorphism h from a Lie group H into G the following diagram commutes:

$$\begin{array}{ccc} H & \xrightarrow{\quad h \quad} & G \\ {\scriptstyle \exp}\uparrow & & \uparrow{\scriptstyle \exp} \\ \mathcal{F} & \xrightarrow{\quad Th(e) \quad} & \mathcal{G} \end{array}$$

$\mathcal{F}$ denotes the Lie algebra of H. In particular we have the commuting diagrams

$$\begin{array}{ccc} G & \xrightarrow{\quad Ad \quad} & GL(\mathcal{G}) \\ {\scriptstyle \exp}\uparrow & & \uparrow{\scriptstyle \exp} \\ \mathcal{G} & \xrightarrow{\quad T\ Ad(e) \quad} & End\ \mathcal{G} \end{array}$$

$$\begin{array}{ccc} G & \xrightarrow{\quad \tau_g \quad} & G \\ {\scriptstyle \exp}\uparrow & & \uparrow{\scriptstyle \exp} \\ \mathcal{G} & \xrightarrow{\quad Ad(g) \quad} & \mathcal{G} \end{array} \quad , g \in G$$

which means for any $t \in \mathbb{R}$ and $\xi \in \mathcal{G}$

$$\exp(t\ Ad(g)(\xi)) = \tau_g(\exp t\xi).$$

A small calculation shows that for all $\xi, \eta \in \mathcal{G}$

$$T\ Ad(e)(\xi)(\eta) = [\xi,\eta].$$

Instead of T Ad(e) write ad, which means more precisely

$$T\ Ad(e)(\xi)(\eta) = ad_\xi\ \eta = [\xi,\eta].$$

The notion ad will be particularly useful if we have to deal with iterations of Lie brackets.

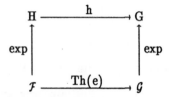

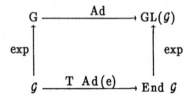

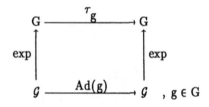

Now we are able to compute the tangent map of exp at an arbitrary point $\xi \in \mathcal{G}$. For any smooth map $f : \mathcal{G} \longrightarrow G$ let $df(\xi) \in \mathrm{End}\ \mathcal{G}$ be defined by

$$df(\xi) = TL_{f(\xi)^{-1}}(f(\xi)) \circ Tf(\xi)$$

(where $T_\xi \mathcal{G}$ is canonically identified with $\mathcal{G}$). Observe that for a second smooth map $g : \mathcal{G} \longrightarrow G$ the tangent map of the pointwise defined product $f \cdot g : \mathcal{G} \longrightarrow G$ is given by

$$T(f\cdot g)(\xi) = TR_{g(\xi)}(f(\xi)) \circ Tf(\xi) + TL_{f(\xi)}(g(\xi)) \circ Tg(\xi)$$

which implies

$$d(f\cdot g)(\xi) = \mathrm{Ad}(g(\xi))^{-1} \circ df(\xi) + dg(\xi)$$

for all $\xi \in \mathcal{G}$. We claim that

$$d\exp(\xi) = \sum_{p=0}^{\infty} \frac{1}{(p+1)!} (-\mathrm{ad}_\xi)^p.$$

To show this consider the two curves c_1 and c_2 in End $\mathcal{G}$ given by

$$c_1(t) = t \cdot d\exp(t\xi)$$

and

$$c_2(t) = \sum_{p=0}^{\infty} \frac{1}{(p+1)!} t^{p+1} (-\mathrm{ad}_\xi)^p$$

for all $t \in \mathbb{R}$. Obviously $c_1(0) = 0 = c_2(0)$. If h_t denotes the linear endomorphism of $\mathcal{G}$ sending each $\xi \in \mathcal{G}$ into $t\xi$, then

$$d(\exp \circ h_t)(\xi) = d\exp(t\xi) \circ h_t = c_1(t)\ .$$

Since

$$\exp \circ h_{t+s} = (\exp \circ h_t) \cdot (\exp \circ h_s)$$

for all $t,s \in \mathbb{R}$ we have

$$\begin{aligned} c_1(t+s) &= d(\exp \circ h_{t+1})(\xi) \\ &= \mathrm{Ad}(\exp(s\xi)^{-1}) \circ c_1(t) + c_1(s) \\ &= \exp(-\mathrm{ad}_{s\xi}) \circ c_1(t) + c_1(s). \end{aligned}$$

Differentiating with respect to s at s=0 yields

$$\begin{aligned} \dot{c}_1(t) &= -\mathrm{ad}_\xi \circ c_1(t) + \dot{c}_1(0) \\ &= -\mathrm{ad}_\xi \circ c_1(t) + \mathrm{id}. \end{aligned}$$

On the other hand the tangent vector of c_2 at t is given by

$$\dot{c}_2(t) = \sum_{p=0}^{\infty} \frac{1}{p!} t^p (-\mathrm{ad}_\xi)^p$$
$$= \mathrm{id} - \mathrm{ad}_\xi \circ c_2(t).$$

Thus both curves satisfy the same differential equation and initial condition and therefore are identical. In particular $c_1(1) = c_2(1)$ which yields the assertion.

Definition 1.3.6 Given a Lie group G and a subgroup $H \subset G$. If H is a submanifold of G, then H is called a Lie subgroup of G.

Certainly the Lie algebra $\mathcal{F}$ of a Lie subgroup H of G is a Lie subalgebra of $\mathcal{G} = T_eG$.

Clearly any open subgroup H of G is a Lie subgroup of G. To show that any closed subgroup of a Lie group is a Lie group again (Cartan's theorem), we study the exponential map somewhat more closely:

Lemma 1.3.7 Given a Lie group G with $\mathcal{G}$ as its Lie algebra. For any $t \in \mathbb{R}$ sufficiently close to zero any pair $\xi,\eta \in \mathcal{G}$ satisfies

$$\exp t\xi \cdot \exp t\eta = \exp\{t(\xi+\eta) + 0(t^2)\}$$

where 0 is a smooth $\mathcal{G}$-valued function for which $\frac{1}{t^2} \cdot 0(t^2)$ is bounded. Thus for any $t \in \mathbb{R}$

$$\lim_{n\to\infty} (\exp \tfrac{t}{n}\, \xi \cdot \exp \tfrac{t}{n}\, \eta)^n = \exp(t(\xi+\eta)).$$

Proof: Since exp is injective in a sufficiently small zero neighbourhood U of $\mathcal{G}$ we find some U-valued curve ζ such that

$$\exp t\xi \cdot \exp t\eta = \exp \zeta(t)$$

for all $t \in \mathbb{R}$ near enough to zero. Thus

$$\frac{d}{dt}(\exp t\xi \cdot \exp t\eta)\Big|_0 = \frac{d}{dt}(\exp \zeta(t))\Big|_0 = \xi + \eta .$$

Thus using the Taylor expansion of ζ we have

$$\zeta(t) = t(\xi+\eta) + 0(t^2)$$

The second half of the lemma follows from

$$(\exp \tfrac{t}{n}\xi \cdot \exp \tfrac{t}{n}\eta)^n = (\exp\{\tfrac{t}{n}(\xi+\eta) + 0((\tfrac{t}{n})^2)\})^n .$$

Now we are ready to prove Cartan's theorem.

Theorem 1.3.8 Any closed subgroup H of a Lie group G is a Lie subgroup of G.

Proof: Consider

$$\mathcal{F} = \{\xi \in \mathcal{G} \mid \exp t\xi \in H \text{ for all } t \in \mathbb{R}\} .$$

By the second half of the above lemma $\mathcal{F} \subset \mathcal{G}$ is a linear subspace of $\mathcal{G}$.

It remains to show the existence of neighbourhoods U of zero in $\mathcal{G}$ and V of e in G related by

$$\exp(U \cap \mathcal{F}) = V \cap H.$$

Suppose the contrary, that no such U exists. Then there is a zero neighbourhood W in $\mathcal{G}$ and a sequence (σ_i) of elements in H converging to e with $\sigma_i \notin \exp(W \cap \mathcal{F})$ for all indices i. Given a complement $\mathcal{K}$ of $\mathcal{F}$ in $\mathcal{G}$, the map

$$\alpha : \mathcal{K} \oplus \mathcal{F} \longrightarrow G$$

sending each pair (ξ,η) into $\exp \xi \cdot \exp \eta$ is invertible in a neighbourhood of zero, since $T\alpha(0,0)$ has maximal rank. W can be choosen to be of the form $W_1 \oplus W_2$ where W_1 and W_2 are zero neighbourhoods in $\mathcal{K}$ and $\mathcal{F}$ respectively for which moreover $\alpha(W)$ is a neighbourhood of $e \in G$ and $\alpha : W \longrightarrow \alpha(W)$ is a diffeomorphism. Let i be large enough such that $\sigma_i \in \alpha(W)$. Then

$$\sigma_i = \exp \xi_i \cdot \exp \eta_i$$

for $\xi_i \in W_1$ and $\eta_i \in W_2$. Since $\sigma_i \in H$ and $\exp \eta_i \in H$, also $\exp \xi_i \in H$. Moreover $\sigma_i \notin \exp(W \cap \mathcal{F})$ implies $\xi_i \neq 0$. Thus $\exp(W_1\backslash\{0\}) \cap H \neq \phi$. Therefore there is a sequence (ζ_i) in $W_1\backslash\{0\}$ converging to zero for which $\exp \zeta_i \in H$ for all indices i. Given a norm $\| \ \|$ on $\mathcal{K}$ there is a subsequence (ζ_j) for which $(\frac{\zeta_j}{\|\zeta_j\|})$ converges to some non zero element $\zeta \in \mathcal{K}$. To any positive real number t there is an integer $n_j(t)$ with

$$\frac{t}{\|\zeta_j\|} - 1 < n_j(t) \leq \frac{t}{\|\zeta_j\|}$$

which yields

$$\lim_{j\to\infty} \|\zeta_j\| \cdot n_j(t) = t.$$

Hence

$$\exp t\zeta = \lim_{j\to\infty} (\exp \zeta_j)^{n_j(t)}$$

is contained in H. Since $\exp(-t\zeta) = (\exp t\zeta)^{-1}$ the element $\exp t\zeta$ belongs to H for all $t \in \mathbb{R}$. Thus $\zeta \in \mathcal{F}$ and hence $\zeta = 0$, a contradiction to $\zeta \neq 0$.

Corollary 1.3.9 The kernel ker h of a Lie group homomorphism h from the Lie group H into the Lie group G is a Lie subgroup of H. The kernel of Th(e) is the Lie algebra of ker h.

Proof: The first assertion follows immediately from Cartan's theorem since $\ker h \subset H$ is a closed subgroup. From the proof of the above theorem we learned that

$$\{\xi \in \mathcal{F} \mid \exp t\xi \in \ker h \text{ for all } t \in \mathbb{R}\}$$

is the Lie algebra of ker h. Since for all $t \in \mathbb{R}$

$$h(\exp t\xi) = \exp t\, Th(e)\xi$$

$T_e \ker h = \ker Th(e)$.

Clearly the kernel of a group homomorphism is a normal subgroup. Corollary 1.3.9 can be sharpened in the following way:

Theorem 1.3.10 Given a connected Lie subgroup H of a connected Lie group G. Then H is normal iff its Lie algebra $\mathcal{F}$ is an ideal of the Lie algebra of G.

Proof: Let $\eta \in \mathcal{F}$ and $\xi \in \mathcal{G}$. Then for $g = \exp \xi$ we have

$$\tau_g(\exp \eta) = \exp(\mathrm{Ad}(g)\eta) = \exp((\exp \mathrm{ad}_\xi)\eta)$$
$$= \exp(\eta + \mathrm{ad}_\xi \eta + \frac{(\mathrm{ad}_\xi)^2}{2!}\eta + \cdots) .$$

Clearly if $\mathcal{F}$ is an ideal the right hand side is in H and then H is normal. Conversely assume H is normal and replace ξ by $t\xi$ for $t \in \mathbb{R}$. Then the left hand side is in H and thus $(\exp(\mathrm{ad}_{t\xi}))\eta \in \mathcal{F}$. Differentiating with respect to t at $t = 0$ yields $\mathrm{ad}_\xi \eta = [\xi,\eta] \in \mathcal{F}$.

Let us apply the results of this section to the adjoint representation of a connected Lie group G. The kernel ker Ad consists of all those $g \in G$ for which $T\tau_g(e) = \mathrm{id}$. Hence

$$\exp \xi = \exp \mathrm{Ad}(g)\xi = \tau_g(\exp \xi)$$

holds for all $\xi \in \mathcal{G}$ and $g \in \ker \mathrm{Ad}$. Thus g is an element of the center of G. The <u>center</u> Z_G of G is defined as

$$Z_G = \{g \in G \mid \tau_g = \mathrm{id}\}.$$

Conversely, if $g \in Z_G$ the above equation holds and thus yields $\mathrm{Ad}(g)\xi = \xi$. Next the <u>center</u> $Z_\mathcal{G}$ of the Lie algebra $\mathcal{G}$ is

$$Z_\mathcal{G} = \{\xi \in \mathcal{G} \mid \mathrm{ad}_\xi = 0\} = \ker \mathrm{ad}.$$

Clearly $Z_G \subset G$ is closed and hence a Lie subgroup. Its Lie algebra is $Z_\mathcal{G}$. We summarize this in:

Proposition 1.3.11 Let G be a connected Lie group. The kernel of the adjoint representation of G is Z_G , a Lie subgroup of G with $Z_\mathcal{G}$ as its Lie algebra.

An immediate consequence is

Corollary 1.3.12 A connected Lie group G is abelian iff $\mathcal{G}$ is abelian.

Proposition 1.3.13 For any connected abelian Lie group G the identity

$$\exp(\xi+\eta) = \exp \xi \cdot \exp \eta$$

holds for any pair $\xi,\eta \in \mathcal{G}$.

Proof: The map α from $\mathbb{R}$ to G sending each $t \in \mathbb{R}$ into

$$\alpha(t) = \exp t\xi \cdot \exp t\eta$$

is a one parameter group with $\xi+\eta$ as tangent vector at e. Thus $\alpha(t) = \exp t(\xi+\eta)$ for all $t \in \mathbb{R}$.

Proposition 1.3.11 describes a special situation of the concept of the isotropy group introduced just below:

Consider a left action

$$\phi : G \times M \longrightarrow M$$

of a Lie group G on a manifold M. Given a point $p \in M$ then the isotropy group G_p consists of all those $g \in G$ for which

$$\phi(g,p) = p.$$

Observe that the isotropy group of $p' \in G \cdot p$ is a conjugate of G_p. Since G_p is closed it is a Lie subgroup and its Lie algebra is the kernel of

$$T\phi_p(e) : \mathcal{G} \longrightarrow T_pG \cdot p$$

where $\phi_p : G \longrightarrow G \cdot p$ maps any $g \in G$ to $g \cdot p$, as introduced at the beginning of this section. Clearly ϕ_p induces a map $\bar{\Phi}_p : G/G_p \longrightarrow G \cdot p$.

This motivates us to investigate the existence of a manifold structure on coset spaces. Therefore, let H be a closed Lie subgroup of a Lie group G. Set $T_eH = \mathcal{F}$.

The left <u>coset space</u> G/H will be equipped with a manifold structure in the following way: The finest topology on G/H among all, for which $\pi : G \longrightarrow G/H$, the canonical projection, is continuous, satisfies the second axiom of countability and the Hausdorff separation axiom. Next consider the involutive distribution $\mathcal{D} \subset TG$ given by $\mathcal{D}_g = TL_g(e)(\mathcal{F})$. By the theorem of Frobenius there is through each $g \in G$ an integral manifold which in this case is a submanifold. Let N_g be such a submanifold containing $g \in G$. Thus g admits an open chart of the form $(W \times V, \varphi)$ modelled in $E \oplus F$ for two finite dimensional vector spaces E and F in which $\varphi(W) \subset E$ as well as $\varphi(V) \subset F$ are open and such that $(V, \varphi|V)$ is a chart of g in N_g. If W is small enough then $(\pi(W), \varphi \circ (\pi|W)^{-1})$ is a chart of $\pi(g) = g \cdot H$ in G/H modelled on E. The maximal atlas defined by all charts of this form defines a manifold structure on G/H, called the <u>final structure</u>.

Throughout the book all such coset spaces are supposed to carry the final structure (cf. [W]). The reader verifies easily the following universal property:

Let f be a map from G/H into a manifold N. Then f is smooth iff $f \circ \pi$ is smooth.

Equipped with these tools we prove next:

<u>Proposition 1.3.14</u> Let ϕ be a smooth left action of a Lie group G on a manifold M, then for each $p \in M$

$$\bar{\Phi}_p : G/G_p \longrightarrow G \cdot p$$

is a smooth diffeomorphism.

Proof: The above universal property of G/G_p guarantees us the smoothness of $\bar{\Phi}_p$. Clearly $\bar{\Phi}_p$ is a bijection. However, a smooth bijection whose tangent map is of maximal rank everywhere, is a smooth diffeomorphism as seen by the inverse function theorem (cf. [D,I]).

Hence given a smooth action of G on M the manifold M splits into a collection of manifolds, called orbits of G. Each one is smooth diffeomorphic to G/G_p where G_p is the isotropy group of a point p in the orbit.

1.4 Immersed Lie groups

By an immersion (resp. submersion) f from a manifold M into N, we mean a smooth map for which Tf(p) is injective (resp. surjective) for any $p \in M$. An immersion f is an embedding if $f : M \longrightarrow f(\ )$ is a homeomorphism, where $f(M) \subset N$ carries the initial topology given by the inclusion.

Let us remark, that an injective immersion does not need to be an embedding at all (cf. [W] and [D,III] p.56).

An immersed Lie group of a Lie group G is defined as the image of an injective immersive Lie group homomorphism h of a Lie group H into G.

The main theorem in this section is

Theorem 1.4.1 Let G be a Lie group with Lie algebra $\mathcal{G}$. Given a Lie subalgebra $\mathcal{F}$ of $\mathcal{G}$. Then there is a connected Lie group H with a Lie algebra isomorphic to $\mathcal{F}$ and an injective Lie group homomorphism from H into G.

Proof: The Lie subalgebra $\mathcal{F}$ defines an involutive distribution $\mathcal{D} \subset TG$ given at $g \in G$ by

$$\mathcal{D}_g = TL_g(e)(\mathcal{F}).$$

Hence there is by the theorem of Frobenius a maximal connected integral manifold H through $e \in G$. Let us show that H is an immersed Lie group. H is a subgroup of G. For if $g_1, g_2 \in H$, then $g_2 \cdot g_1^{-1} \cdot H$ is a maximal connected integral manifold containing $g_2 \in H$. Hence $g_2 \cdot g_1^{-1} \cdot H = H$ and then $g_2 \cdot g_1^{-1} \in H$. Clearly an open neighbourhood of e in H is a submanifold of G, out of which we conclude the smoothness of the operations. The inclusion $j : H \longrightarrow G$ is an immersive Lie group homomorphism.

The structure of the image of a Lie group homomorphism is described next as an easily derivable consequence of the above theorem:

Corollary 1.4.2 Given any two Lie groups H,G and a Lie group homomorphism $h : H \longrightarrow G$. If H is connected then h(H) is an immersed Lie group of G.

The relation between Lie group homomorphisms and Lie algebra homomorphisms is given in the next theorem. First let us introduce some notions. By a local homomorphism h from a Lie group H into another one G we mean a smooth map defined on a neighbourhood U of the neutral element $e \in H$ with range in a neighbourhood V of the neutral element $e \in G$ such that

$$h(g_1 \cdot g_2) = h(g_1) \cdot h(g_2)$$

whenever $g_1 \cdot g_2 \in U$ for $g_1, g_2 \in U$. Call a connected manifold M simply connected if the fundamental group $\pi_1(M)$ of M vanishes. (As a reference for algebraic topology we suggest [G] and [Sp]). By construction the universal covering $\tilde{M}$ of a connected manifold M is simply connected [G].

Theorem 1.4.3 Let H and G be two Lie groups with Lie algebra $\mathcal{F}$ and $\mathcal{G}$ respectively. Given a Lie algebra homomorphism $u : \mathcal{F} \longrightarrow \mathcal{G}$, there is a local homomorphism h from H to G with $Th(e) = u$. Any other local homomorphism h′ from H to G with $Th'(e) = u$ coincides with h on some neighbourhood of $e \in H$. Moreover if H is connected and simply connected, there is a unique Lie group homomorphism h from H to G with $Th(e) = u$.

Proof: Consider $H \times G$, the product Lie group. The graph of u in $\mathcal{F} \times \mathcal{G}$ is a Lie algebra, thus defines by theorem 1.4.1 a Lie group K immersed in $H \times G$. Let $j : K \longrightarrow H \times G$ be the injection. Denote by $pr_H: H \times G \longrightarrow H$ the canonical projection onto H. Clearly $pr_H \circ j : K \longrightarrow H$ is a Lie group homomorphism. $T(pr_H \circ j)(e)$ maps the Lie algebra of K isomorphically onto $\mathcal{F}$. Thus $pr_H \circ j$ is a diffeomorphism of a neighbourhood U of e in K onto a neighbourhood V of e in H. Hence $pr_G \circ j \circ a^{-1}: V \longrightarrow G$ is a local Lie group homomorphism h where pr_G is the canonical projection of $H \times G$ onto G and a is the restriction of $pr_H \circ j$ to U. Due to the theorem of Frobenius any local homomorphism h′ of H into G with $Th'(e) = u$ has to coincide with h on some neighbourhood of e.

Let H be connected and simply connected. The map $pr_H \circ j: K \longrightarrow H$ is a covering map (cf. [W]), which due to the simply connectedness of H has to be injective and open, hence $pr_H \circ j : K \longrightarrow H$ is an isomorphism. Thus $h = pr_G \circ j \circ (pr_H \circ j)^{-1}$ is the proposed Lie group homomorphism.

1.5 Examples

We begin our selection of examples of Lie groups by characterizing all connected commutative Lie groups of a given dimension.

Example 1

Assume therefore, that G is a connected commutative Lie group. Then $\mathcal{G} = \mathbb{R}^n$ for some integer n. Since exp: $\mathbb{R}^n \longrightarrow G$ is a Lie group homomorphism and since Texp has maximal rank everywhere, exp is a covering map, meaning that $G = \mathbb{R}^n/D$ for some discrete subgroup $D \subset \mathbb{R}^n$.

The crucial observation is now the following characterization of discrete subgroups of $\mathbb{R}^n$.

Proposition 1.5.1 For each closed subgroup F of $\mathbb{R}^n$, there exists an automorphism u of the vector space $\mathbb{R}^n$ such that $u(F) = \mathbb{Z}^p \times \mathbb{R}^r$ with $p,r \geq 0$ and $p+r \leq n$. $\mathbb{Z}^p$ denotes the discrete $\mathbb{Z}$–module with basis $e_1,\dots,e_p$ and $\mathbb{R}^r$ is the vector space spanned by $e_{p+1},\dots,e_{p+r}$, where $e_1,\dots,e_n$ is the canonical basis of $\mathbb{R}^n$.

We refer to chapter XIX in [D,IV] for a detailed proof. This leads to the characterization mentioned above.

Proposition 1.5.2 Every connected commutative Lie group of dimension n is isomorphic to one of the products $(\mathbb{R}/\mathbb{Z})^p \times \mathbb{R}^{n-p}$ where $0 \leq p \leq n$.

Proof: Without loss of generality we may assume $D = \mathbb{Z}^p$. The canonical isomorphisms $\mathbb{R}^n/\mathbb{Z}^p \cong \mathbb{R}^p/\mathbb{Z}^p \times \mathbb{R}^{n-p}$ and $\mathbb{R}^p/\mathbb{Z}^p \cong (\mathbb{R}/\mathbb{Z})^p$ are both Lie group isomorphisms.

Example 2: GL(E), SL(E)

Let E be either a real or complex linear space of dimension n and End E the vector space of all linear maps from E to E. The collection of all invertible endomorphisms of E is denoted by GL(E). We already know, that $GL(E) \subset \text{End } E$ is an open subset and thus a manifold of dimension n^2 if E is real and of dimension $2n^2$ if E is complex. Since the operations in GL(E) – namely the composition and passage to the inverse – are both smooth GL(E) is a Lie group. The subgroup $SL(E) \subset GL(E)$ formed by all $l \in GL(E)$ with $\det l = 1$ is closed and hence by theorem 1.3.8 a Lie subgroup of GL(E).

Let us find the Lie algebras of the Lie groups GL(E) and SL(E). Since $GL(E) \subset \text{End } E$ is open $T_{id}GL(E) = \text{End } E$. To determine the Lie bracket consider two vector fields X,Y on GL(E). Since

$$T\,GL(E) = GL(E) \times \text{End } E$$

any vector field Z on GL(E) is characterized by its principal part, called Z again. Next assume

$$X(g) = DL_g(id)(l_1) \text{ and } Y(g) = DL_g(id)(l_2)$$

for each $g \in GL(E)$ and $l_1,l_2 \in \text{End } E$. Since the left translation

$$L_g : GL(E) \longrightarrow GL(E)$$

maps any h into $g \circ h$ we find that

$$X(g) = g \circ l_1 \text{ and } Y(g) = g \circ l_2$$

for all $g \in GL(E)$. Observe that X and Y depend linearly on g. The Lie bracket [X,Y] at id is thus

$$\begin{aligned}[X,Y](id) &= DY(id)(X(id)) - DX(id)(Y(id)) \\ &= l_1 \circ l_2 - l_2 \circ l_1 .\end{aligned}$$

The Lie algebra of GL(E) is hence End E with the commutator as Lie product. If $H \subset GL(E)$ is a Lie subgroup the Lie algebra $\mathcal{F}$ of H is certainly a subalgebra of End E. Thus the Lie bracket in $\mathcal{F}$ is again the commutator. The Lie algebra sl(E) of SL(E) consists of all traceless endomorphisms in End E. This is due to the fact that for $l \in$ End E

$$\det \exp l = e^{\operatorname{tr} l}$$

as easily seen. Thus dim SL(E) is n^2-1 or $2n^2-2$ accordingly as to whether E is real or complex.

In this section we will prove that GL(E) has two connected components if E is real. Those are $GL^+(E)$ and $GL^-(E)$ containing those transformations with positive and negative determinants respectively.

The situation is different if E is a complex linear space. We will see that GL(E) is connected.

<u>Example 3: 0(p,n–p), the Lorentz group</u>

Our next example again arises from linear algebra. We consider $\mathbb{R}^n$ with its canonical basis $e_1,\dots,e_n$, where e_j is $(0,0,\dots,0,1,0,\dots,0)$ with 1 at the j–th entry. We replace $GL(\mathbb{R}^n)$ by GL(n). Let

$$b : \mathbb{R}^n \times \mathbb{R}^n \longrightarrow \mathbb{R}$$

be the bilinear form mapping any two vectors

$$x = \sum_1^n \alpha^i e_i \text{ and } y = \sum_1^n \beta^j e_j \text{ to}$$

$$b(x,y) = \sum_1^p \alpha^i \cdot \beta^i - \sum_{p+1}^n \alpha^r \cdot \beta^r ,$$

where $p \leq n$ is a fixed natural number. If $p = n$ the bilinear form is just the usual positive definite scalar product on $\mathbb{R}^n$.

By $0(p,n-p)$ we mean the closed subgroup of all linear maps l in $GL(n)$ satisfying

$$b(l(x),l(y)) = b(x,y) .$$

This group carries the topology inherited from $GL(n)$. By theorem 1.3.8 $0(p,n-p)$ is a Lie subgroup of $GL(n)$, called the orthogonal group of b.

The Lie algebra $o(p,n-p)$ consists of all those $l \in \mathrm{End}\, \mathbb{R}^n$ satisfying

$$b(lx,y) + b(x,ly) = 0.$$

If $p = n$ we write $0(n)$ instead of $0(n,n-0)$. Writing any $l \in 0(n)$ as a matrix $M(l)$ with respect to the given basis, we observe that

$$\{M(l) \mid l \in 0(n)\}$$

forms a closed subspace of

$$\{M(l) \mid l \in \mathrm{End}\, \mathbb{R}^n\},$$

where the topology of the latter space is the finest one among all those for which

$$\mathrm{End}\, \mathbb{R}^n \xrightarrow{M} \{M(l) \mid l \in \mathrm{End}\, \mathbb{R}^n\}$$

is continuous. $\mathrm{End}\, \mathbb{R}^n$ carries a natural positive definite scalar product

$$\mathrm{End}\, \mathbb{R}^n \times \mathrm{End}\, \mathbb{R}^n \xrightarrow{< >} \mathbb{R} ,$$

sending (a,c) into

$$<a,c> = \mathrm{tr}\, a \cdot \tilde{c} ,$$

where $\tilde{c}$ is the adjoint of c with respect to b. Since the set

$$\{M(l) \mid l \in 0(n)\}$$

is bounded as well, $0(n)$ is a compact subgroup of $GL(n)$. In general $0(p,n-p)$ is however not compact. In fact we have:

Theorem 1.5.3 $0(m)$ is a compact subgroup of $GL(m)$ for all integers m. Any compact subgroup of $0(p,n-p)$ is contained in a conjugate to $0(p) \times 0(n-p)$.

Proof: Let $H \subset O(p,n-p)$ be a compact subgroup. Assume for a moment the existence of a positive definite bilinear map

$$\bar{b} : \mathbb{R}^n \times \mathbb{R}^n \longrightarrow \mathbb{R}$$

which is left invariant by H. Both b and $\bar{b}$ diagonalize with respect to a common basis $\bar{e}_1,\ldots,\bar{e}_p,\bar{e}_{p+1},\ldots,\bar{e}_n$ as shown in [Gr,1]. Let

$$g : \mathbb{R}^n \longrightarrow \mathbb{R}^n$$

be the linear map sending e_i to $\bar{e}_i$ for all i. Then $g \in O(p,n-p)$ and

$$g\, H\, g^{-1} \subset O(p) \times O(n-p).$$

The proof is completed by the following lemma:

Lemma 1.5.4 Given any compact subgroup $H \subset O(p,n-p)$ there is a positive definite bilinear form

$$\bar{b} : \mathbb{R}^n \times \mathbb{R}^n \longrightarrow \mathbb{R}$$

which is left invariant by H.

Proof: Let $H \subset O(p,n-p)$ and ξ be an element in $\mathcal{T}$, the Lie algebra of H. Since H operates on $\mathbb{R}^n$ by applying each $1 \in H$ to each vector $x \in \mathbb{R}^n$, the set K given by

$$\{(\exp t\xi(x) \mid t \in \mathbb{R}\}$$

is a compact subset of $\mathbb{R}^n$ for each $x \in \mathbb{R}^n$. This subset is contained in the unit ball of b if x is identical with one of the vectors $e_1,\ldots,e_p$ mentioned at the beginning of this section. Since ξ is skew symmetric relative to b each eigenspace of the self–adjoint linear map $\xi^2 \in \operatorname{End} \mathbb{R}^n$ is at least two–dimensional. Moreover each of the vectors $e_1,\ldots,e_p,e_{p+1},\ldots,e_n$ is an eigenvector of ξ^2 (cf. [Gr,1]). Let $b(e_i,e_i) = 1$. Assume (without loss of generality), that $\xi e_i \neq 0$. Then e_i and ξe_i span an invariant plane E of ξ. Clearly $\exp t\xi$ leaves E invariant as well for each $t \in \mathbb{R}$. Since K is compact and contained in the unit ball of b restricted to E, the bilinear map b has to be positive definite on E. Thus H maps $\operatorname{span}(e_1,\ldots,e_p)$ into itself and in analogy we also conclude that it maps $\text{span}(e_{p+1},\dots,e_n)$ into itself. Obviously H leaves $\bar{b} : \mathbb{R}^n \times \mathbb{R}^n \longrightarrow \mathbb{R}$ given by

$$\bar{b}(e_i,e_i) = b(e_i,e_i),\ i = 1,\dots,p$$

$$\bar{b}(e_i,e_i) = -b(e_i,e_i),\ i = p+1,\dots,n$$

invariant.

A special example of a group of type $0(p,n-p)$ is the Lorentz group. Let b be a non degenerate symmetric bilinear form on $\mathbb{R}^4$. Assume that for a suitable basis e_1,e_2,e_3,e_4 b maps $x = \sum_1^4 \alpha^i e_i$ and $y = \sum_1^4 \beta^i e_i$ into

$$b(x,y) = \sum_{i=1}^{3} \alpha^i \cdot \beta^i - \alpha^4 \cdot \beta^4,$$

i.e. b is of index 3 (cf. [Gr,1]) and $(\mathbb{R}^4,b)$ is a <u>Minkowski space</u>. The group $0(3,1)$ is called the <u>Lorentz group</u>.

There is a different description of $0(3,1)$, bringing much more light into the nature of the Lorentz group: Consider the two dimensional complex plane $\mathbb{C}^2$ equipped with a positive definite Hermitian inner product $<,>$ (cf. example 5). On the four dimensional real vector space S, the space of all self–adjoint (complex) linear transformations of $\mathbb{C}^2$ (cf. example 5), we use the following real scalar product: Any two $a,c \in S$ are send into $<a,c>$ given by

$$<a,c> := \frac{1}{2}(\text{tr}(a \circ c) - \text{tr}\ a \cdot \text{tr}\ c).$$

Thus $<,>$ is a non degenerate symmetric bilinear form of index 3. Note that tr is the complex trace, which is real valued on the space of all self–adjoint transformations. Associated with an orthonormal basis $z_1,z_2 \in \mathbb{C}^2$ are the linear maps a_1,a_2,a_3 in S whose matrices are

$$\begin{bmatrix} 1 & 0 \\ 0 & -1 \end{bmatrix}, \begin{bmatrix} 0 & 1 \\ 1 & 0 \end{bmatrix} \text{ and } \begin{bmatrix} 0 & -i \\ i & 0 \end{bmatrix} \text{ respectively.}$$

These matrices are called the Pauli matrices. The maps a_1, a_2, a_3 and id form a basis of S and satisfy

$$\langle a_i, a_j \rangle = \delta_{ij}\,, \text{ for } i,j = 1,2,3$$
$$\langle a_i, id \rangle = 0$$
$$\langle id, id \rangle = -1,$$

where δ_{ij} denotes the Kronecker symbol and id is the identity on $\mathbb{C}^2$. These statements are immediate from the relations

$$\langle a,a \rangle = \tfrac{1}{2}(\text{tr } a^2 - (\text{tr } a)^2) = -\det a$$

and
$$\langle a, id \rangle = -\tfrac{1}{2} \text{ tr } a$$

for all $a \in S$. Clearly $(S, \langle\ \rangle)$ is a Minkowski space [Gr,1].

A Lorentz transformation a, i.e. a linear map on S preserving $\langle , \rangle$, is called orthochroneous if it maps fore–cone to fore–cone and past–cone to past–cone [Gr,1]. Such a transformation still may have a positive or negative determinant (of absolute value one). Call a proper if det a = 1. Such a transformation is of the following form. Let $c : \mathbb{C}^2 \longrightarrow \mathbb{C}^2$ be any linear transformation with det c = 1, then

$$T(c) : S \longrightarrow S$$

given for all $s \in S$ by

$$T(c)\, s := c \circ s \circ \tilde{c}\,,$$

where $\tilde{c}$ denotes the adjoint of c with respect to $\langle , \rangle$, is a proper orthochroneous Lorentz transformation on $(S, \langle , \rangle)$. Vice versa any such transformation is of the form T(c) (cf. [Gr,1]). The group L of all Lorentz transformations of $(S, \langle , \rangle)$ is isomorphic to 0(3,1) and hence a Lie group. Call the collection of all proper orthochroneous Lorentz transformations of S into itself by $L_+^{\uparrow}$. Then observe that for the complex linear space $\mathbb{C}^2$

$$T : SL(2,\mathbb{C}) \longrightarrow L_+^{\uparrow}$$

is a smooth surjective group homomorphism. We use $SL(2,\mathbb{C})$ and $SL(\mathbb{C}^2)$

synonymously. Clearly $L_+^{\uparrow}$, being a closed subgroup of the group L, is a Lie group. The kernel of T consists – as shown in [Gr,1] – of $\{\pm \mathrm{id}\}$. The value of the map T is the following: $SL(2,\mathbb{C})$ is simply connected. This can be seen as follows: Consider the submersion

$$P : SL(2,\mathbb{C}) \longrightarrow \mathbb{C}^2 - \{0,0\}$$

which associates to any $a \in SL(2,\mathbb{C})$ with matrix

$$\begin{bmatrix} \alpha & \beta \\ \gamma & \delta \end{bmatrix}$$

the vector $\alpha z_1 + \gamma z_2$. The fibre $P^{-1}(z)$ is diffeomorphic to $\mathbb{C}$ for each $z \in \mathbb{C}^2 - \{0,0\}$. The map

$$Q : \mathbb{C}^2 - \{0,0\} \longrightarrow SL(2,\mathbb{C})$$

which sends $\alpha^1 z_1 + \gamma^1 z_2$ to the automorphism with matrix

$$\begin{bmatrix} \alpha^1 & -\overline{\gamma}^1/(\alpha\cdot\overline{\alpha} + \beta\cdot\overline{\beta}) \\ \gamma^1 & \overline{\alpha}^1/(\alpha\cdot\overline{\alpha} + \beta\cdot\overline{\beta}) \end{bmatrix}$$

satisfies $P \circ Q = \mathrm{id}$. Since P and Q are smooth $SL(2,\mathbb{C})$ is diffeomorphic to $(\mathbb{C}^2 - \{0,0\}) \times \mathbb{C}$, which itself is diffeomorphic to $S^3 \times \mathbb{R}^3$, where S^3 is the unit sphere of $\mathbb{R}^4$. Hence $SL(2,\mathbb{C})$ is simply connected. This means now, that $L_+^{\uparrow}$ is a connected Lie group with $SL(2,\mathbb{C})$ as its universal covering. On the other hand $L_+^{\uparrow}$ is open in L. Thus $L_+^{\uparrow}$ is the connected component of $\mathrm{id} \in L$. Define furthermore the following three sets

$$L_-^{\uparrow} := \{a \in L \mid \det a = -1, \text{ fore–cone to fore–cone}\}$$

$$L_+^{\downarrow} := \{a \in L \mid \det a = 1, \text{ fore–cone to past–cone}\}$$

$$L_-^{\downarrow} := \{a \in L \mid \det a = -1, \text{ fore–cone to past–cone}\}$$

Thus $L = L_+^{\uparrow} \cup L_-^{\uparrow} \cup L_+^{\downarrow} \cup L_-^{\downarrow}$ is a disjoint union of open sets. Moreover $L_-^{\uparrow}$, $L_+^{\downarrow}$ and $L_-^{\downarrow}$ are images of left translations of $L_+^{\uparrow}$ with the matrices (given with respect to a_1, a_2, a_3 and id)

$$\begin{bmatrix} +1 & & & \\ & -1 & & \\ & & -1 & \\ & & & -1 \end{bmatrix}, \quad \begin{bmatrix} -1 & & & \\ & -1 & & \\ & & -1 & \\ & & & -1 \end{bmatrix} \text{ and } \begin{bmatrix} -1 & & & \\ & +1 & & \\ & & +1 & \\ & & & +1 \end{bmatrix}$$

respectively. Thus we have the following:

Theorem 1.5.5 The Lorentz group $0(3,1)$ being six dimensional has four connected components. The universal covering of $0_{id}(3,1)$, the component of id $\in 0(3,1)$, is diffeomorphic to $SL(2,\mathbb{C})$ which itself is diffeomorphic to $S^3 \times \mathbb{R}^3$. The kernel of the covering map is isomorphic to $\mathbb{Z}_2$.

The orthogonal group $0(n)$ determines the homotopy type of $GL(n)$. The reason of this i:

Proposition 1.5.6 Let $b : \mathbb{R}^n \times \mathbb{R}^n \longrightarrow \mathbb{R}$ be a non degenerate positive definite bilinear form. Denote by $P(n) \subset GL(n)$ the collection of all self–adjoint positive linear isomorphisms of $\mathbb{R}^n$. Then $P(n)$ is contractible to $id_{\mathbb{R}^n}$ and is open within the linear space F of all self–adjoint maps of $\mathbb{R}^n$.

$$\Pi : 0(n) \times P(n) \longrightarrow GL(n)$$

sending each pair (g,f) with $g \in 0(n)$ and $f \in P(n)$ into $g \circ f$ is a diffeomorphism. Thus the decomposition of $1 \in GL(n)$ into $1 = g \circ f$, called the polar decomposition, is unique.

Proof: Clearly $P(n)$ is open in F. Moreover it is connected. This is easily seen as follows: Let $f \in P(n)$. Join the maps $id_{\mathbb{R}^n}$ and f by the line segment $\{(1-t)\, id + t\, f \mid 0 \le t \le 1\}$ in End $\mathbb{R}^n$. The whole segment is in fact in $P(n)$ and thus $P(n)$ is contractible to $id_{\mathbb{R}^n}$. The map Π is surjective: To verify it choose any $a \in GL(n)$. Then $a \circ \tilde{a}$, where $\sim$ denotes the adjoint, is a self–adjoint positive isomorphism. Thus there exists a unique self–adjoint positive square root $p \in P(n)$ of $a \circ \tilde{a}$. Then

$$a = g \circ p$$

with $g = a \circ p^{-1} \in O(n)$. The decomposition of a into $g \circ p$ is unique as easily shown. Since the derivative of Π is invertible everywhere, Π is a diffeomorphism.

The next goal is to show that $GL(n)$ has two components only. Let $GL^+(n)$ denote the set of all automorphisms of $\mathbb{R}^n$ with positive determinant and $SO(n) := O(n) \cap GL^+(n)$. Consider $1 \in GL^+(n)$ and its polar decomposition into $1 = g \circ f$, where $g \in O(n)$ and $f \in P(n)$. Since f has positive determinant, $g \in SO(n)$. Given id and 1 both in $GL^+(n)$ we have a smooth arc

$$\sigma : [0,1] \longrightarrow GL^+(n)$$

with $\sigma(0) = id$ and $\sigma(1) = 1$. The arc σ is given by composing some smooth curve

$$\beta : [0,1] \longrightarrow SO(n)$$

satisfying $\beta(0) = id$ and $\beta(1) = g$ ($SO(n)$ is compact and connected as seen below) and

$$\gamma : [0,1] \longrightarrow P(n).$$

γ sends any $t \in [0,1]$ to

$$\gamma(t) = (1-t)id + tf.$$

As mentioned $\sigma(t) = \beta(t) \circ \gamma(t)$ for all $t \in [0,1]$. This means that $GL^+(n)$ and in turn $GL^-(n) := \{1 \in GL(n) \mid \det 1 < 0\}$ are connected.

Example 4: SO(n)

Again consider $\mathbb{R}^n$ and a non degenerate symmetric bilinear form

$$b : \mathbb{R}^n \times \mathbb{R}^n \longrightarrow \mathbb{R} .$$

of type $(p,n-p)$. Write $SL(n)$ instead of $SL(\mathbb{R}^n)$. The intersection $O(p,n-p) \cap SL(n)$ is a closed subgroup of $O(p,n-p)$ and thus a Lie group. It is called $SO(p,n-p)$.

If b is positive definite $SO(n)$ is, as a closed Lie subgroup of $O(n)$, compact (theorem 1.5.3). Let us determine its connected component. To this end we introduce the concept of a maximal torus. By proposition 1.5.2 any connected abelian Lie subgroup of a compact Lie group is a torus. Such a torus is called maximal if it is not properly contained in any other one. To characterize maximal tori in SO(n), choose an orthonormal basis $e_1,\dots,e_n$ in $\mathbb{R}^n$. If n is even then the set consisting of all linear maps in SO(n) whose matrices with respect to the given basis are of the form

$$\begin{bmatrix} A_1 & & 0 \\ & \ddots & \\ 0 & & A_{\frac{n}{2}} \end{bmatrix},$$

is a torus. Here A_j is a matrix of the type

$$\begin{bmatrix} \cos\theta_j & \sin\theta_j \\ -\sin\theta_j & \cos\theta_j \end{bmatrix}.$$

In case n is odd the matrix of a linear map in T looks like

$$\begin{bmatrix} A_1 & & 0 & 0 \\ & \ddots & & \\ 0 & & A_{\frac{n-1}{2}} & 0 \\ 0 & & 0 & 1 \end{bmatrix}$$

where A_j again is as above. We will show now, that for a maximal torus T

$$SO(n) = \bigcup_{g \in SO(n)} g \cdot T \cdot g^{-1}$$

holds. As shown in [Gr,1] to every element $g \in O(n)$ there exists an orthogonal decomposition of $\mathbb{R}^n$ into stable subspaces F_i of dimension one and two. A subspace $F \subset \mathbb{R}^n$ is stable under $g \in \operatorname{End} \mathbb{R}^n$ if $g(F) \subset F$. Choose now an orthonormal basis $v_1,\dots,v_n$ of $\mathbb{R}^n$ such that each of the vectors is contained in one of the stable subspaces F_i. Thus the matrix of $g \in O(n)$ looks like

$$\begin{bmatrix} A_1 & & 0 & & & 0 \\ & \ddots & & & & \\ 0 & & A_p & & & \\ & & & \epsilon_{2p+1} & & 0 \\ & & & & \ddots & \\ 0 & & & 0 & & \epsilon_n \end{bmatrix}$$

where A_j is of the form

$$\begin{bmatrix} \cos\theta_j & \sin\theta_j \\ -\sin\theta_j & \cos\theta_j \end{bmatrix}$$

and $\epsilon_i = \pm 1$ for $i = 2p+1,\ldots,n$. Now given $g \in SO(n)$ it is easy to find $g_1 \in SO(n)$ such that

$$g \in g_1 \circ T \circ g_1^{-1}.$$

Since $\bigcup_{g \in SO(n)} g \circ T \circ g^{-1}$ is connected we have:

Proposition 1.5.7 $SO(n)$ is connected and hence is the connected component of $O(n)$ containing id.

Example 5: Pin(p,n–p) and Spin(p,n–p)

Here we follow [Gr,2]. Suppose, $b : E \times E \longrightarrow \mathbb{R}$ is a non–degenerate symmetric bilinear map of index p defined on some n–dimensional real vector space E. The tensor algebra $\otimes E$ is given by

$$\otimes E = \sum_{k=0}^{\infty} \otimes^k E,$$

where $\otimes^k E$ is the k–fold tensor product of E with itself and $\otimes^0 E$ stands for $\mathbb{R}$. Each summand $\otimes^k E$ is an $\mathbb{R}$–linear space, thus $\otimes E$ has a $\mathbb{R}$–linear structure. The multiplication is $\otimes$. Thus $\otimes E$ is an associative algebra over $\mathbb{R}$ with a unit element 1. The set

$$\{x \otimes x - b(x,x) \cdot 1 \mid x \in E\}$$

generates an ideal. $\otimes E$ modulo this ideal is an associative 2^n–dimensional $\mathbb{R}$–algebra with a unit element 1, called the Clifford–algebra of b. It is denoted by C_E. This algebra has a characteristic universal property: Given an associative $\mathbb{R}$–algebra A with a unit element e_A. To a linear map

$$a : E \longrightarrow A$$

satisfying $(a(x))^2 = b(x,x) \cdot e_A$ exists a unique $\mathbb{R}$–algebra homomorphism $\bar{a} : C_E \longrightarrow A$, which coincides with a on the generating set E. We will introduce now a few notions and state without proofs some of their properties. As a general reference we suggest [Gr,2].

The map

$$w : E \longrightarrow E,$$

given by $w(x) = -x$ is an isometry of b, meaning

$$b(w(x),w(y)) = b(x,y)$$

for all $x,y \in E$. It induces a homomorphism

$$w_E : C_E \longrightarrow C_E$$

with the property

$$w_E^2 = id_{C_E} .$$

This is the reason why w_E is called the degree involution. It induces a $\mathbb{Z}_2$–gradation of C_E, meaning that

$$C_E = C_E^o \oplus C_E^1$$

with $C_E^o = \ker(w_E - id_{C_E})$ and $C_E^1 = \ker(w_E + id_{C_E})$. The multiplicative group C_E^* of all invertible elements in C_E, i.e. of all elements $c \in C_E$ which allow an inverse c^{-1} satisfying $c \cdot c^{-1} = 1$, admits a natural representation

$$ad : C_E^* \longrightarrow GL(C_E).$$

By $GL(C_E)$ we mean the group of all invertible $\mathbb{R}$–linear automorphisms of C_E. This map ad, called the twisted adjoint representation, is then given by

$$ad(a)(u) = w_E(a) \cdot u \cdot a^{-1}$$

where a varies in C_E^* and u in C_E. Within C_E^* there is a subgroup, denoted by Γ_E, consisting of all those elements $a \in C_E^*$ with

$$ad(a)(E) \subset E.$$

The group Γ_E is called the Clifford group. It is generated by all those $e \in E$ yielding $b(e,e) \neq 0$. Each element $a \in \Gamma_E$ defines an isometry on E by

$$\tau_a = ad(a) \mid E.$$

Thus we have a homomorphism

$$\phi_E : \Gamma_E \longrightarrow 0(p,n-p)$$

where $0(p,n-p)$ means the group of all isometries of b on E. This homomorphism is surjective. Let

$$\lambda_E : \Gamma_E \longrightarrow \mathbb{R}\backslash\{0\}$$

be the map which is determined on the generating set $\{e \in E \mid b(e,e) \neq 0\}$ of Γ_E by

$$\lambda_E(e) = -b(e,e).$$

Now ϕ_E is still surjective on

$$\mathrm{Pin}(p,n-p) := \{a \in \Gamma_E \mid \lambda_E(a) = +1\}$$

and its kernel consists of 1 and -1, i.e. of a group isomorphic to $\mathbb{Z}_2$. Thus we have the exact sequence

$$1 \longrightarrow \mathbb{Z}_2 \longrightarrow \mathrm{Pin}(p,n-p) \xrightarrow{\varphi} 0(p,n-p) \longrightarrow 1,$$

where $\varphi = \phi_E \mid \mathrm{Pin}(p,n-p)$. Since $\mathrm{Pin}(p,n-p)$ covers $0(p,n-p)$ two fold, it has a unique Lie group structure making φ to a local diffeomorphism. $\mathrm{Pin}(p,n-p)$ is compact if $p = n$ By $\mathrm{Spin}(p,n-p)$ we mean $\varphi^{-1}(SO(p,n-p))$. Thus we again have

$$1 \longrightarrow \mathbb{Z}_2 \longrightarrow \mathrm{Spin}(p,n-p) \xrightarrow{\varphi} SO(p,n-p) \longrightarrow 1.$$

Clearly $\mathrm{Spin}(p,n-p)$ is a Lie group and compact if $p = n$. Moreover as shown in [A,B,S] $\mathrm{Spin}(p,n-p)$ is simply connected.

Example 6: U(n), SU(n)

On a complex linear space E we have given a positive–definite Hermitian inner product, i.e. a bilinear map

$$h : E \times E \longrightarrow E$$

such that

$$h(\lambda x_1 + \mu x_2, y) = \lambda h(x_1,y) + \mu h(x_2,y)$$
$$h(x, \lambda y_1 + \mu y_2) = \bar{\lambda} h(x,y_1) + \bar{\mu} h(x,y_2)$$
$$h(x,y) = \overline{h(y,x)} \text{ (hence } h(x,x) \in \mathbb{R})$$

$$h(z,z) > 0 \text{ for all } z \in E \text{ with } z \neq 0$$

for all $x,x_1,x_2,y,y_1,y_2 \in E$ and for all $\lambda,\mu \in \mathbb{C}$. Clearly $(x,y) := \Sigma\, \xi^i \cdot \bar{\eta}^i$ where $x = (\xi^1,\ldots,\xi^n)$ and $y = (\eta^1,\ldots,\eta^n)$ is a positive–definite Hermitian inner product on $\mathbb{C}^n$. Call a linear map l from E into itself unitary if

$$h(lx,ly) = h(x,y) \text{ for all } x,y \in E.$$

The collection $U_h(E)$ of all unitary endomorphisms on E forms a closed subgroup of GL(E) and is hence a Lie subgroup of GL(E).

Observe that for any two positive–definite Hermitian inner products h_1 and h_2 the Lie groups $U_{h_1}(E)$ and $U_{h_2}(E)$ are isomorphic. By U(n) we mean $U_{(,)}(\mathbb{C}^n)$. It is called the unitary group of $\mathbb{C}^n$.

The collection of all unitary maps in U(n) with determinant equal to one, the special unitary group of $\mathbb{C}^n$, is denoted by SU(n). Clearly SU(n) is a Lie subgroup of U(n).

The Lie algebra u(n) of U(n) consists of the collection of all skew Hermitian endomorphisms of $\mathbb{C}^n$ with respect to (,), i.e.:

$$u(n) = \{l \in \text{End } \mathbb{C}^n \mid (lx,y)+(x,ly) = 0 \text{ for all } x,y \in \mathbb{C}^n\}.$$

The Lie product is the commutator.

A map $l \in \text{End } \mathbb{C}^n$ is called self–adjoint with respect to (,), if $(lx,y) = (x,ly)$ for all $x,y \in \mathbb{C}^n$. Thus $l \in u(n)$ iff $(i\cdot l)^\sim = -i\cdot\tilde{l}$. Here ~ denotes the adjoint again. We denote by $P(\mathbb{C}^n)$ the collection of all self–adjoint maps in End $\mathbb{C}^n$. Then End $\mathbb{C}^n = u(n) \oplus P(\mathbb{C}^n)$.

Now we easily deduce, that U(n) is n^2– and SU(n) is (n^2-1)–dimensional. In addition both U(n) and SU(n) are compact, since every eigenvalue of a unitary mapping has norm one.

Next we will see that U(n) and SU(n) both are connected. Again we proceed by covering both groups by conjugates of their respective maximal tori. The collection of all maps in U(n), whose matrices with respect to the canonical basis in $\mathbb{C}^n$ have the form

$$\begin{bmatrix} e^{i\theta_1} & & 0 \\ & \ddots & \\ 0 & & e^{i\theta_n} \end{bmatrix}$$

with $\theta_1,\dots,\theta_n \in \mathbb{R}$, forms a maximal torus T. $T \cap SU(n)$ consists of these maps which in addition satisfy $\sum_{s=1}^{n} \theta_s = 0$. It is left as a (simple) exercise to show that

$$U(n) = \bigcup_{g \in U(n)} g \cdot T \cdot g^{-1}$$

and

$$SU(n) = \bigcup_{g \in SU(n)} g \cdot T \cdot g^{-1} .$$

Hence both groups U(n) and SU(n) are connected.

<u>Remark 1</u> The analogue of proposition 1.5.6 is also valid for $GL(\mathbb{C}^n)$, U(n) and $\mathbb{P}(\mathbb{C}^n)$ where $\mathbb{P}(\mathbb{C}^n)$ denotes the collection of all positive Hermitian maps on $\mathbb{C}^n$. Thus any $l \in GL(\mathbb{C}^n)$ can uniquely be written as

$$l = g \circ f,$$

where $g \in U(n)$ and $f \in \mathbb{P}(\mathbb{C}^n)$.

Now we proceed as in the case of a real vector space E to show that $GL(\mathbb{C}^n)$ and $SL(\mathbb{C}^n)$ are connected.

<u>Example 7: Sp(n), the real symplectic group</u>

Regard $\mathbb{R}^{2n}$ as $\mathbb{R}^n \times \mathbb{R}^n$. Any vector $x \in \mathbb{R}^{2n}$ can be represented as a pair (x_1,x_2) where $x_1,x_2 \in \mathbb{R}^n$. A non degenerate skew symmetric bilinear form

$$w_0: \mathbb{R}^{2n} \times \mathbb{R}^{2n} \longrightarrow \mathbb{R}$$

is given by

$$w_0((x_1,x_2),(y_1,y_2)) = (x_1,y_2) - (y_1,x_2),$$

where (,) is a positive definite scalar product on $\mathbb{R}^n$. Clearly (,) can be extended to $\mathbb{R}^n \times \mathbb{R}^n$ by choosing (,) on each factor. Given vice versa any non degenerate skew symmetric bilinear form

$$w : E \times E \longrightarrow \mathbb{R}$$

and a positive definite scalar product $<,>$ on a 2n–dimensional real vector space E, we find a w–adapted splitting $E = E_1 \oplus E_2$ as follows: Represent w via $<,>$ by an isomorphism $J : E \longrightarrow E$, which means, that

$$w(x,y) = <Jx,y>$$

for all $x,y \in E$. Since J is a skew symmetric isomorphism we have

$$E = \bigoplus_{j=1}^{n} E_{\aleph_j}$$

where $E_{\aleph_1},\ldots,E_{\aleph_n}$ are the invariant two planes of J, in which J^2 has eigenvalue $-(\aleph_j^2)$ (cf. [Gr, 1]).

Moreover, the planes are mutually orthogonal. Each of these invariant planes $E_{\aleph_j}$ splits into

$$E_{\aleph_j} = \mathbb{R}\cdot e_j \oplus \mathbb{R}\cdot f_j$$

where $e_j = \aleph_j^{-\frac{1}{2}}\cdot\bar{e}_j$ and $f_j = \aleph_j^{-\frac{1}{2}}\cdot\bar{f}_j$, and $\bar{e}_j$ and $\bar{f}_j$ are orthonormal with respect to $<,>$. Then $Je_j = f_j$ and $Jf_j = -e_j$ where $J = \aleph_j^{-1}J$ on $E_{\aleph_j}$. Let next E_1 be the span of $e_1,\ldots,e_n$ and E_2 be the span of $f_1,\ldots,f_n$. Then on $E = E_1 \oplus E_2$, where

$\dim E_1 = \dim E_2 = n$

$$w : E_1 \times E_2 \times E_1 \times E_2 \longrightarrow \mathbb{R}$$

has the property that

$$w(e_i,e_j) = \aleph_i<Je_i,e_j> = \aleph_i<f_i,e_j> = 0$$

$$w(f_i,f_j) = \aleph_i<Jf_i,f_j> = -\aleph_i<e_i,f_j> = 0$$

and

$$w(e_i,f_j) = \aleph_i<je_i,f_j> = \aleph_i<f_i,g_j> = \delta_{ij}$$

$$w(f_i,e_j) = \aleph_i<Jf_i,e_j> = -\aleph_i<e_i,e_j> = -\delta_{ij}.$$

The basis $e_1,\ldots,e_n,f_1,\ldots,f_n$ is called a symplectic basis on E. The two form w applied to $x = \Sigma\, \alpha^i e_i + \Sigma \beta^i f_i$ and $y = \Sigma\, \varphi^j e_j + \Sigma\, \Psi^j f_j$ yields

$$w(x,y) = w((x_1,x_2),(y_1,y_2)) = \Sigma\, \alpha^i \Psi^i - \Sigma\, \varphi^i \beta^i$$

with $x_1 = \Sigma\, \alpha^i e_i$, $x_2 = \Sigma\, \beta^i f_i$, $y_1 = \Sigma\, \varphi^j e_j$ and $y_2 = \Sigma\, \Psi^j f_j$. If E_1 is identified with E_2 by sending e_j into f_j for each j then

$$w(x,y) = (x_1,y_2) - (y_1,x_2)$$

where (,) is the positive definite scalar product on $E_1 = E_2$ given by $(e_i,e_j) = \delta_{ij}$ for all i,j.

Thus any non–degenerate skew symmetric bilinear map on E is of the form w_0 if E is identified with $\mathbb{R}^{2n}$. The collection of all linear transformations $a \in \text{End}\, \mathbb{R}^{2n}$ which preserve w_0 forms a group, denoted by Sp(n). Certainly $Sp(n) \subset GL(2n)$ is closed and thus a Lie group.

The Lie algebra sp(n) consists of all linear maps $l \in \text{End}\, \mathbb{R}^n$ which satisfy

$$w_0(lx,y) + w_0(x,ly) = 0.$$

If $J \in \text{End}\, \mathbb{R}^{2n}$ again yields

$$w_0(x,y) = (Jx,y)$$

for all $x,y \in \mathbb{R}^{2n}$, then $a \in Sp(n)$ iff

$$\tilde{a}\, J\, a = J$$

where ~ denotes the adjoint with respect to (,). Clearly

$$J^2 = -\mathrm{id}\ .$$

Hence

$$w_0(Jx,Jy) = -(x,Jy) = (Jx,y) = w_0(x,y).$$

Thus $J \in Sp(n)$. On the other hand

$$w_0(Jx,y) + w_0(x,Jy) = -\langle x,y\rangle + \langle Jx,Jy\rangle = 0.$$

Hence $J \in sp(n)$ also. Furthermore by [Gu,St]

$$J = \exp(\tfrac{\pi}{2} J).$$

An element $l \in sp(n)$ has to fulfil

$$((J \circ l)x,y) + (Jx,ly) = 0$$

for all $x,y \in \mathbb{R}^{2n}$. Thus it satisfies

$$J \circ l = -\tilde{l} \circ J,$$

from this one concludes

$$\dim Sp(n) = n\cdot(2n+1).$$

1.6 Aut G for a connected G

Let G be a connected Lie group. By Aut G we mean the collection of all (smooth) Lie group isomorphisms of G onto itself. We will show that Aut G carries a natural Lie group structure.

To this end assume G to be simply connected. Each $h \in \mathrm{Aut}\, G$ defines Th(e), a Lie algebra isomorphism of the Lie algebra $\mathcal{G}$ onto itself. Call the collection of all Lie algebra isomorphisms of $\mathcal{G}$ onto itself, i.e. the collection of all Lie algebra automorphisms of $\mathcal{G}$ by Aut $\mathcal{G}$. The map

$$a\colon \mathrm{Aut}\, G \longrightarrow \mathrm{Aut}\, \mathcal{G}$$

sending each h to Th(e) is bijective by theorem 1.4.3. Aut $\mathcal{G}$ being a closed subgroup of GL($\mathcal{G}$), the group of all linear isomorphisms of $\mathcal{G}$ onto itself, carries a Lie group structure. Thus we have on Aut G a unique Lie group structure such that a is a (smooth) Lie group isomorphism.

Next we drop the assumption that G is simply connected. The universal covering $\tilde{G}$ of G is certainly a Lie group and the covering homomorphism $\pi: \tilde{G} \longrightarrow G$ a Lie group homomorphism. Since $\tilde{G}$ is connected and simply connected Aut $\tilde{G}$ is a Lie group. Now π induces a homomorphism

$$\pi^* : \text{Aut } G \longrightarrow \text{Aut } \tilde{G}$$

in the following way. Given h $\in$ Aut G, we compose it with π to $h \circ \pi : \tilde{G} \longrightarrow G$. By the lifting property of $\tilde{G}$ we can lift $h \circ \pi$ uniquely to a homomorphism $\pi^*(h) : \tilde{G} \longrightarrow \tilde{G}$. By theorem 1.4.1 $\pi^*(h)$ is in fact a member of Aut $\tilde{G}$. By the same theorem again we conclude that π^* is injective. π^*(Aut G) consists of all those elements in Aut $\tilde{G}$ which preserve the kernel of π. Since $\pi^*(\text{Aut } G) \subset \text{Aut } \tilde{G}$ is closed, we have

<u>Proposition 1.6.1</u> Aut G is a Lie group. π^* is a smooth diffeomorphism of Aut G onto $\pi^*(\text{Aut } G) \subset \text{Aut } \tilde{G}$.

Finally we determine the Lie algebra of Aut G in case G is connected and simply connected. The Lie algebra of Aut $\mathcal{G}$ has to be a subalgebra of End $\mathcal{G}$, the collection of all linear maps of $\mathcal{G}$ with the commutator as Lie product. Thus the Lie algebra of Aut $\mathcal{G}$ consists of all those elements $l \in \text{End } \mathcal{G}$ for which $\exp tl \in \text{Aut } \mathcal{G}$ for all $t \in \mathbb{R}$. This means, that for each $\xi,\eta \in \mathcal{G}$.

$$\exp tl[\xi,\eta] = [(\exp tl)\xi, (\exp tl)\eta].$$

Differentiating this with respect to t at zero yields

$$l[\xi,\eta] = [l\xi,\eta] + [\xi,l\eta].$$

Thus l is, by definition, a derivation of $\mathcal{G}$. Conversely, let $l \in \text{End}\,\mathcal{G}$ be a derivation of $\mathcal{G}$. Then consider for each t the vector $\nu(t)$ given by

$$\nu(t) = (\exp tl)[\xi,\eta] - [(\exp tl)\xi,(\exp tl)\eta].$$

Differentiating both terms at t yields due to $\frac{d}{dt}\exp tl|_t = l \circ \exp tl$

$$l(\exp tl)[\xi,\eta] - l[\exp tl)\xi,(\exp tl)\eta] = \frac{d\ \nu(t)}{dt}|_t .$$

Since $\nu(0) = 0$ the differential equation

$$l\,\nu(t) = \frac{d\ \nu(t)}{dt}|_t$$

has the unique solution $\nu(t) = 0$ for all $t \in \mathbb{R}$. Thus

$$(\exp tl)[\xi,\eta] = [(\exp tl)\xi,(\exp tl)\eta],$$

meaning that $\exp tl \in \text{Aut}\,\mathcal{G}$ for all t. If $D\mathcal{G}$ denotes the Lie algebra of all derivations of $\mathcal{G}$ with the commutator as its Lie product we have

Proposition 1.6.2 Let $\mathcal{G}$ be a finite dimensional Lie algebra. The Lie algebra of Aut $\mathcal{G}$ is $D\mathcal{G}$.

This immediately yields via $a : \text{Aut}\, G \longrightarrow \text{Aut}\,\mathcal{G}$

Corollary 1.6.3 The Lie algebra of Aut G for a connected and simply connected Lie group G is isomorphic to $D\mathcal{G}$.

1.7 The semidirect product

We recall the definition of a semidirect product of groups. Let G be a group. Given any two subgroups H and N of which N is normal in G, then G is said to be the semidirect product of H and N, if

$$G = H \cdot N$$

and

$$H \cap N = \{e\}.$$

Observe that $H \cdot N = N \cdot H$ since N is normal in G. Assume now that G is the semidirect product of H and N. Each $g \in H$ defines an inner automorphism τ_g of G by $\tau_g(x) = gxg^{-1}$. Clearly $\tau_g(N) \subset N$. Since $\tau_g \cdot \tau_{g'} = \tau_{gg'}$ for $g,g' \in H$, the map $\sigma : H \longrightarrow \mathcal{A}(N)$, assigning to each $g \in H$ the automorphism τ_g is a homomorphism of groups. By $\mathcal{A}(N)$ we mean the group of all automorphisms of N.

Next let H and N be any groups and

$$\sigma : H \longrightarrow \mathcal{A}(N)$$

a homomorphism. This means that H acts on N. On the cartesian product of the sets N and H we introduce a group structure as follows

$$(x,y)\cdot(x',y') = (x\cdot\sigma(y)(x'),yy')$$

with $(x,y),(x',y')$ being any two tuples in $N \times H$. It is an easy exercise to check the group axioms. The group just introduced is denoted by $N \times_\sigma H$. If we introduce the subgroups $\bar{N}$ and $\bar{H}$ given by

$$\bar{N} = \{(x,e_H) \mid x \in N\} \text{ and } \bar{H} = \{(e_N,y) \mid y \in H\}$$

with e_H and e_N the neutral elements of H and N respectively, then

$$N \times_\sigma H = \bar{N} \cdot \bar{H} .$$

Moreover $\bar{N} \cap \bar{H} = \{(e_N,e_H)\}$. Therefore we call $N \times_\sigma H$ the semidirect product of N and H (with respect to σ). If σ is the constant map, assigning the identity as value, then $N \times_\sigma H$ is denoted by $N \times H$ and is called the direct product of N and H.

As an example of a semidirect product consider the group G of all Euclidean motions in an Euclidean space $(\mathbb{R}^n,<,>)$. By $<,>$ we mean a positive–definite scalar product on $\mathbb{R}^n$. The scalar product $<,>$ yields a metric on $\mathbb{R}^n$ (cf. [K]), namely

$$\varphi(x,y) = ||x-y||$$

for all $x,y \in \mathbb{R}^n$. A map $b : \mathbb{R}^n \longrightarrow \mathbb{R}^n$ is said to preserve φ if

$$\varphi(b(x),b(y)) = \varphi(x,y)$$

for all $x,y \in \mathbb{R}^n$.

An Euclidean motion on $(\mathbb{R}^n,\varphi)$ is a φ preserving map $a : \mathbb{R}^n \longrightarrow \mathbb{R}^n$, which is of the form

$$a(x) = l(x) + x'$$

where $l \in \text{End}\, \mathbb{R}^n$ and x' is a fixed element in $\mathbb{R}^n$. Certainly $l : \mathbb{R}^n \longrightarrow \mathbb{R}^n$ has to be a linear isomorphism. In fact it has to be orthogonal, meaning that

$$<lx,ly> = <x,y>$$

for all $x,y \in \mathbb{R}^n$ as easily checked. Call for a fixed $x' \in \mathbb{R}^n$

$$t_{x'} : \mathbb{R}^n \longrightarrow \mathbb{R}^n$$

mapping any $x \in \mathbb{R}^n$ to $t_{x'}(x) = x + x'$ a translation; it is uniquely determined by $x' \in \mathbb{R}^n$. Thus any Euclidean motion a splits uniquely into $a = t_{x'} \circ l$. Observe, that l itself is an Euclidean motion. Call E(n) the group of all Euclidean motions, 0(n) the group of all orthogonal transformations of $(\mathbb{R}^n,<,>)$ and denote the group of all translations by $\mathbb{R}^n$. Then

$$E(n) = \mathbb{R}^n \cdot 0(n)$$

$\mathbb{R}^n \subset E(n)$ is a normal subgroup and id is the only element in $0(n) \cap \mathbb{R}^n$. Thus E(n) is the semidirect product of 0(n) and $\mathbb{R}^n$. The semidirect product of $\mathbb{R}^4$ and the Lorentz group (cf. examples) is called the <u>Poincaré group</u>.

Now we turn over to the semidirect product of Lie groups. Assume therefore, that G is a connected Lie group and H a connected Lie group embedded in G. Moreover let $N \subset G$ be a closed, connected, normal subgroup. Finally assume G to be the semidirect product of H and N. Clearly the canonical projection $\pi: G \longrightarrow G/N$ provides a Lie group isomorphism of H onto G/N. σ: H $\longrightarrow$ Aut N given by $\sigma(h) = \tau_h$ is a Lie group homomorphism. It yields the semidirect product $N \times_\sigma H$ of groups. This group carries a manifold structure, namely the cartesian product of the manifolds N and H. Since the map $\tilde{\sigma} : N \times H \longrightarrow N$ sending each tuple (g,h) with g $\in$ N and h $\in$ H into $\sigma(h)(g)$ is smooth in h and g the group operations on $N \times_\sigma H$ are smooth. Thus $N \times_\sigma H$ is a Lie group. It is easy to verify:

Proposition 1.7.1 If G is the semidirect product of N and H where H is identified with the image of the embedding the map

$$c : N \times_\sigma H \longrightarrow G$$

sending $(g,h) \in N \times_\sigma H$ into $g \cdot h$ is a Lie group isomorphism (and hence is a diffeomorphism).

In conclusion of this section we turn to the structure of the Lie algebra of the semidirect product. Under the hypothesis made above, n, the Lie algebra of N, is an ideal in $\mathcal{G}$ as assured by theorem 1.3.10. By corollary 1.6.3 the tangent map $T\sigma(e)$ of σ at e can be thought as having values in Dn, thus

$$T\sigma(e): \mathcal{F} \longrightarrow Dn$$

where $\mathcal{F}$ denotes the Lie algebra of H. If $\xi, \xi' \in n$ and $\eta, \eta' \in \mathcal{F}$, then

$$[\xi+\eta, \xi'+\eta'] = [\xi,\xi'] + (T\sigma(e)\eta)\xi' - (T\sigma(e)\eta')\xi + [\eta,\eta'].$$

If we introduce on the vector space $n \times \mathcal{F}$ the Lie product defined by the equation

$$[(\xi,\eta),(\xi',\eta')] = ([\xi,\xi'] + (T\sigma(e)\eta)(\xi') - (T\sigma(e)\eta')(\xi), [\eta,\eta'])$$

we call the resulting Lie algebra by $n \times_{T\sigma(e)} \mathcal{F}$, the semidirect product of the Lie algebras n and $\mathcal{F}$ (cf. [D,IV]).

The Euclidean group E(n) is certainly a Lie group. We have it as an exercise to determine its Lie algebra.

References, Chapter 1

[D,I] Dieudonné, J. Foundations of Modern Analysis (vol I), Academic Press, New York and London, (1969).

[D,III] Dieudonné, J. Treatise on Analysis (vol III), Academic Press, New York and London, (1972).

[D,IV] Dieudonné, J. Treatise on Analysis (vol IV), Academic Press, New York and London, (1974).

[G] Greenberg, M. Lectures on Algebraic Topology, W.A. Benjamin, Inc., Menlo Park, California, (1977).

[Gr,1] Greub, W. Linear Algebra, (4 ed.), Springer Verlag, New York, Heidelberg, Berlin, (1975).

[Gr,2] Greub, W. Multilinear Algebra, (2nd ed.), Springer–Verlag, New York, Heidelberg, Berlin, (1978).

[Gu,St] Guillemin, V. Sternberg, S. Geometric Asymptotics Mathematical Surveys, Number 14, AMS, Providence, Rhode Island, (1977).

[Hi] Hirsch, M.W. Differential Topology, Springer Verlag, New York, Heidelberg, Berlin, (1976).

[K] Kelley, J.L. General Topology, Springer Verlag, New York, Heidelberg, Berlin, (1975).

[L] Lang, S. Differential Manifolds, Addison Wesley Publishing Company, Reading, Massachusetts, Menlo Park, California, (1972).

[Sp] Spanier, E.H. Algebraic Topology, Springer Verlag, New York, Heidelberg, Berlin, (1981).

[W] Warner, F.W. Foundations of Differentiable Manifolds and Lie Groups, Springer–Verlag, New York, Heidelberg, Berlin, (1983).

Chapter 2

VECTOR BUNDLES

2.1 Fibre bundles

Let M,N and F be smooth manifolds together with a smooth (surjective) projection $\pi: N \longrightarrow M$. The map π has the local product property in $p \in M$ with respect to F if there is an open set $U \in M$ containing p and a smooth diffeomorphism

$$\pi^{-1}(U) \xrightarrow{\Psi} U \times F$$

such that

$$\begin{array}{ccc} \pi^{-1}(U) & \xrightarrow{\Psi} & U \times F \\ & {\scriptstyle \pi} \searrow \quad \swarrow {\scriptstyle pr_1} & \\ & U & \end{array}$$

commutes. pr_1 means the projection onto the first factor U. Such a chart U is called a trivializing chart and Ψ a trivializing diffeomorphism for U. The pair (U,Ψ) is called a local trivializing representation.

(N,π,M,F) is called a (smooth) fibre bundle over M if π has the local product property with respect to F in any point $p \in M$. We call N the total space, M the base space and F the typical fibre of the bundle. The set $\pi^{-1}(p)$, denoted by N_p, is a closed smooth submanifold of N diffeomorphic to F for any $p \in M$ and is called the fibre over p. As a short hand for (N,π,M,F) we use N if no misunderstanding will arise.

Clearly the notion of a fibre bundle can be generalized to manifolds with boundary.

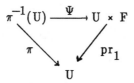

As an example consider a Lie group G and a closed subgroup H. As shown in section 1.3 the coset space G/H carries a natural manifold structure. We will show that $(G,\pi,G/H,H)$ is a fibre bundle. Given any $p \in G/H$ we will establish the existence of a (local) smooth section $s : V \longrightarrow G$, where $V \subset G/H$ is a suitable open neighbourhood of p. "<u>Section</u>" means $\pi \circ s = id_V$. Choose some $g \in G$ with $\pi(g) = p$. Then by the implicite function theorem there is an open neighbourhood of g of the form $V \times W$ where V is an open neighbourhood of g in the coset gH and W is a subset of G which is mapped by π diffeomorphically onto an open neighbourhood U of p. Thus $s = \pi_{|W}^{-1} : U \longrightarrow W \subset G$ is a smooth section. In fact for any $h \in H$

$$s\cdot h : U \longrightarrow G$$

given by $(s\cdot h)(v) = s(v)\cdot h$ is also a smooth section. Thus we have a smooth map

$$\pi^{-1}(U) \xrightarrow{\Psi} U \times H$$

constructed as follows. Let $g' \in \pi^{-1}(U)$. Then there is a unique $h \in H$ with $g'\cdot h \in W$. Then set

$$\Psi(g') = (\pi(g'),h).$$

The rest of the definition of a fibre bundle is easily checked for G/H.

A fibre bundle (N,π,M,F) is called <u>trivial</u> if there is a smooth diffeomorphism φ : $N \longrightarrow M \times F$ for which

$$\begin{array}{ccc} N & \xrightarrow{\varphi} & M \times F \\ & {\scriptstyle\pi}\searrow \quad \swarrow{\scriptstyle pr_1} & \\ & M & \end{array}$$

commutes. This notion clarifies the term "trivializing representation".

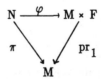

2.2 Vector bundles

A smooth fibre bundle $(\mathbb{E},\pi,M,E)$ is called a vector bundle, if E and $\mathbb{E}_p = \pi^{-1}(p)$ are real vector spaces for all $p \in M$ and moreover, if there is a covering collection of trivializing representations (U_α,Ψ_α) such that each trivializing diffeomorphism

$$\Psi_\alpha : \pi^{-1}(U_\alpha) \longrightarrow U_\alpha \times E$$

is fibrewise linear, i.e. the map

$$\Psi_{\alpha,p} : \mathbb{E}_p \longrightarrow E\,,$$

given by $\Psi_{\alpha,p} = pr_2 \circ \Psi_{\alpha|\mathbb{E}_p}$, is a linear isomorphism for every $p \in U_\alpha$. Here α varies in an index set J. By "covering" we mean

$$\bigcup_{\alpha \in J} U_\alpha = M.$$

Call $(\mathbb{E},\pi,M,E)$ trivial if it is trivial as a smooth fibre bundle in which $\varphi_p : \mathbb{E}_p \longrightarrow E$ is a linear isomorphism for each $p \in M$, where $\varphi : \mathbb{E} \longrightarrow M \times E$ denotes the trivializing diffeomorphism.

By a bundle map between two vector bundles $(\mathbb{E},\pi_{\mathbb{E}},M,E)$ and $(\mathbb{F},\pi_{\mathbb{F}},M',F)$ we mean a pair (Ψ,f) of smooth maps

$$\Psi : \mathbb{E} \longrightarrow \mathbb{F}$$

and

$$f : M \longrightarrow M'$$

such that

$$\begin{array}{ccc} \mathbb{E} & \longrightarrow & \mathbb{F} \\ {\scriptstyle \pi_{\mathbb{E}}}\downarrow & & \downarrow{\scriptstyle \pi_{\mathbb{F}}} \\ M & \longrightarrow & M' \end{array}$$

commutes and $\Psi|_{\mathbb{E}_p}$ is a linear homomorphism $\mathbb{E}_p \longrightarrow \mathbb{F}_{f(p)}$ for all $p \in M$. If $M = M'$ and f is the identity then (Ψ,f) is called a strong bundle map. In this case we refer to it just as Ψ.

Clearly the composition of two bundle maps is a bundle map again. A bundle map (Ψ,f) is called an isomorphism if Ψ (and hence f) is a diffeomorphism.

Let M be a smooth manifold where all chart maps have their values in $\mathbb{R}^n$. As it is easily verified $(TM,\pi_M,M,\mathbb{R}^n)$ is a vector bundle. Also $(TTM,\pi_{TM},TM,\mathbb{R}^{2n})$ is a vector bundle. However observe that TTM, in the sequel denoted by T^2M, carries a second bundle structure, namely

$$(T^2M,T\pi_M,TM,\mathbb{R}^{2n}).$$

On a chart (U,φ) of M the bundle $(T^2M,\pi_{TM},TM,\mathbb{R}^{2n})$ trivializes on TU as

$$\begin{array}{ccc} T^2U & \xrightarrow{T(T\varphi)} & U \times \mathbb{R}^n \times \mathbb{R}^n \times \mathbb{R}^n \\ \downarrow{\scriptstyle \pi_{TU}} & & \downarrow{\scriptstyle pr_{1,2}} \\ TU & \xrightarrow{T\varphi} & U \times \mathbb{R}^n \end{array}$$

where $pr_{1,2}$ denotes the projection onto the first two factors, while $(T^2M,T\pi_M,TM,\mathbb{R}^{2n})$ trivializes on TU as

$$\begin{array}{ccc} T^2U & \xrightarrow{T(T\varphi)} & U \times \mathbb{R}^n \times \mathbb{R}^n \times \mathbb{R}^n \\ \downarrow{\scriptstyle T\pi_U} & & \downarrow{\scriptstyle pr_{1,3}} \\ TU & \xrightarrow{T\varphi} & U \times \mathbb{R}^n \end{array}$$

where $pr_{1,3}$ is the projection onto the first and third factor. Observe that the last diagram arises from

$$\begin{array}{ccc} TU & \xrightarrow{T\varphi} & U \times \mathbb{R}^n \\ \downarrow{\scriptstyle \pi_U} & & \downarrow{\scriptstyle pr_2} \\ U & \xrightarrow{\varphi} & \mathbb{R}^n \end{array}$$

This two bundle structures of T^2M are carried into each other by a unique strong bundle map ι, which in a local representation amounts to the smooth map which interchanges the middle two factors in $U \times \mathbb{R}^n \times \mathbb{R}^n \times \mathbb{R}^n$. Clearly $\iota^2 = \mathrm{id}$. ι is called the <u>canonical involution</u> [Be].

2.3 Construction of vector bundles, the pull back

In this section we follow partly [G,H,V]. Assume that to any point p in a smooth manifold M there is associated an m–dimensional real vector space $\mathbb{E}_p$. Call its disjoint union $\bigcup_{p\in M} \mathbb{E}_p$ by $\mathbb{E}$. Furthermore let E be an m–dimensional $\mathbb{R}$–linear space. Next assume the existence of a covering $\{U_\alpha | \alpha\in J\}$ of M of open sets, where J is an index set, and isomorphisms

$$\Psi_{\alpha,p} : \mathbb{E}_p \longrightarrow E$$

for each $\alpha \in J$ and $p \in U_\alpha$, which fulfil the following condition:

If $\alpha,\beta \in J$ then

$$g_{\alpha,\beta}: U_\alpha \cap U_\beta \longrightarrow GL(E)$$

given by

$$g_{\alpha,\beta}(p) = \Psi_{\beta,p} \circ \Psi_{\alpha,p}^{-1}$$

is smooth. Then

$$\Psi_\alpha : \bigcup_{p\in U_\alpha} \mathbb{E}_p \longrightarrow U_\alpha \times E$$

given by

$$\Psi_\alpha(x_p) = (p, \Psi_{\alpha,p}(x_p))$$

for all $x_p \in \mathbb{E}_p$ and $p \in U_\alpha$, is smooth.

Now it is easy to verify that there is a unique smooth manifold structure on $\mathbb{E}$, with respect to which

$$(\mathbb{E},\pi,M,E)$$

is a vector bundle and (U_α,Ψ_α) are local trivializing representations. Here $\pi(x_p) = p$ for all $x_p \in \mathbb{E}_p$ and all $p \in M$.

Let $(\mathbb{E},\pi_{\mathbb{E}},N,E)$ be any vector bundle over N and $f : M \longrightarrow N$ a smooth map from the smooth manifold M into N. We will use the above mechanism to construct the pull back bundle of $\mathbb{E}$ over M. To each $p \in M$ we associate the vector space $\mathbb{E}_{f(p)}$ and let $f^*\mathbb{E}$ be $\bigcup_{p\in M} \mathbb{E}_{f(p)}$. Let $\{(V_\alpha,\varphi_\alpha) \mid \alpha \in J\}$ be a covering collection of trivializing representations for $\mathbb{E}$. Then $\{f^{-1}(V_\alpha) \mid \alpha \in J\}$ is an open covering of M. Denote $f^{-1}(V_\alpha)$ by U_α and let

$$\Psi_{\alpha,p} : \mathbb{E}_{f(p)} \longrightarrow E$$

be given by

$$\Psi_{\alpha,p} = \varphi_{\alpha,f(p)}$$

for all $p \in U_\alpha$. Now define $\Psi_\alpha : \pi_f^{-1}(U_\alpha) \longrightarrow U_\alpha \times E$ as above, where $\pi_f : f^*\mathbb{E} \longrightarrow M$ is given by $\pi_f(x_{f(p)}) = p$ for all $x_{f(p)} \in \mathbb{E}_{f(p)}$ and all $p \in M$. Clearly for every $\alpha,\beta \in J$ the map

$$g_{\alpha,\beta} : U_\alpha \cap U_\beta \longrightarrow GL(E)$$

defined by

$$g_{\alpha,\beta}(p) = \Psi_{\beta,p} \circ \Psi_{\alpha,p}^{-1}$$

is smooth. Hence there is a unique smooth manifold structure on $f^*\mathbb{E}$ such that

$$(f^*\mathbb{E},\ \pi_f,M,E)$$

is a vector bundle which trivializes over each U_α by $\Psi_\alpha : \pi_f^{-1}(U_\alpha) \longrightarrow U_\alpha \times E$. This bundle is called the <u>pull back</u> of $\mathbb{E}$ by f. The pull back $f^*\mathbb{E}$ has the following universal property. Let $(\mathbb{F},\pi_{\mathbb{F}},M,F)$ be any vector bundle over M and (Ψ,f) a bundle map from $\mathbb{F}$ to $\mathbb{E}$. Then there is a unique strong bundle map $\varphi : \mathbb{F} \longrightarrow f^*\mathbb{E}$ for which

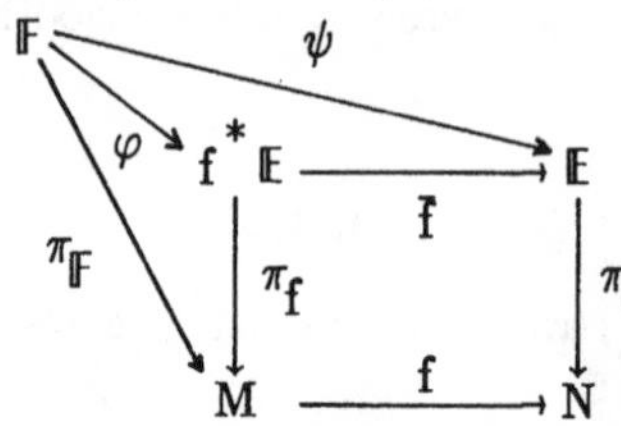

commutes. The map $\bar{f}$ sends any $x_{f(p)} \in \pi_f^{-1}(p)$ to $x_{f(p)} \in \pi_{\mathbb{E}}^{-1}(f(p))$. The verification of this universal property is left to the reader as a simple exercise.

Let M' be another smooth manifold and $g : M' \longrightarrow M$ be smooth. Then the vector bundles $g^* f^* \mathbb{E}$ and $(f \circ g)^* \mathbb{E}$ are strong isomorphic. Finally define the <u>restriction</u> $\mathbb{E}|M$ of a vector bundle $\mathbb{E}$ over N to a submanifold $M \subset N$ as $i^* \mathbb{E}$, where $i : M \longrightarrow N$ denotes the inclusion.

2.4 <u>Homotopy</u>

Let $\mathbb{E}$ be a vector bundle over M with projection π and typical fibre E. A map $s : M \longrightarrow \mathbb{E}$ is called a <u>section</u>, if $\pi \circ s = id_M$. First we prove

<u>Lemma 2.4.1</u> Let M be compact, $K \subset M$ a closed submanifold and $s : K \longrightarrow \mathbb{E}$ be a smooth map with $\pi \circ s = id_K$. Then s extends to a smooth section on all of M.

<u>Proof</u>: Let $U \subset M$ be a trivializing chart with $U \cap K \neq \emptyset$. Then by Tietze's extension theorem (cf. [K]) s regarded as an $\mathbb{E}$–valued function on $U \cap K$ can be extended to U. Since finitely many of such charts cover K we can extend s to an open neighbourhood V of K. There is an open neighbourhood $V' \subset V$ of K whose closure $\overline{V'}$ is contained in $\overline{M \setminus V}$. Now we find a smooth function τ on M which is zero on $M \setminus V$ and identical to one on K. Hence $\tau \cdot s$ is a smooth section defined on V which agrees with s on K. It can be extended to M by putting it to zero outside of V. This yields an extension $\bar{s}$ of s. Certainly $\bar{s}$ is smooth.

<u>Proposition 2.4.2</u> Let $\mathbb{E}$ be a vector bundle over a smooth manifold N. If M is compact and the smooth maps $f_0, f_1 : M \longrightarrow N$ are homotopic then $f_0^* \mathbb{E}$ and $f_1^* \mathbb{E}$ are strong isomorphic.

Proof: Let $H : M \times [0,1] \longrightarrow N$ be a smooth homotopy of f_0 and f_1 and $pr: M \times [0,1] \longrightarrow M$ the projection onto the first factor. Let $t \in [0,1]$ and consider the vector bundles $H^* \mathbb{E}$ and $(H_t \circ pr)^* \mathbb{E}$ over the manifold with boundary $M \times [0,1]$. $H_t: M \longrightarrow N$ is given by $H_t(p) = H(p,t)$ for all $p \in M$. Clearly the restrictions of the above two bundles to $M \times \{t\}$ are strong isomorphic. Thus we have a smooth section $s_t : M\times\{t\} \longrightarrow L(H^* \mathbb{E}|M\times\{t\}, (H_t \circ pr)^* \mathbb{E}|M\times\{t\})$. (The bundle $L(\mathbb{E},\mathbb{F})$ for two vector bundles $\mathbb{E}$ and $\mathbb{F}$ over the same base space is defined in the next section, its fibre over p is the linear space of all linear mappings from $\mathbb{E}_p$ to $\mathbb{F}_p$). By the above lemma s_t can be extended to a smooth section

$$s: M\times[0,1] \longrightarrow L(H^* E,(H_t \circ pr)^* \mathbb{E}).$$

Let U be the set of all those points $(p,\lambda) \in M\times[0,1]$ such that $s(p,\lambda)$ is an isomorphism. Then U is open in $M \times [0,1]$ and contains $M \times \{t\}$. Thus $H^* \mathbb{E}|M\times I_t$ and $(H_t \circ pr)^* \mathbb{E}|M\times I_t$ are isomorphic for some open neighbourhood I_t of t in [0,1]. Therefore the strong isomorphism class of $H_t^* \mathbb{E}$ is a locally constant function of t. This implies that $H_0^* \mathbb{E}$ and $H_1^* \mathbb{E}$ are strong isomorphic, since [0,1] is connected. We followed [A].

Corollary 2.4.3 Any vector bundle $\mathbb{E}$ over a compact contractible manifold M is trivial.

This is simply true because id and the contraction to a point are assumed to be homotopic.

2.5 $\mathbb{E} \oplus \mathbb{F}$, $\mathbb{E} \otimes \mathbb{F}$, $L(\mathbb{E},\mathbb{F})$

Let $\mathbb{E}$ and $\mathbb{F}$ be two vector bundles over M with respective projections $\pi_{\mathbb{E}}$ and $\pi_{\mathbb{F}}$ and typical fibres E and F. We associate to each $p \in M$ the vector space $\mathbb{E}_p \oplus \mathbb{F}_p$. Consider $E \oplus F$ and $GL(E \oplus F)$, as well as an open covering $\{U_\alpha | \alpha \in J\}$ of M for some index set J over which both $\mathbb{E}$ and $\mathbb{F}$ trivialize simultaneously, with respective local trivializing representations

$$\Psi_\alpha : \pi_{\mathbb{E}}^{-1}(U_\alpha) \longrightarrow U_\alpha \times E$$

and

$$\varphi_\alpha : \pi_{\mathbb{F}}^{-1}(U_\alpha) \longrightarrow U_\alpha \times F.$$

Let $g_{\alpha,\beta} : U_\alpha \cap U_\beta \longrightarrow GL(E \oplus F)$ be given by

$$g_{\alpha,\beta}(p) = \Psi_{\beta,p} \circ \Psi_{\alpha,p}^{-1} \oplus \varphi_{\beta,p} \circ \varphi_{\alpha,p}^{-1}$$

for all $p \in U_\alpha \cap U_\beta$. For each choice of $\alpha,\beta \in J$ the map $g_{\alpha,\beta}$ is smooth. Thus $\mathbb{E} \oplus \mathbb{F} = \bigcup_{p \in M} \mathbb{E}_p \oplus \mathbb{F}_p$ carries a unique bundle structure such that $\mathbb{E} \oplus \mathbb{F}$ trivializes on $\{U_\alpha | \alpha \in J\}$ via $\Psi_\alpha \oplus \varphi_\alpha : \pi_{\mathbb{E}\oplus\mathbb{F}}^{-1}(U_\alpha) \longrightarrow U_\alpha \times E \oplus F$, mapping $x_p \oplus y_p$ into $(p, \Psi_{\alpha,p}(x_p) \oplus \varphi_{\alpha,p}(y_p))$. By $\pi_{\mathbb{E}\oplus\mathbb{F}}$ we mean the obvious projection from $\mathbb{E} \oplus \mathbb{F}$ to M. The bundle $\mathbb{E} \oplus \mathbb{F}$ is called the Whitney sum of $\mathbb{E}$ and $\mathbb{F}$.

In a similar manner we define $\mathbb{E} \otimes \mathbb{F}$ and $L(\mathbb{E},\mathbb{F})$, typical fibres are $E \otimes F$ and $L(E,F)$ respectively. By $L(E,F)$ we mean the $\mathbb{R}$–linear space of all linear maps from E into F. The functions into the respective general linear groups

$$g_{\otimes_{\alpha,\beta}} : U_\alpha \cap U_\beta \longrightarrow GL(E \otimes F)$$

and

$$g_{L_{\alpha,\beta}} : U_\alpha \cap U_\beta \longrightarrow GL(L(E,F))$$

are respectively given by

$$g_{\otimes_{\alpha,\beta}}(p) = (\Psi_{\beta,p} \circ \Psi_{\alpha,p}^{-1}) \otimes (\varphi_{\beta,p} \circ \varphi_{\alpha,p}^{-1})$$

and $$g_{L_{\alpha,\beta}}(p)(l) = \varphi_{\beta,p} \circ \varphi_{\alpha,p} \circ l \circ \Psi_{\beta,p} \circ \Psi_{\alpha,p}^{-1}$$

for all $l \in L(E,F)$. If $F = \mathbb{R}$, the bundle $\mathbb{F}$ is called a line bundle.

The bundle $L(\mathbb{E}, M \times \mathbb{R})$ is called the dual bundle of $\mathbb{E}$ and is simply denoted by $\mathbb{E}'$. Here $M \times \mathbb{R}$ is regarded as a trivial line bundle over M.

Lemma 2.5.1 Let $\mathbb{E}$ and $\mathbb{F}$ be two vector bundles over M with typical fibres E and F respectively. Then the canonical isomorphism

$$L(E,F) \cong E' \otimes F$$

where $'$ denotes the dual space, yields a strong bundle isomorphism

$$L(\mathbb{E},\mathbb{F}) \cong \mathbb{E}' \otimes \mathbb{F} .$$

This can be verified as follows:

Clearly $L(\mathbb{E}_p,\mathbb{F}_p) \cong L(\mathbb{E}_p \otimes \mathbb{F}_p', \mathbb{R})$ and thus $(\mathbb{E}_p \otimes \mathbb{F}_p')' \cong \mathbb{E}_p' \otimes \mathbb{F}_p$ for all $p \in M$. Let $\iota_p : L(\mathbb{E}_p,\mathbb{F}_p) \longrightarrow \mathbb{E}_p' \otimes \mathbb{F}_p$ be the isomorphism just mentioned. It induces a map

$$\iota : L(\mathbb{E},\mathbb{F}) \longrightarrow \mathbb{E}' \otimes \mathbb{F}$$

by

$$\iota | L(\mathbb{E},\mathbb{F})_p = \iota_p$$

for all $p \in M$. Using $g_{\otimes_{\alpha,\beta}}$ and $g_{L_{\alpha,\beta}}$ one easily demonstrates the smoothness of ι.

The above consideration can be applied in the same way to all canonical linear isomorphisms to obtain canonical strong bundle isomorphisms of the corresponding vector bundles, e.g.:

$$\mathbb{E} \oplus \mathbb{F} \cong \mathbb{F} \oplus \mathbb{E}$$

$$(\mathbb{E} \oplus \mathbb{F}) \oplus G \cong E \oplus (\mathbb{F} \oplus G)$$

$$\mathbb{E} \otimes \mathbb{F} \cong \mathbb{F} \otimes \mathbb{E}$$

$$(\mathbb{E} \otimes \mathbb{F}) \otimes G \cong \mathbb{E} \otimes (\mathbb{F} \otimes G)$$

$$\mathbb{E} \otimes (\mathbb{F} \oplus G) \cong (\mathbb{E} \otimes \mathbb{F}) \oplus (\mathbb{E} \otimes G)$$

$$\mathbb{E}'' \cong \mathbb{E}$$

$$(\mathbb{E} \otimes \mathbb{F})' \cong \mathbb{E}' \otimes \mathbb{F}' .$$

Clearly $\mathbb{E}$,$\mathbb{F}$ and G denote vector bundles over the same base space. In the following $(\mathbb{E} \oplus \mathbb{F}) \oplus G$ is written as $\mathbb{E} \oplus \mathbb{F} \oplus G$. Moreover $\bigoplus_{i=1}^{m} \mathbb{E}_i$ denotes $\mathbb{E}_1 \oplus \cdots \oplus \mathbb{E}_m$. In case $\mathbb{E}_1 = \cdots = \mathbb{E}_m = \mathbb{E}$ we write $\oplus^m\mathbb{E}$. The symbols $\mathbb{E} \otimes \mathbb{F} \otimes G$, $\bigotimes_{i=1}^{m} \mathbb{E}_i$ and $\otimes^m\mathbb{E}$ are used analogously for the tensor product.

For vector bundles $\mathbb{E}_1,\ldots,\mathbb{E}_m,\mathbb{F}$ over M with typical fibres $E_1,\ldots,E_m,F$ respectively define the vector bundle $L^m(\mathbb{E}_1,\ldots,\mathbb{E}_m,\mathbb{F})$ similarly as above. The typical fibre is the vector space of all m–linear maps from $E_1 \times \cdots \times E_m$ into F, denoted by $L^m(E_1,\ldots,E_m,F)$. If $\mathbb{E}_1 = \mathbb{E}_2 = \cdots = \mathbb{E}_m = \mathbb{E}$, we simply write $L^m(\mathbb{E},\mathbb{F})$ and $L^m(E,F)$. There are canonical strong isomorphisms

$$L^m(\mathbb{E}_1,\ldots,\mathbb{E}_m,\mathbb{F}) \cong L(\mathbb{E}_1 \otimes \cdots \otimes \mathbb{E}_m,\mathbb{F})$$
$$\cong \mathbb{E}_1' \otimes \cdots \otimes \mathbb{E}_m' \otimes \mathbb{F} .$$

In particular in case $\mathbb{E}_1 = \mathbb{E}_2 = \cdots = \mathbb{E}_m = \mathbb{E}$ and $\mathbb{F} = M \times \mathbb{R}$ we have

$$L^m(\mathbb{E},M\times\mathbb{R}) \cong \otimes^m \mathbb{E}' \cong (\otimes^m \mathbb{E})' .$$

2.6 $\Lambda^m\mathbb{E}$

In this section $\mathbb{E}$ is a vector bundle over M with typical fibre E and projection $\pi_{\mathbb{E}}$. Let Λ^mE be the mth exterior power of E. It is characterized as follows (cf. [W]). There is a natural m–linear map π from the m–fold cartesian product $\times_m E$ into Λ^mE such that any m–linear totally skew symmetric map b from $\times_m E$ into some $\mathbb{R}$–linear space F uniquely factors to a linear map $\bar{b}$ such that

$$\begin{array}{ccc} \underset{m}{\times} E & \xrightarrow{\quad b \quad} & F \\ \pi \big\downarrow & \nearrow \bar{b} & \\ \Lambda^m E & & \end{array}$$

commutes. Any map $l \in GL(E)$ induces a unique linear map $\Lambda^m l \in GL(\Lambda^m E)$ such that

$$\begin{array}{ccc} \underset{m}{\times} E & \xrightarrow{\quad \underset{m}{\times} l \quad} & \underset{m}{\times} E \\ \pi \big\downarrow & & \big\downarrow \pi \\ \Lambda^m E & \xrightarrow{\quad \Lambda^m l \quad} & \Lambda^m E \end{array}$$

commutes. By $\underset{m}{\times} l$ we mean the map sending any element $(x_1,\dots,x_m)$ into $(lx_1,\dots,lx_m)$, which composed with π yields an m–linear totally skew symmetric map to which the above universal property applies. To construct the bundle $\Lambda^m \mathbb{E}$ we proceed as usual: We choose an open covering $\{U_\alpha | \alpha \in J\}$ of M for some index set J over which $\mathbb{E}$ trivializes via

$$\Psi_\alpha : \pi_{\mathbb{E}}^{-1}(U_\alpha) \longrightarrow U_\alpha \times E.$$

Then define

$$g_{\alpha,\beta} : U_\alpha \cap U_\beta \longrightarrow GL(\Lambda^m E)$$

by

$$g_{\alpha,\beta}(p) = \Lambda^m(\Psi_{\beta,p} \circ \Psi_{\alpha,p}^{-1})$$

for any $p \in U_\alpha \cap U_\beta$. By the mechanism described in section 2.3 we obtain a natural bundle structure on $\bigcup_{p \in M} \Lambda^m \mathbb{E}_p$ over M. This bundle will be referred to as $\Lambda^m \mathbb{E}$ and is called the m–fold exterior product of $\mathbb{E}$.

If $\mathbb{F}$ is another vector bundle over M with typical fibre F, then we can construct the bundle $A^m(\mathbb{E},\mathbb{F})$ with typical fibre $A^m(E,F)$, the vector space of all m–linear totally skew symmetric maps from $\underset{m}{\times} E$ to F, in an analogous manner as in section 2.5. Clearly the vector bundles $A^m(\mathbb{E}, M \times \mathbb{R})$ and $(\Lambda^m \mathbb{E})' \cong \Lambda^m \mathbb{E}'$ are canonically strong isomorphic.

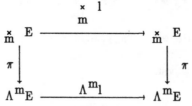

2.7. Section modules of $\mathbb{E}$, orientation in $\mathbb{E}$

Given a vector bundle $(\mathbb{E},\pi,M,E)$, a smooth map $s:M \longrightarrow \mathbb{E}$ is called a smooth section of $\mathbb{E}$, iff

$$\pi \circ s = id_M .$$

Certainly the space $\Gamma\mathbb{E}$ of all smooth sections of $\mathbb{E}$ forms a $\mathbb{R}$–linear space. However, each τ in the $\mathbb{R}$–algebra $C^\infty(M)$ of all real valued smooth functions of M yields together with s $\in \Gamma\mathbb{E}$ the smooth section

$$\tau \cdot s : M \longrightarrow \mathbb{E}$$

given by $\tau \cdot s(p) = \tau(p) \cdot s(p)$ for all $p \in M$. Hence $\Gamma\mathbb{E}$ is module over $C^\infty(M)$.

It is a projective module. This means that $\Gamma\mathbb{E}$ as a module is a direct summand of a free module (cf. [L,2]). In addition $\Gamma\mathbb{E}$ is finitely generated (cf. [L,2]).

Vice versa, the theorem of Swan asserts, that any projective, finitely generated module over $C^\infty(M)$ is a module of sections into some vector bundle $\mathbb{E}$ over M.

For later purposes we state a technical lemma:

Lemma 2.7.1 For any two vector bundles $\mathbb{E}$ and $\mathbb{F}$ over M there are $C^\infty(M)$–modul isomorphisms

$$\Gamma(\mathbb{E} \oplus \mathbb{F}) \cong \Gamma\mathbb{E} \oplus \Gamma\mathbb{F}$$

and

$$\Gamma(\mathbb{E} \otimes \mathbb{F}) \cong \Gamma\mathbb{E} \otimes_{C^\infty(M)} \Gamma\mathbb{F} ,$$

where $\otimes_{C^\infty(M)}$ denotes the tensor product over $C^\infty(M)$. The proof is straight forward.

Let E be the typical fibre of $\mathbb{E}$. Certainly if m > dim E, then $\Lambda^m\mathbb{E} = M \times \{0\}$. An orientation of $\mathbb{E}$ is a nowhere vanishing section in $\Lambda^{\dim E}\mathbb{E}$. The smooth sections $s : M \longrightarrow \Lambda^m T^* M$ are called (smooth) $\mathbb{R}$–valued m–forms of M. Here T^*M denotes the dual bundle of TM. Observe that $s \in \Gamma\Lambda^m T^* M$ can naturally be identified with a section of $A^m(TM, M\times\mathbb{R}) \subset L^m(TM, M\times\mathbb{R})$.

In the following we will extend the notion of an $\mathbb{R}$–valued m–form to a bundle valued m–form.

First consider an m–linear totally skew symmetric map

$$b : \times_m E \longrightarrow F$$

where E and F are finite dimensional real vector spaces. b factors uniquely to a linear map

$$\bar{b}: \Lambda^m E \longrightarrow F$$

and hence b can be viewed as an element of $(\Lambda^m E)' \otimes F$, since

$$L(\Lambda^m E, F) \cong (\Lambda^m E)' \otimes F.$$

Moreover $(\Lambda^m E)'$ is naturally strong isomorphic to $\Lambda^m(E')$. If $\mathbb{E}$ and $\mathbb{F}$ are vector bundles over M with typical fibres E and F respectively, then by lemma 2.5.1 we have

$$L(\Lambda^m\mathbb{E}, \mathbb{F}) \cong \Lambda^m\mathbb{E}' \otimes \mathbb{F}.$$

Thus we obtain by lemma 2.7.1

$$\Gamma L(\Lambda^m\mathbb{E}, \mathbb{F}) = \Gamma(\Lambda^m\mathbb{E}') \otimes_{C^\infty(M)} \Gamma\mathbb{F}\ .$$

If $\mathbb{E} = TM$, then

$$\Gamma L(\Lambda^m TM, \mathbb{F}) \cong \Gamma(\Lambda^m T^* M) \otimes_{C^\infty(M)} \Gamma\mathbb{F}\ .$$

This isomorphism motivates the following definition. An element

$$s \in \Gamma L(\Lambda^m TM, \mathbb{F}) \cong \Gamma(\Lambda^m T^* M) \otimes_{C^\infty(M)} \Gamma\mathbb{F}$$

is called a $\mathbb{F}$–valued (smooth) m–form on M.

Let $s_1 \in \Gamma\Lambda^m T^* M$ and $s_2 \in \Gamma L(\Lambda^n TM, \mathbb{F})$. We define $s_1 = s_2 \in \Gamma L(\Lambda^{m+n} TM, \mathbb{F})$ by

$$s_1 = s_2(p)(v_1, \dots, v_m, v_{m+1}, \dots, v_{m+n}) =$$
$$= \frac{1}{m!n!} \sum_{\sigma} \epsilon(\sigma) \cdot s_1(p)(v_{\sigma(1)}, \dots, v_{\sigma(m)}) \cdot s_2(p)(v_{\sigma(m+1)}, \dots, v_{\sigma(m+n)}).$$

Here σ varies over all permutations of m+n elements. $\epsilon(\sigma)$ is 1 if the permutation is even and −1 otherwise. It is a matter of simple routine to show the smoothness of $s_1 = s_2$.

2.8 The jet bundle

In the following we proceed closely to [Go,Gui]. Let M,N be two smooth manifolds and $p \in M$ some point. Suppose for $f,g \in C^\infty(M,N)$ we have $f(p) = g(p) = q$. f has a first order contact at p with g if $Tf(p) = Tg(p)$ and a <u>contact of order</u> n if $Tf, Tg: NM \longrightarrow TN$ have a $(n-1)^{th}$–order contact at any $v_p \in T_pM$. It is easily seen, that having k^{th} order contact at p is an equivalence relation in $C^\infty(M,N)$.

Denote the set of equivalence classes by $J^k(M,N)_{p,q}$. The elements are called <u>k–jets</u> from M to N, p is referred to as the <u>source</u> and q as the <u>target</u> of the jets. The k–jet of f at p is denoted by $j^k(f)(p)$. By $J^k(M,N)$ we mean $\bigcup_{(p,q)\in M\times N} J^k(M,N)_{p,q}$. Let $J^0(M,N)$ be $M \times N$.

The map $\alpha : J^k(M,N) \longrightarrow M$ assigning to any k–jet its source is the <u>source map</u>. In an analogous way we define the <u>target map</u> $\beta : J^k(M,N) \longrightarrow N$.

Certainly the jet $j^k(f)(p)$ depends only on the values of the functions in some neighbourhood of p. Therefore using charts (U,φ) and (V,Ψ) of p and q respectively we easily observe that $g \in j^k(f)(p)$ iff

$$T^k(\Psi \circ f \circ \varphi^{-1})(\varphi(p)) = T^k(\Psi \circ g \circ \varphi^{-1})(\varphi(p))$$

where $T^k(h)(x)$ for a smooth $\mathbb{R}^n$–valued function $h : \varphi(U) \longrightarrow \mathbb{R}^n$ means the Taylor polynomial of h around $x \in \varphi(U)$ up to the kth total derivative of h in x. This observation now allows us to introduce a topology on $J^k(M,N)$. Consider two open charts (U,φ) and (V,Ψ) in M and N respectively. A k–jet $j^k(h)(x)$ of a smooth function $h : \varphi(U) \longrightarrow \Psi(V)$ at $x \in \varphi(U)$, identified with $T^k(h)(x)$ is an element of

$$\mathbb{R}^m \oplus \mathbb{R}^n \oplus \bigoplus_{s=1}^{k} S^s(\mathbb{R}^m,\mathbb{R}^n) ,$$

where m = dim M, n = dim N and $S^s(\mathbb{R}^m,\mathbb{R}^n)$, denotes the symmetric s–linear maps from $\times_s \mathbb{R}^m$ to $\mathbb{R}^n$. Vice versa $T^k(h)(x)$ uniquely determines $j^k(h)(x)$. The set

$$J^k(\varphi(U),\Psi(V))$$

is open in $\mathbb{R}^m \oplus \mathbb{R}^n \oplus \bigoplus_{s=1}^{k} S^s(\mathbb{R}^m,\mathbb{R}^n)$. Assigning to any jet $j^k(f)(p) \in J^k(U,V)$ its Euclidean representation

$$\theta_{U,V}(j^k(f)(p)) = j^k(\Psi \circ f \circ \varphi^{-1})(\varphi(p))$$

yields a bijection from $J^k(U,V)$ onto $J^k(\varphi(U),\Psi(V))$. Thus $J^k(M,N)$ inherits a unique topology under which it is a smooth manifold having $J^k(U,V)$ as charts. Given a map $f \in C^\infty(M,N)$ we have its <u>k–jet prolongation</u>

$$j^k(f) : M \longrightarrow J^k(M,N),$$

which associates to any $p \in M$ the k–jet $j^k(f)(p)$ in $J^k(M,N)_{p,f(p)}$.

It is a simple exercise to show (cf. [Go,Gui],[Hi])

Proposition 2.8.1 For any pair of smooth manifolds M and N, $J^k(M,N)$ is a smooth manifold for which the source map

$$J^k(M,N) \xrightarrow{\alpha} M$$

is a smooth submersion, in fact a fibre bundle whose typical fibre is an affine space. Moreover

$$J^k(M,N) \xrightarrow{\beta} N$$

and

$$j^k(f) : M \longrightarrow J^k(M,N)$$

are smooth for all $f \in C^\infty(M,N)$.

If $k > r$, then we have a natural smooth map

$$\pi_{k,r} : J^k(M,N) \longrightarrow J^r(M,N)$$

which associates to each jet $j^k(f)(p)$, regarded in its local representation as the Taylor polynomial T^k of degree k, the r–jet $j^r(f)(p)$ corresponding to the Taylor polynomial T^r of degree r obtained by taking the first $r + 1$ terms of T^k only. The fibre of $\pi_{k,r}$ thus is a vector space. In fact

$$(J^k(M,N), \pi_{k,r}, J^r(M,N), \bigoplus_{s=r+1}^{k} S^s(\mathbb{R}^m,\mathbb{R}^n))$$

is a vector bundle.

For the rest of this section assume that $N = \mathbb{E}$ is a vector bundle over M. The collection of k–jets of sections in $\Gamma\mathbb{E}$ will be denoted by $J^k\mathbb{E}$. The projection $\pi : \mathbb{E} \longrightarrow M$ yields

$$\pi_* : J^k(M,\mathbb{E}) \longrightarrow J^k(M,M)$$

which assigns to $j^k(f)(p)$ the jet $j^k(\pi \circ f)(p)$ for each $f \in C^\infty(M,\mathbb{E})$ and $p \in M$. This is a smooth submersion. Let $I \subset J^k(M,M)$ be all those k–jets which are represented by id_M. Since I is a smooth submanifold of $J^k(M,M)$ (diffeomorphic to M) the total preimage $\pi_*^{-1}(I)$ is a smooth submanifold of $J^k(M,\mathbb{E})$. However

$$\pi_*^{-1}(I) = J^k\mathbb{E} .$$

Now it is simple to check

Proposition 2.8.2 For any vector bundle $(\mathbb{E},\pi,M,E)$ over an m–dimensional smooth manifold M

$$(J^k\mathbb{E},\alpha,M,E \oplus \bigoplus_{s=1}^{k} S^s(\mathbb{R}^m,E))$$

is a vector bundle.

2.9 The canonical 1–form on J^kN

In this section we assume N to be a smooth fibred manifold over a smooth manifold M, that is we require the existence of a smooth map $\pi : N \longrightarrow M$ which is a surjective submersion.

By ΓN we denote the collection of all smooth maps $s : M \longrightarrow N$ with $\pi \circ s = id_M$, that is the collection of all (smooth) sections. By a local section we mean a smooth map $s : U \longrightarrow N$ satisfying $\pi \circ s = id_U$ where U is some open subset of M.

We follow closely [G,St] in order to construct the canonical TJ^kN–valued 1–form which will be used in the calculus of variations in later chapters. We begin by defining J^kN.

Any local section $s : U \longrightarrow N$ defines

$$j^k(s)(p)$$

for all $p \in U$. The collection J^kN of all k–jets of all local smooth sections is a smooth manifold again. This can be shown as in the case of $N = \mathbb{E}$. All we really used in the previous section to show that $J^k\Gamma\mathbb{E}$ is a smooth manifold is the fact that $\pi : \mathbb{E} \longrightarrow M$ is a smooth surjective submersion. The source map

$$\alpha : J^kN \longrightarrow M$$

is a surjective submersion again. Any local section s defined on U determines a local section

$$j^k(s) : U \longrightarrow J^kN$$

assigning to any $p \in U$ the k–jet $j^k(s)(p)$. Moreover

$$\pi_k : J^kN \longrightarrow J^{k-1}N$$

which when represented in charts assigns to the kth–order Taylor polynomial the $(k-1)^{th}$–order one, is clearly a smooth surjective submersion.

The smooth map $\pi : N \longrightarrow M$ defines the vector bundle VN over N of <u>vertical</u> vectors as follows. Given any $q \in N$ we set

$$V_qN = \ker T\pi(q) \subset T_qN$$

and let

$$VN = \bigcup_{p\in N} V_qN \subset TN.$$

Obviously any $v \in V_qN$ can be obtained as $v = \dot{\sigma}(0)$ with some smooth curve σ assuming its values in $\pi^{-1}(\pi(q))$ and satisfying $\sigma(0) = q$. It is an easy exercise to show that VN is a vector bundle over N with π_N restricted to VN as its projection. VN is called the <u>vertical bundle</u> of the fibred manifold N. Consider the map $\pi \circ \pi_N$: $VN \longrightarrow M$. Clearly VN is fibred over M with respect to $\pi \circ \pi_N$.

Next we have to extend the notion of a vector bundle–valued form. In section 2.7. we required that the range bundle is defined over the domain of the form. In the following extension this condition will be dropped. Let $\mathbb{E} \longrightarrow M_1$ be a smooth vector bundle over a manifold M_1 and f a smooth map from the smooth manifold M to M_1. By an <u>$\mathbb{E}$–valued r–form</u> on M we mean an $f^*\mathbb{E}$–valued r–form. Clearly $f^*\mathbb{E}$ is a vector bundle over M.

Now we are in a position to show the existence and uniqueness of the canonical 1–form ω_k, as shown in [G,St].

Proposition 2.9.1 There exists a unique $TJ^{k-1}N$–valued 1–form

$$\omega_k : TJ^kN \longrightarrow TJ^{k-1}N$$

on J^kN such that

1)
$$\omega_k(T_j^k(s)(p)(v)) = 0$$

for any smooth local section $s:U \longrightarrow N$, $v \in T_pU$ and $p \in U$,

2)
$$\omega_k(\xi) = T\pi_k(\xi)$$

for all $\xi \in VJ^kN := \ker T\alpha$.

Proof: Following [G,St] we first prove the uniqueness of such a 1–form ω_k. Suppose there are two 1–forms ω_k and $\bar{\omega}_k$ satisfying (1) and (2). Let us consider the difference $\omega_k - \bar{\omega}_k$. By (2) we get

$$(\omega_k - \bar{\omega}_k)(\xi) = T\pi_k(\xi) - T\pi_k(\xi)$$

for all $\xi \in VJ^kN$. Thus $(\omega_k - \bar{\omega}_k)\, VJ^kN = 0$, our first observation. Clearly by condition (1) we have our second observation

$$(\omega_k - \bar{\omega}_k)(Tj^k(s)(p)(v)) =$$
$$= \omega_k(Tj^k(s)(p)(v)) - \bar{\omega}_k(Tj^k(s)(p)(v)) = 0.$$

To show that $\omega_k = \bar{\omega}_k$ we represent $q \in J^kN$ by $q = j^k(s)(p)$ for some local section $s : U \longrightarrow N$ defined on a neighbourhood U of $p = \alpha(q)$. Thus given $\xi \in T_qJ^kN$

$$\xi^v := \xi - Tj^k(s)(p)(T\alpha(q)(\xi)) \in VJ^kN.$$

Therefore by the two observations made above

$$(\omega_k - \bar{\omega}_k)(\xi) = (\omega_k - \bar{\omega}_k)(\xi^v) + (\omega_k - \bar{\omega}_k)(\xi - \xi^v) = 0.$$

Thus $\omega_k = \bar{\omega}_k$.

Let us construct ω_k. We represent any $q \in J^kN$ by $q = j^k(s)(p)$ as above. Consider the map

$$T\, j^{k-1}(s)(p) : T_qM \longrightarrow T_{j^{k-1}(s)(p)} J^{k-1}N = T_{\pi_k(q)} J^{k-1}N.$$

This map depends only on $q \in J^kN$. We set

$$\omega_k(\xi) = T\pi_k(q)(\xi) - Tj^{k-1}(s)(p)(T\,\alpha(q)(\xi))$$

for all $\xi \in T_qJ^kN$. Conditions (1) and (2) are trivially satisfied.

The following proposition states a converse of the above condition (1). In order to formulate it we introduce

$$\pi_0 : J^kN \longrightarrow J^0N = N$$

which when represented in charts amounts to truncate the Taylor polynomial after the 0th–order derivative.

Proposition 2.9.2 If $\mathscr{s}: U \longrightarrow J^kN$ is a local section satisfying

$$\omega_k(T\mathscr{s}(p)(v)) = 0$$

for all $p \in U$ and $v \in T_pU$, then if $s = \pi_0 \circ \mathscr{s}$

$$\mathscr{s} = j^k(s).$$

Proof: We proceed by induction. Let $k = 1$ and $\mathscr{s}: U \longrightarrow J^1N$ satisfying the required condition. Let us represent $\mathscr{s}(p)$ by

$$\mathscr{s}(p) = j^1(\bar{s})(p)$$

for some local section $\bar{s}$. Clearly

$$(\pi_0 \circ \mathscr{s})(p) = \pi_0(j^1(\bar{s})(p)) = \bar{s}(p).$$

Then by definition of ω_1 we have

$$\begin{aligned} 0 &= \omega_1(T\mathscr{s}(p)(v)) \\ &= (T\pi_0(\mathscr{s}(p)) - T\, j^0(\bar{s})(p) \circ T\alpha(\mathscr{s}(p)))(T\mathscr{s}(p)(v)) \\ &= (T(\pi_0 \circ \mathscr{s})(p) - T\,\bar{s}(p))(v) \end{aligned}$$

for all $v \in T_pU$. Thus we obtain

$$T(\pi_0 \circ s)(p) = T\,\bar{s}(p)$$

and we deduce

$$s(p) = j^1(\pi_0 \circ s)(p).$$

Assume next that the proposition holds for $k-1$. Again we represent $s(p)$ by

$$s(p) = j^k(\bar{s})(p)$$

for some local section $\bar{s}$. Obviously

$$(\pi_k \circ s)(p) = \pi_k(j^k(\bar{s})(p)) = \bar{s}(p).$$

Then by definition

$$\begin{aligned} 0 &= \omega_k(T\,s(p)(v)) \\ &= (T\,\pi_k(s(p)) - T\,j^{k-1}(\bar{s})(p) \circ T\alpha(s(p)))(T\,s(p)(v)) \\ &= (T(\pi_k \circ s)(p) - T\,j^{k-1}(\bar{s})(p))(v). \end{aligned}$$

Now observe that

$$\pi_k \circ s = j^{k-1}(\pi_0 \circ s).$$

This is due to the following:

For any smooth section $s\colon U \longrightarrow J^kN$ we have

$$T\,\pi_{k-1} \circ \omega_k \circ T s' = \omega_{k-1} \circ T(\pi_k \circ s')$$

as is easily verified. Hence $s' = s$ yields $\omega_{k-1} \circ T(\pi_k \circ s) = 0$. Now using the induction hypothesis

$$\pi_k \circ s = j^{k-1}(\pi_0 \circ \pi_k \circ s) = j^{k-1}(\pi_0 \circ s)$$

we conclude

$$T\,j^{k-1}(\pi_0 \circ s)(p) = T\,j^{k-1}(\bar{s})(p)$$

saying that

$$j^k(\pi_0 \circ s)(p) = s(p).$$

2.10 Vertical and horizontal bundles, connnections

Let $\mathbb{E}$ be a smooth vector bundle over the smooth manifold M with typical fibre E. The projection from $\mathbb{E}$ to M is denoted by π.

A submanifold $\mathbb{F} \subset \mathbb{E}$ which is a vector bundle over M with typical fibre $F \subset E$ and projection $\pi|\mathbb{F}$ is called a vector subbundle or just a subbundle of $\mathbb{E}$.

As in section 2.9. we denote the vertical bundle over $\mathbb{E}$ by $V\mathbb{E} = \ker T\pi$. Clearly $V\mathbb{E}$ is a subbundle of $T\mathbb{E}$ with $\pi_{\mathbb{E}}$, the restraiction of $\pi_{\mathbb{E}} : T\mathbb{E} \longrightarrow \mathbb{E}$ to $V\mathbb{E}$, as its projection. If $U \times E$ is a local trivialization of $\mathbb{E}$ over some open set $U \subset M$ then the restriction of $T\mathbb{E}$ to $\pi^{-1}(U) \subset \mathbb{E}$ trivializes as

$$U \times E \times \mathbb{R}^n \times E$$

where n is the dimension of M. Locally $T\pi$ is represented as

$$pr_{1,3} : U \times E \times \mathbb{R}^n \times E \longrightarrow U \times \mathbb{R}^n,$$

mapping any quadruple (p,e,v,e') into (p,v). Thus $V\mathbb{E}|\pi^{-1}(U)$ is diffeomorphic to $U \times E \times E$. Clearly the dimension of $V\mathbb{E}$ is $n + 2\cdot\dim E$. Call the fibre of $V\mathbb{E}$ over $e \in \mathbb{E}$ by $V_e\mathbb{E}$. Since $V_e\mathbb{E}$ is the tangent space to the fibre $\mathbb{E}_{\pi(e)} = \pi^{-1}(\pi(e))$ at e, the following lemma is evidently true:

Lemma 2.10.1 For all $e \in \mathbb{E}$ the vertical space $V_e\mathbb{E}$ is canonically isomorphic to $\mathbb{E}_{\pi(e)}$.

A subbundle $H\mathbb{E}$ of $T\mathbb{E}$ is called a horizontal bundle over $\mathbb{E}$ if the Whitney sum of $V\mathbb{E}$ and $H\mathbb{E}$ is strongly isomorphic to $T\mathbb{E}$:

$$V\mathbb{E} \oplus H\mathbb{E} = T\mathbb{E}.$$

In this case $\dim H\mathbb{E} = 2n + \dim E$. The vectors in $H\mathbb{E}$ are called horizontal. Let $H_e\mathbb{E}$ denote the fibre of $H\mathbb{E}$ over $e \in \mathbb{E}$. Obviously we have

Lemma 2.10.2 If $\mathbb{H}\mathbb{E}$ is a horizontal bundle over $\mathbb{E}$, then

$$T\pi(e) : H_e\mathbb{E} \longrightarrow T_{\pi(e)}M$$

is an isomorphism for all $e \in \mathbb{E}$.

Since $\mathbb{V}\mathbb{E}$ consists of all tangent vectors along all fibres of $\mathbb{E}$ we easily show that $\mathbb{V}\mathbb{E}$ is an involutive distribution of tangent subspaces of $T\mathbb{E}$. However a horizontal bundle is usually not involutive. This will be shown later by means of the tool introduced next.

$T\mathbb{E}$ is not only a vector bundle over $\mathbb{E}$ via the projection $\pi_{\mathbb{E}}$, but also a vector bundle over TM via the projection $T\pi$. To define a vector space structure on the fibre $T\pi^{-1}(v)$ for $v \in TM$ we represent $T\pi$ locally as

$$pr_{1,3}: U \times E \times \mathbb{R}^n \times E \longrightarrow U \times \mathbb{R}^n$$

as above. This induces for any $v \in TM|U \cong U \times \mathbb{R}^n$ a diffeomorphism from $T\pi^{-1}(v)$ onto $E \times E$ and thus a vector space structure on $T\pi^{-1}(v)$. It is straightforward to verify that this structure is independent of the chosen trivialisation and that $T\mathbb{E}$ is a vector bundle over TM with projection $T\pi$ and typical fibre $E \times E$. Moreover the projections satisfy

$$\pi_M \circ T\pi = \pi \circ \pi_{\mathbb{E}} ,$$

where π_M denotes the projection of TM to M.

Following [D,III], by a connection in $\mathbb{E}$ we mean a smooth choice of horizontal vectors given by a smooth map

$$C : TM \oplus \mathbb{E} \longrightarrow T\mathbb{E}$$

with the following properties:

1) $T\pi(e)(C((v,e)) = v$ for all $v \in TM$ and $e \in \mathbb{E}$ with $\pi_M(v) = \pi(e)$.

2) $\pi_{\mathbb{E}}(C(v,e)) = e$ for all $v \in TM$ and $e \in \mathbb{E}$ with $\pi_M(v) = \pi(e)$.

3) The map $T_pM \longrightarrow T_e\mathbb{E}$ sending $v \in T_pM$ into $C(v,e)$ is linear for all $e \in \mathbb{E}$ and $p = \pi(e)$.

4) The map $\mathbb{E}_p \longrightarrow T\pi^{-1}(v)$ assigning to each $e \in \mathbb{E}_p$ the vector $C(v,e)$ is linear for all $v \in T_pM$.

The conditions (1), (2) and (3) show that $\{C(v,e) \mid v \in T_pM\}$ for fixed $p \in M$ and $e \in \mathbb{E}_p$ constitute a horizontal subspace of $T_e\mathbb{E}$. The collection of all these subspaces form a horizontal bundle $H^C\mathbb{E}$.

To construct a connection in $\mathbb{E}$ let $\{U_\alpha | \alpha \in I\}$ be a local finite covering of open sets of M over which $\mathbb{E}$ trivializes. Specify C_α on $TU_\alpha \oplus E|U_\alpha$ such that the conditions (1) – (4) hold. This can be done e.g. locally by the map

$$U_\alpha \times \mathbb{R}^n \times E \longrightarrow U_\alpha \times E \times \mathbb{R}^n \times E$$

sending (p,v,e) into (p,e,v,o). Then patch the maps C_α together by a partition of unity. This also shows that each vector bundle $\mathbb{E}$ over M admits a horizontal bundle.
Associated to a connection C there is a differential operator, the covariant derivative:

Let N be a smooth manifold and $f : N \longrightarrow \mathbb{E}$ a smooth map. Then for any $w \in T_qN$

$$Tf(q)(w) - C(T(\pi \circ f)(q)(w),f(q) \in V_{f(q)}\mathbb{E}$$

and thus by Lemma 2.10.1 can be viewed as a vector in $\mathbb{E}_{\pi(f(q))}$, which we denote by $\nabla_w f$. Assigning to each $w \in TN$ the vector $\nabla_w f$ yields a smooth map $\nabla f : TN \longrightarrow \mathbb{E}$. In particular any $x \in \Gamma TM$ and any $s \in \Gamma\mathbb{E}$ yield $\nabla_X s \in \Gamma\mathbb{E}$, defined by

$$(\nabla_X s)(p) = \nabla_{X(p)} s \in \mathbb{E}_p$$

for all $p \in M$. The map

$$\nabla : \Gamma TM \times \Gamma\mathbb{E} \longrightarrow \Gamma\mathbb{E}$$

sending $(X,s) \in \Gamma TM \times \Gamma\mathbb{E}$ into $\nabla_X s$ has the following properties:

1) $\nabla_X(s_1+s_2) = \nabla_X s_1 + \nabla_X s_2$
2) $\nabla_X(\tau\cdot s) = X(\tau)\cdot s + \tau\cdot\nabla_X s$
3) $\nabla_{X_1+X_2} s = \nabla_{X_1} s + \nabla_{X_2} s$
4) $\nabla_{\tau X} s = \tau\cdot\nabla_X s$

for all $X,X_1,X_2 \in \Gamma TM$, $s,s_1,s_2 \in \Gamma\mathbb{E}$ and $\tau \in C^\infty(M,\mathbb{R})$. A covariant derivative for $\mathbb{E}$ is a map $\nabla : \Gamma TM \times \Gamma\mathbb{E} \longrightarrow \Gamma\mathbb{E}$ satisfying the above four conditions. It defines a curvature operator R assigning to each pair $X,Y \in \Gamma TM$ and each $s \in \Gamma\mathbb{E}$ the section

$$R(X,Y)s := \nabla_X\nabla_Y s - \nabla_Y\nabla_X s - \nabla_{[X,Y]}s \in \Gamma\mathbb{E}.$$

Suppose now that $s \in \Gamma L^r(TM,\mathbb{E})$. Furthermore assume that both TM and $\mathbb{E}$ have covariant derivatives $\tilde{\nabla}$ and ∇ respectively. These two covariant derivatives define a covariant derivative ∇^L on $L^r(TM,\mathbb{E})$ as follows:

$$\nabla^L_Z(s)(X_1,\ldots,X_r) = \nabla_Z(s(X_1,\ldots,X_r)) - \sum_{j=1}^{r} \nabla_Z(s(X_1,\ldots,X_{j-1},\tilde{\nabla}_Z X_j,X_{j+1},\ldots,X_r))$$

for all $Z,X_1,\ldots,X_r \in \Gamma TM$. One easily verifies the conditions (1) – (4) of a covariant derivative for ∇^L.

2.11 Riemannian structures on vector bundles

Given an $\mathbb{R}$–vector space E. We will construct the two–fold symmetric power V^2E. Consider $E \otimes E$ and the subspace A spanned by all tensors of the form

$$\{\tfrac{1}{2}(x \otimes y - y \otimes x) \mid x,y \in E\}.$$

Then $V^2E = E \vee E$ denotes $E \otimes E_{/A}$. The purpose of $E \vee E$ is the following. Any symmetric bilinear map

$$b : E \times E \longrightarrow F$$

into some $\mathbb{R}$–vector space F can be factored uniquely over a linear map $\bar{b} : E \vee E \longrightarrow F$,

meaning that

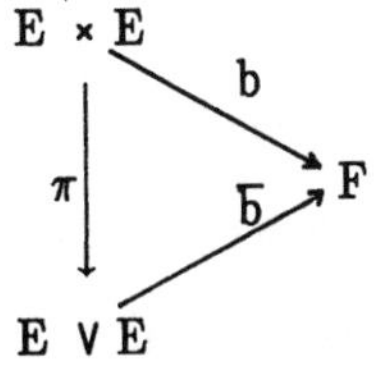

commutes, where π assigns to any $(x,y) \in E \times E$ the equivalence class of $x \otimes y$ in $E \vee E$.

The space $S^2(E,\mathbb{R})$ of all symmetric bilinear maps is then naturally isomorphic to $(E \vee E)'$ which itself is canonically isomorphic to $E' \vee E'$.

By the mechanism described in 2.3. we associate to any vector bundle $\mathbb{E}$ over a smooth manifold M with typical fibre E the vector bundle

$$V^2\mathbb{E}'$$

whose typical fibre is V^2E'.

A section in $\Gamma V^2E'$ evaluated at a point $p \in M$ can be identified with a symmetric $\mathbb{R}$–valued bilinear form on $\mathbb{E}_p$, the fibre of $\mathbb{E}$ over p.

By a <u>pseudo–Riemannian structure</u> on $\mathbb{E}$ we mean a smooth section in $\Gamma V^2\mathbb{E}'$ which at any point $p \in M$ is non degenerate. The section is called a <u>Riemannian structure</u> if moreover every value is positive definite.

Since on each trivial bundle $U_\alpha \times E$ where U_α is a chart of M we have a Riemannian structure, called $\langle , \rangle_\alpha$ say, we obtain a Riemannian structure on $\mathbb{E}$ as follows. Subordinated to a local finite covering $\{U_\alpha | \alpha \in J\}$ over which $\mathbb{E}$ trivializes and J is the index set of the atlas, there is a partition of unity $\{\varphi_\alpha | \alpha \in J\}$. Then

$$\langle , \rangle := \sum_{\alpha \in J} \varphi_\alpha \langle , \rangle_\alpha$$

is the desired structure. We have the following theorem [Hi].

Theorem 2.11.1 Let $\Psi: \mathbb{E} \longrightarrow \mathbb{F}$ be a strong bundle isomorphism between two vector bundles $\mathbb{E}$ and $\mathbb{F}$ over M, each of which is equipped with a Riemannian structure. Then Ψ is homotopic to a strong bundle isomorphism from $\mathbb{E}$ into $\mathbb{F}$, which fibrewise is an isometry.

Proof: Assume $\mathbb{E} = M \times E$ and $\mathbb{F} = M \times F$. Since by Proposition 1.5.6 the space of all self–adjoint positive linear automorphisms of an Euclidean vector space is contractible to the identity map we may assume without loss of generality by using the theorem of Fischer–Riesz [Gr,1], that $E = F$ and that the Riemannian structures on $\mathbb{E}$ and $\mathbb{F}$, which are nothing else but maps from M into V^2E', are identical and constant, say $<,>$. Now Ψ yields a map

$$\Psi_M : M \longrightarrow GL(E)$$

by setting $\Psi_M(p) = \Psi | \mathbb{E}_p$. Again by Proposition 1.5.6 Ψ_M is homotopic to a smooth map $h_M : M \longrightarrow 0(E)$, where $0(E)$ is the orthogonal group of $<,>$ on E. Hence

$$h : \mathbb{E} \longrightarrow \mathbb{F}$$

given by $h | \mathbb{E}_p = h_M(p)$ is a strong fibrewise isometric bundle isomorphism.

Applying this procedure to $U_\alpha \times E$ and $U_\alpha \times F$, where $\{U_\alpha | \alpha \in J\}$ is a local finite covering of M over which $\mathbb{E}$ and $\mathbb{F}$ both trivialize simultaneously we obtain the above theorem.

References, Chapter 2

[A]	Atiyah, M.	K–Theory, W.A. Benjamin, Inc., Amsterdam, New York, (1967).
[Be]	Berger, M.	Lectures in Riemannian Geometry, Tata Institute of Fundamental Research, Bombay, (1965).
[D,III]	Dieudonné, J.	Treatise on Analysis, (Vol III), Academic Press, New York and London, (1972).
[Go,Gui]	Golubitski, M. Guillemin, V.	Stable Mappings and their Singularities, Springer Verlag, New York, Heidelberg, Berlin, (1980).
[G,St]	Goldschmidt, H. Sternberg, S.	The Hamilton–Cartan Formalism in the Calculus of Variations, Annals Inst. Fourier (Grenoble), (vol.23), pp.203–267, (1973).
[Gr,1]	Greub, W.	Linear Algebra, (4 ed.), Springer Verlag, New York, Heidelberg, Berlin, (1975).
[G,H,V]	Greub, W. Halperin, S. Vanstone, R.	Connections, Curvature and Cohomology, (vol I and II), Academic Press, New York and London, (1972).
[Hi]	Hirsch, M.W.	Differential Topology, Springer Verlag, New York, Heidelberg, Berlin, (1976).
[K]	Kelley, J.L.	General Topology, Springer Verlag, New York, Heidelberg, Berlin, (1975).
[L,2]	Lang, S.	Algebra, Addison–Wesley Publishing Company, Reading, Massachusetts, Menlo Park, California, (1965).
[W]	Warner, F. W.	Foundations of Differentiable Manifolds and Lie Groups, Springer–Verlag, New York, Heidelberg, Berlin,(1983).

Chapter 3

ELEMENTARY DIFFERENTIAL GEOMETRY

We begin this small chapter by a lemma from analysis.

3.1 The Lemma of Poincaré

Let M be a smooth manifold. Instead of $\Gamma\Lambda^k T^* M$ we just write $A^k(M)$. The exterior derivative

$$\partial : A^k(M) \longrightarrow A^{k+1}(M)$$

is given by

$$\partial\omega(X_0,\ldots,X_k) = \sum_{s=0}^{k} (-1)^s X_s(\omega(X_0,\ldots,\hat{X}_s,\ldots,X_k))$$
$$+ \sum_{0\leq s<r\leq k} (-1)^{s+r}\omega([X_s,X_r],X_0,\ldots,\hat{X}_s,\ldots,\hat{X}_r,\ldots,X_k)$$

for all $X_0,\ldots,X_k \in \Gamma TM$. Here $\hat{X}_s$ means that X_s has to be deleted from $\omega(X_0,\ldots,X_k)$.

Let us give a local expression of this formula. We choose any chart U of M. As usual we identify domain and range of the chart map. The k–form ω restricted to the k–fold cartesian product $T_pU \times \cdots \times T_pU$ is denoted by $\omega(p)$. Clearly $\omega(p)$ is a real–valued skew symmetric k–linear map. Since ω is smooth we thus can view its restriction to U as a smooth map

$$\omega_U\colon U \longrightarrow (\Lambda^k\mathbb{R}^n)' = \Lambda^k(\mathbb{R}^{n'})$$

where n is the dimension of M. This map assigns to each $p \in U$ the map $\omega(p)$. Then we immediately verify

$$(\partial\omega)_U(p)(v_0,\ldots,v_k) = \sum_{k=0}^{s} (-1)^s D\omega_U(p)(v_s)(v_0,\ldots,\hat{v}_s,\ldots,v_k).$$

Here D $\omega_U(p)(v_s)$ denotes the Fréchet derivative of ω_U at p evaluated at v_s.

As an example consider $f \in A^0(M) = C^\infty(M,\mathbb{R})$. Let df: $TM \longrightarrow \mathbb{R}$ be given by splitting

$$Tf : TM \longrightarrow T\mathbb{R} = \mathbb{R} \times \mathbb{R}$$

into Tf = (f,df), as mentioned in section 1.1. Then

$$\partial f(X) = X(f) = df(X)$$

for all $X \in \Gamma TM$. Here we have written df(X) instead of df $\circ$ X. Moreover we have

$$\begin{aligned}\partial(\partial f)(X,Y) &= X(df(Y)) - Y(df(X)) - df([X,Y]) \\ &= df([X,Y]) - df([X,Y]) = 0\end{aligned}$$

for all $X,Y \in \Gamma TM$. This is a special case of the following

<u>Lemma 3.1.1</u> For any $\omega \in A^k(M)$ and any integer $k \geq 0$

$$\partial(\partial\omega) = 0.$$

The proof is standard and can be found in [L] e.g.

The lemma of Poincaré states the converse for very special manifolds M. A subset $0 \subset \mathbb{R}^n$ is called <u>star–shaped</u> with respect to a point $p_0 \in 0$, if the segment

$$[p_0,q] := \{p_0 + t(q-p_0) \mid t \in [0,1]\}$$

is contained in 0 for any $q \in 0$.

The <u>lemma of Poincaré</u> reads now:

<u>Lemma 3.1.2</u> Let $0 \subset \mathbb{R}^n$ be an open set which is star–shaped with respect to zero. If $\omega \in A^k(0)$ satisfies

$$\partial\omega = 0$$

then there is $\alpha \in A^{k-1}(0)$ such that

$$\omega = \partial\alpha.$$

This statement holds for any integer $k \geq 1$.

Proof: We will construct a linear map

$$h : A^k(0) \longrightarrow A^{k-1}(0)$$

such that for any $\omega \in A^k(0)$

$$\partial(h\omega) + h(\partial\omega) = \omega$$

is satisfied. Evidently $\partial\omega = 0$ implies

$$\omega = \partial(h\omega).$$

The map h is defined pointwise by

$$(h\omega)(x)(v_1,\ldots,v_{k-1}) = \int_0^1 t^{k-1}\omega(tx)(x,v_1,\ldots,v_{k-1})dt$$

for any $\omega \in A^k(0)$, $x \in 0$ and $v_1,\ldots,v_{k-1} \in \mathbb{R}^n$. As usual here we have identified T_y0 and $\mathbb{R}^n$ for all $y \in 0$. Using the definition of h a straightforward calculation shows that $\partial(h\omega) + h(\partial\omega) = \omega$.

Given a smooth manifold M it is not necessarily a subset of $\mathbb{R}^n$ and star–shaped as defined above makes no sense then.

Therefore we define

$Z^k : = \{\omega \in A^k(M) \mid \partial\omega = 0\}$ for $k \geq 0$,

$B^k : = \{\omega \in A^k(M) \mid \omega = \partial\alpha$ for some $\alpha \in A^{k-1}(M)\}$ for $k \geq 1$

and $B^0 = 0$. Both Z^k and B^k are vector spaces. Obviously $B^k \subset Z^k$. We set

$$H^k(M) : = Z^k/B^k.$$

This vector space $H^k(M)$ is called the k–th cohomology group of M with coefficients in $\mathbb{R}$. Evidently lemma 3.1.2 holds on all of M iff $H^k(M) = 0$.

In fact $H^k(M) \neq 0$ in general. But trivially $H^k(M) = 0$ for all $k > \dim M$. We refer to [D,IX] for the relation of $H^k(M)$ with topologically defined cohomology groups, as expressed by the theorem of de Rham.

Next we extend the <u>exterior derivative</u> ∂ from $\mathbb{R}$–valued forms to E–valued forms for any finite dimensional real vector space E.

Let us choose a basis $e_1,\ldots,e_n$ in E. Then any $\alpha \in \Gamma L(\Lambda^k TM, M \times E)$ is of the form

$$\alpha = \sum_{i=1}^{n} \omega^i e_i$$

with $\omega^i \in A^k(M)$. Then we define

$$\partial\alpha := \sum_{i=1}^{n} (\partial\omega^i)e_i.$$

It is left to the reader to verify that $\partial\alpha$ is independent of the particular basis chosen. Clearly

$$\partial(\partial\alpha) = 0$$

and in particular, if $f \in C^\infty(M,E)$ then

$$\partial(df) = 0.$$

Here df is the E–valued 1–form given by $Tf = (f,df)$ as in case of $E = \mathbb{R}$.

As a generalization of lemma 3.1.2 we easily obtain

<u>Lemma 3.1.3</u> Let α be an E–valued k–form defined on some open set $0 \subset \mathbb{R}^n$ which is star–shaped with respect to $0 \in \mathbb{R}^n$. If $k \geq 1$ and $\partial\alpha = 0$ then there is an E–valued (k–1)–form β on 0 such that $\partial\beta = \alpha$.

In concluding this section we define the <u>pull back of a form</u>. Let M,N be smooth manifolds and

$$f : M \longrightarrow N$$

smooth map. Any E–valued k–form $\alpha \in \Gamma L(\Lambda^k TN, N \times E)$, where E is a finite dimensional real vector space, gives rise to the pull back $f^* \alpha \in \Gamma L(\Lambda^k TM, M \times E)$ given by

$$f^* \alpha(v_1,\dots,v_k) = \alpha(Tf(v_1),\dots,Tf(v_k))$$

for all $v_1,\dots,v_k \in T_pM$ and $p \in M$. If L is another smooth manifold, then the chain rule implies

$$(f \circ g)^* = g^* \circ f^*$$

for any smooth map $g : L \longrightarrow M$. In particular, if f is a diffeomorphism, then

$$(f^{-1})^* = (f^*)^{-1} .$$

<u>Lemma 3.1.4</u> $\partial f^* = f^* \partial.$

The proof is done e.g. via a simple calculation in charts. This is left to the reader.

As a particular consequence of the above lemma the k–th cohomology groups of diffeomorphic manifolds are isomorphic. In particular lemma 3.1.3 is valid for any vector valued k–form, $k \geq 1$, defined on a smooth manifold for which $H^k(M) = 0$.

3.2 Induced Riemannian metrics, covariant derivatives and second fundamental tensors on submanifolds of Euclidean spaces

By a Riemannian metric on a smooth manifold M we mean a smooth map

$$g : TM \oplus TM \longrightarrow \mathbb{R}$$

such that $g|T_pM \times T_pM$ is a symmetric positive definite bilinear map for all $p \in M$. Clearly, any Riemannian metric on M can be regarded as a smooth section in ΓV^2T^*M. Thus a Riemannian metric on M is nothing but a Riemannian structure on TM as defined in section 2.11. As shown there, any smooth manifold admits a Riemannian metric.

Another way of constructing a Riemannian metric on M is as follows. Choose some smooth embedding of M into $\mathbb{R}^n$, i.e. a smooth map

$$j : M \longrightarrow \mathbb{R}^n$$

for which $Tj(p)$ is injective for all $p \in M$ and which maps M homeomorphically onto $j(M) \subset \mathbb{R}^n$. We refer to [Hi] for Whitney's embedding theorem, which guarantees the existence of such an embedding if n is sufficiently large. As mentioned in the previous section we have the $\mathbb{R}^n$–valued 1–form

$$dj: TM \longrightarrow \mathbb{R}^n$$

given by $Tj = (j,dj)$. Next let $<,>$ be a scalar product on $\mathbb{R}^n$. Then define a Riemannian metric on M by

$$g(X,Y) := <djX,djY>$$

for all $X,Y \in \Gamma TM$, where

$$<djX,djY>(p) = <dj(X(p)),dj(Y(p))>$$

for all $p \in M$. Instead of $g(X,Y)$ one usually writes in this situation $j^*<X,Y>$.

Associated with $Tj(TM) \subset \mathbb{R}^n \times \mathbb{R}^n$ we have for each $p \in M$

$$\nu_p := \{w \in \mathbb{R}^n \mid \langle w, dj(v) \rangle = 0 \text{ for all } v \in T_pM\}.$$

ν_p is called the <u>normal space</u> of $dj(T_pM) \subset \mathbb{R}^n$. Let

$$\nu := \bigcup_{p \in M} \{p\} \times \nu_p \subset M \times \mathbb{R}^n$$

be the disjoint collection of all normal spaces. It is a vector bundle over M with typical fibre $\mathbb{R}^{n-m}$, where m is the dimension of M. To verify it, let

$$\pi_\nu : \nu \longrightarrow M$$

denote the projection onto the first factor. If U is any chart of M for which TU trivializes as $TU = U \times \mathbb{R}^m$, then for all $p \in U$ we can identify $dj(T_pU)$ with $\mathbb{R}^m$ and hence ν_p with $\mathbb{R}^{n-m}$, the orthogonal complement of $\mathbb{R}^m \subset \mathbb{R}^n$. Thus we get a map

$$\pi_\nu^{-1}(U) \xrightarrow{\ \Psi\ } U \times \mathbb{R}^{n-m}$$

which is linear on the fibres. If $\{U_\alpha \mid \alpha \in J\}$ is an atlas of M, then $\{(\pi_\nu^{-1}(U_\alpha), \Psi_\alpha) \mid \alpha \in J\}$ is an atlas on ν, which turns ν into a smooth vector bundle, in fact into a subbundle of $M \times \mathbb{R}^n$.

ν is called the <u>normal bundle</u> of M inside $\mathbb{R}^n$. It is a simple exercise to show that there exists a strong bundle isomorphism

$$TM \oplus \nu \cong M \times \mathbb{R}^n .$$

Now let us examine dj. For any $X, Y \in \Gamma TM$ we have smooth maps

$$dj(Y) : M \longrightarrow \mathbb{R}^n$$

and

$$d(dj(Y))(X) : M \longrightarrow \mathbb{R}^n .$$

For each $p \in M$ we split $d(dj(Y))(X)(p) \in \mathbb{R}^n$ into the components A(p) and B(p) with respect to $dj(T_pM) \subset \mathbb{R}^n$ and $\nu_p \subset \mathbb{R}^n$ respectively. Thus

$$d(dj(Y))(X)(p) = A(p) + B(p).$$

Call A(p) and B(p) the tangential and the normal component of $d(dj(Y))(X)(p)$ respectively. Evidently there is a unique vector field $Z \in \Gamma TM$ such that

$$dj(Z)(p) = A(p)$$

for all $p \in M$. Since Z depends on X and Y, we replace Z by $\nabla_X Y$. Here ∇ expresses that $\nabla_X Y \in \Gamma TM$ is obtained via a differentiation.

Moreover B(p) also depends on X and Y. Let us replace it by S(X,Y)(p) for all $p \in M$. The following formulas are easy to verify. For all $X, X_1, X_2, Y, Y_1, Y_2 \in \Gamma TM$ and $f \in C^\infty(M,\mathbb{R})$ we have

1) $\nabla_X(Y_1+Y_2) = \nabla_X Y_1 + \nabla_X Y_2$
2) $\nabla_X(fY) = df(X)\cdot Y + f\cdot \nabla_X Y$
3) $\nabla_{X_1+X_2} Y = \nabla_{X_1} Y + \nabla_{X_2} Y$
4) $\nabla_{fX} Y = f\,\nabla_X Y$
5) $S(X_1 + X_2, Y) = S(X_1,Y) + S(X_2,Y)$
6) $S(X,Y_1+Y_2) = S(X,Y_1) + S(X,Y_2)$
7) $S(fX,Y) = S(X,fY) = f\,S(X,Y)$.

These formulas justify to call ∇ the <u>covariant derivative associated with j</u>. S is referred to as the <u>second fundamental tensor</u> of j.

Our next observations are based on

$$\partial\, dj(X,Y) = 0$$

for all $X,Y \in \Gamma TM$. We have

$$\begin{aligned} 0 &= \partial\, dj(X,Y) \\ &= dj(\nabla_X Y) - dj(\nabla_Y X) - dj([X,Y]) + S(X,Y) - S(Y,X). \end{aligned}$$

Thus we obtain:

Proposition 3.2.1 Any smooth embedding $j : M \longrightarrow \mathbb{R}^n$ defines a covariant derivative ∇ for TM and a second fundamental tensor S for which

$$\nabla_X Y - \nabla_Y X = [X,Y]$$

and

$$S(X,Y) = S(Y,X)$$

for all $X,Y \in \Gamma TM$.

Next let us consider the smooth function

$$j^* \langle X,Y \rangle : M \longrightarrow \mathbb{R}$$

for any choice of $X,Y \in \Gamma TM$. Given $Z \in \Gamma TM$ we have

$$\begin{aligned} d(j^* \langle X,Y \rangle)(Z) &= d(\langle dj(X),dj(Y) \rangle)(Z) = \\ &= \langle d(dj(X))(Z),dj(Y) \rangle + \langle dj(X),d(dj(Y))(Z) \rangle \\ &= j^* \langle \nabla_Z X,Y \rangle + j^* \langle X,\nabla_Z Y \rangle. \end{aligned}$$

We will prove:

Proposition 3.2.2 The covariant derivative associated with any smooth embedding $j : M \longrightarrow \mathbb{R}^n$ depends only on $j^* \langle , \rangle$.

Proof: Since for any triplet $X,Y,Z \in \Gamma TM$

$$d(j^* \langle X,Y \rangle)(Z) = j^* \langle \nabla_Z X,Y \rangle + j^* \langle X,\nabla_Z Y \rangle$$

we deduce by switching the roles of X and Z as well as of Y and Z

$$d(j^* \langle Y,Z \rangle)(X) = j^* \langle \nabla_X Y,Z \rangle + j^* \langle Y,\nabla_X Z \rangle$$

and

$$d(j^* \langle Z,X \rangle)(Y) = j^* \langle \nabla_Y X,Z \rangle + j^* \langle Z,\nabla_Y X \rangle.$$

Thus we find

$$\begin{aligned} d(j^* \langle X,Y \rangle)(Z) &= j^* \langle \nabla_X Z,Y \rangle + j^* \langle [Z,X],Y \rangle \\ &+ j^* \langle X,\nabla_Y Z \rangle - j^* \langle X,[Y,Z] \rangle \end{aligned}$$

and

$$d(j^*\langle Z,X\rangle)(Y) = j^*\langle \nabla_Y Z,X\rangle + j^*\langle Z,\nabla_X Y\rangle$$
$$-j^*\langle Z,[X,Y]\rangle.$$

Then by forming

$$d(j^*\langle Y,Z\rangle)(X) - d(j^*\langle X,Y\rangle)(Z) + d(j^*\langle Z,X\rangle)(Y)$$

we find

$$\begin{aligned} &j^*\langle \nabla_X Y,Z\rangle \\ &= \tfrac{1}{2}(d(j^*\langle Y,Z\rangle)(X) + d(j^*\langle Z,X\rangle)(Y) \\ &\quad - d(j^*\langle X,Y\rangle)(Z) + j^*\langle Z,[X,Y]\rangle \\ &\quad + j^*\langle Y,[Z,X]\rangle - j^*\langle X,[Y,Z]\rangle), \end{aligned}$$

an expression of which the right hand side does not depend on ∇ any more. Thus given any embedding $i : M \longrightarrow \mathbb{R}^n$ for which

$$j^*\langle,\rangle = i^*\langle,\rangle,$$

its associated covariant derivative, denoted by $\tilde{\nabla}$, satisfies

$$j^*\langle \nabla_X Y,Z\rangle = j^*\langle \tilde{\nabla}_X Y,Z\rangle.$$

Hence $\nabla_X Y = \tilde{\nabla}_X Y$ since $j^*\langle,\rangle$ is non degenerated everywhere.

The covariant derivative associated with j is called the <u>covariant derivative of Levi–Cività</u> associated with the Riemannian metric $j^*\langle,\rangle$.

A celebrated theorem of Nash states, that any Riemannian metric g on a smooth manifold M is of the form $g = j^*\langle,\rangle$ for some embedding j mapping M into an $\mathbb{R}^n$ of very high dimension n. Thus any Riemannian metric yields the covariant derivative of Levi–Cività. This will be seen in this chapter on a much more elementary way then by using Nash's theorem.

In the next section we take up the concept of a connection introduced in chapter 2 and return to embeddings later again.

3.3 Linear connections, sprays, geodesics and the exponential map

Let M be a smooth manifold of dimension n and

$$C : TM \oplus TM \longrightarrow T^2M$$

a connection in TM. As shown in section 2.10 C induces a covariant derivative ∇ on TM. Conversely, given a covariant derivative ∇ on TM, we redefine a connection in TM based on the defining equation of the covariant derivative of a connection given in section 2.10 as follows:

Set

$$C(X(p),Y(p)) = TY(p)(X(p)) - (\nabla_X Y)(p) \in T_{Y(p)}TM$$

for all $X,Y \in \Gamma TM$, where $(\nabla_X Y)(p) \in T_pM$ is considered as an element of $T_{Y(p)}(T_pM) \subset T^2M$. Observe that C(X(p),Y(p)) depends only on X(p) and Y(p) and not on X and Y. Thus C is a map defined on $TM \oplus TM$. Moreover C satisfies the conditions of a connection, and its associated covariant derivative is ∇. Therefore we have a canonical one–to–one correspondence between connections and covariant derivatives on TM. This justifies that both notions are often used simultaneously for ∇ as well as for C.

Let us describe C and ∇ locally. Given any chart U of M, the representation of C in U is a map

$$C_U : U \times \mathbb{R}^n \times \mathbb{R}^n \longrightarrow U \times \mathbb{R}^n \times \mathbb{R}^n \times \mathbb{R}^n$$

which is due to the conditions (1) and (2) in the definition of connections given in section 2.10 of the form

$$C_U(p,v,w) = (p,w,v, -\Gamma(p)(v,w))$$

for all $p \in U$ and $v,w \in \mathbb{R}^n$. Moreover for each $p \in U$

$$\Gamma(p) : \mathbb{R}^n \times \mathbb{R}^n \longrightarrow \mathbb{R}^n$$

is bilinear by (3) and (4) of the definition of a connection in section 2.10.

To compute ∇ in local charts, let X and Y be two smooth vector fields on M. Their local representations read as

$$X_U = (\mathrm{id},x) : U \longrightarrow U \times \mathbb{R}^n,$$

$$Y_U = (\mathrm{id},y) : U \longrightarrow U \times \mathbb{R}^n.$$

Then

$$(TY \circ X)_U : U \longrightarrow U \times \mathbb{R}^n \times \mathbb{R}^n \times \mathbb{R}^n$$

maps each $p \in U$ into $(p,y(p),x(p),Dy(p)(x(p)))$ and thus

$$(TY \circ X)_U(p) - C_U(p,x(p),y(p))$$
$$= (p,y(p),O,Dy(p)(x(p)) + \Gamma(p)(x(p),y(p)))$$

since the difference has to be taken in the vector space

$$T_{y(p)}TU = p \times y(p) \times \mathbb{R}^n \times \mathbb{R}^n.$$

Identifying the vertical space $T_{y(p)}(T_pU) = p \times y(p) \times O \times \mathbb{R}^n$ with $T_pU = p \times \mathbb{R}^n$ yields

$$(\nabla_X Y)_U(p) = (p,Dy(p)(x(p)) + \Gamma(p)(x(p),y(p))).$$

If $e_1,\dots,e_n$ is a basis of $\mathbb{R}^n$, we can write

$$\Gamma(p)(e_i,e_j) = \sum_{k=1}^{n} \Gamma^k_{ij}(p)\, e_k$$

where the Γ^k_{ij} are smooth real valued functions defined on U. They are called the <u>Christoffel symbols of second kind</u>. Thus we obtain

$$(\nabla_X Y)_U(p) = (p,\sum_k(\sum_i x^i(p)\cdot D_i y^k(p) + \sum_{i,j} x^i(p) y^j(p) \Gamma^k_{ij}(p)) e_k)$$

where

$$x = \sum_i x^i e_i \text{ and } y = \sum_i y^i e_i$$

and D_i denotes the partial derivative $\frac{\partial}{\partial e_i}$. In particular, if we consider D_i as a vector field on U, i.e. $D_i(p) = (p,e_i)$ for all $p \in U$, then we have

$$(\nabla_{D_i} D_j)(p) = (p, \sum_k \Gamma^k_{ij}(p) e_k) = \sum_k \Gamma^k_{ij}(p)\, D_k(p).$$

Next we define the notion of a symmetric connection. To do so recall from section 1.1 that any $a_v \in T_vTM$ can be considered as a linear map

$$a_v : C^\infty(TM,\mathbb{R}) \longrightarrow \mathbb{R}$$

satisfying

$$a_v(fg) = a_v(f)g(v) + f(v)a_v(g)$$

for all $f,g \in C^\infty(TM,\mathbb{R})$. Conversely, any such map can be viewed as a tangent vector of T^2M at $v \in TM$. Moreover, if $a_v(d\tau) = 0$ for all $\tau \in C^\infty(M,\mathbb{R})$, then $a_v = 0$. Thus a_v is uniquely determined by its values $a_v(d\tau)$ for all $\tau \in C^\infty(M,\mathbb{R})$.

Now consider the <u>canonical involution</u>

$$\iota : T^2M \longrightarrow T^2M$$

as introduced in section 2.2. This map is uniquely determined by the following two conditions:

(i) $T\pi_M \circ \iota = \pi_{TM}$

(ii) $\iota(a)(d\tau) = a(d\tau)$ for all $a \in T^2M$ and all $\tau \in C^\infty(M,\mathbb{R})$.

If ι is represented locally as

$$\iota_U : U \times \mathbb{R}^n \times \mathbb{R}^n \times \mathbb{R}^n \longrightarrow U \times \mathbb{R}^n \times \mathbb{R}^n \times \mathbb{R}^n,$$

then

$$\iota_U(p,v,w,z) = (p,w,v,z)$$

for all $p \in U$ and $v,w,z \in \mathbb{R}^n$.

A connection C in TM is called symmetric, if

$$C(v_p, w_p) = \iota \circ C(w_p, v_p)$$

for all $v_p, w_p \in T_pM$ and all $p \in M$. If C is written locally in the form

$$C_U(p,v,w) = (p,w,v,-\Gamma(p)(v,w))$$

as above, then symmetry of C implies

$$\Gamma(p)(v,w) = \Gamma(p)(w,v)$$

for all $p \in U$ and $v,w \in \mathbb{R}^n$ and thus

$$\Gamma^k_{ij} = \Gamma^k_{ji} \quad \text{for all } i,j,k = 1,\dots,n.$$

The torsion tensor T of a connection C is defined by

$$T(X,Y) = \nabla_X Y - \nabla_Y X - [X,Y]$$

for all $X,Y \in \Gamma TM$. T is bilinear over $C^\infty(M,\mathbb{R})$ and thus $T(X,Y)(p)$ depends only on $X(p)$ and $Y(p)$. Locally T is given by

$$T_U(D_i,D_j) = \sum_k (\Gamma^k_{ij} - \Gamma^k_{ji}) D_k$$

since $[D_i,D_j] = 0$. Therefore we have

Lemma 3.3.1 A connection on TM is symmetric iff its torsion tensor vanishes.

We define the spray $S : TM \longrightarrow T^2M$ of a connection C by

$$S(v) := C(v,v)$$

for all $v \in TM$. The map S is certainly a smooth vector field on TM by (2), (here we refer again to the conditions (1) – (4) in the definition of connections given in section 2.10). In fact it is a vector field of second order, meaning that

$$T\pi_M S = id_{TM} .$$

This is due to (1).

Moreover let $\lambda \in \mathbb{R}$ and denote the multiplication of every vector $v \in TM$ by λ with the symbol μ_λ. Clearly $\mu_\lambda : TM \longrightarrow TM$ is smooth. Then we find by (3) and (4)

$$S(\lambda v) = \lambda \cdot T\mu_\lambda\, S(v)$$

for each $\lambda \in \mathbb{R}$, saying that S is <u>quadratically homogeneous</u>.

Given a chart U of M, the local representation of a spray S is given as

$$S_U : U \times \mathbb{R}^n \longrightarrow U \times \mathbb{R}^n \times \mathbb{R}^n \times \mathbb{R}^n$$

sending each tuple (p,v) into $(p,v,v,\phi(p)(v))$, where for all $p \in U$

$$\phi(p) : \mathbb{R}^n \longrightarrow \mathbb{R}^n$$

is a smooth homogeneous map of order two and thus a quadratic form. Clearly, if S is the spray of a connection C, then

$$\phi(p)(v) = -\Gamma(p)(v,v)$$

for all $p \in U$ and $v \in \mathbb{R}^n$.

<u>Lemma 3.3.2</u> Let S be a spray on M. Then there exists a unique symmetric connection C on TM, such that S is the spray of C.

<u>Proof</u>: Let first C be a symmetric connection with spray S. If C and S are represented locally by Γ and ϕ as above, then

$$\Gamma(p)(v,w) = -\tfrac{1}{2}(\phi(p)(v+w) - \phi(p)(v) - \phi(p)(w)) \qquad (*)$$

for all $p \in U$ and $v,w \in \mathbb{R}^n$. Thus Γ is determined by ϕ, which shows the uniqueness of C.

Now let S be a spray on M. Again we represent S locally by

$$S_U(p,v) = (p,v,v,\phi(p)(v)).$$

For each $p \in U$ let

$$-\Gamma(p) : \mathbb{R}^n \times \mathbb{R}^n \longrightarrow \mathbb{R}^n$$

be the symmetric bilinear form associated with the quadratic from $\phi(p)$, i.e. $\Gamma(p)$ is defined by equation (*). Setting

$$C_U(p,v,w) = (p,w,v,-\Gamma(p)(v,w))$$

we thus get a symmetric connection on TU with S_U as its spray. If $V \subset M$ is another chart, then C_U and C_V coincide over $U \cap V$ due to the uniqueness. Therefore we get a globally defined symmetric connection with spray S.

An explicite formula for the symmetric connection C with spray S is given by

$$C(v_p,w_p)(d\tau) = \tfrac{1}{2}(S(v_p+w_p)(d\tau)-S(v_p)(d\tau)-S(w_p)(d\tau))$$

for all $v_p,w_p \in T_pM$, $p \in M$ and $\tau \in C^\infty(M,\mathbb{R})$. As remarked above, the vector $C(v_p,w_p) \in T_{w_p}TM$ is determined by its values on $d\tau$ for all $\tau \in C^\infty(M,\mathbb{R})$.

Finally we introduce the notion of geodesics and define the exponential map of a spray S. A smooth curve $\sigma : I \longrightarrow M$ defined on an open interval $I \subset \mathbb{R}$ is called a <u>geodesic</u> if

$$\dot{\sigma} : I \longrightarrow TM$$

is an integral curve of S. If on the other hand $\beta : I \longrightarrow TM$ is some integral curve of S, then due to the fact that S is a vector field of second order

$$\pi_M \circ \beta : I \longrightarrow M$$

is a geodesic. More precisely

$$(\pi_M \circ \beta)^{\cdot} = \beta .$$

Thus for any $v \in T_pM$ there exists a geodesic σ_v defined on an open interval containing zero which satisfies the initial conditions $\sigma_v(0) = p$ and $\dot{\sigma}_v(0) = v$. It is unique up to the domain of definition. Therefore we have a unique geodesic of which the domain can not be enlarged any further and which satisfies the above initial conditions.

Given a connection C on TM with spray S and a smooth curve $\sigma : I \longrightarrow M$ we can consider

$$\nabla_{D_t}\dot{\sigma} \in T_{\sigma(t)}M$$

as defined in section 2.10, where $D_t \in T_tI$ is equal to $1 \in \mathbb{R}$ under the canonical identification $T_tI = \mathbb{R}$. This vector is also denoted by $\nabla_{\dot{\sigma}(t)}\dot{\sigma}$ or $\nabla_{\dot{\sigma}}\dot{\sigma}(t)$, thus $\nabla_{\dot{\sigma}}\dot{\sigma}$ is a smooth curve in TM. Recalling the definition of $\nabla_{D_t}\dot{\sigma}$ given in section 2.10, we have

$$\begin{aligned}\nabla_{\dot{\sigma}}\dot{\sigma}(t) &= T\,\dot{\sigma}(t)(D_t) - C(T\sigma(t)(D_t),\dot{\sigma}(t))\\ &= \ddot{\sigma}(t) - S(\dot{\sigma}(t))\end{aligned}$$

where $\nabla_{\dot{\sigma}}\dot{\sigma}(t)$ is considered as an element in $T_{\dot{\sigma}(t)}(T_{\sigma(t)}M)$. This shows that σ is a geodesic iff

$$\nabla_{\dot{\sigma}}\dot{\sigma} = 0.$$

The local representation of a geodesic σ is immediately obtained by writing the defining equation

$$\ddot{\sigma}(t) = C(\dot{\sigma}(t),\dot{\sigma}(t))$$

in local form. It reads as

$$\ddot{\sigma}(t) = -\sum_{k,i,j} \Gamma^k_{ij}(\sigma(t))\dot{\sigma}^i(t)\,\dot{\sigma}^j(t)\,e_k$$

or

$$\ddot{\sigma}^k(t) = -\sum_{i,j} \Gamma^k_{ij}(\sigma(t))\dot{\sigma}^i(t)\dot{\sigma}^j(t)$$

where

$$\dot{\sigma}(t) = \sum_i \dot{\sigma}^i(t)\,e_i\,.$$

To define the <u>exponential map</u> at a point $p \in M$, let $W \subset T_pM$ be the set of all tangent vectors v at p such that the maximal geodesic σ_v with initial conditions $\sigma_v(0) = p$ and $\dot{\sigma}_v(0) = v$ is defined at $1 \in \mathbb{R}$. Then W is an open zero neighbourhood in T_pM due to the quadratic homogeneity of S. Define

$$\exp : W \longrightarrow M$$

by $\exp(v) = \sigma_v(1)$ for all $v \in W$. This map, called the exponential map at p, is smooth. Moreover

$$T_0\exp : T_0W \longrightarrow T_pM$$

is the identity map when T_0W is identified with T_pM in the canonical way. Thus there is some open neighbourhood $V \subset W$ of zero on which

$$\exp : V \longrightarrow M$$

is a diffeomorphism onto the open neighbourhood $\exp(V)$ of $p \in M$.

3.4 The canonical one– and two–form on T^*M, Riemannian spray and the Levi–Cività connection

Next we construct the connection of Levi–Cività of a Riemannian metric g on M. In order to do so we first introduce the canonical one– and two–form on T^*M. Let M be a smooth manifold of dimension n. The canonical projection $\pi^* : T^*M \longrightarrow M$ induces a smooth map

$$T\pi^* : T\,T^*M \longrightarrow TM.$$

Given $l_p \in T_p^*M$ and $v_{l_p} \in T_{l_p}T^*M$ then $T\pi^*(l_p)(v_{l_p}) \in T_pM$. Now we define the canonical 1–form

$$\theta(v_{l_p}) = -l_p(T\pi^*(l_p)(v_{l_p})).$$

By computing θ in local coordinates one immediately deduces that θ is smooth. Let therefore $U \subset M$ be a chart over which TU, T^*U and $T\,T^*U$ trivialize as

$$TU = U \times \mathbb{R}^n$$
$$T^*U = U \times \mathbb{R}^{n'}$$

and

$$T\,T^*U = U \times \mathbb{R}^{n'} \times \mathbb{R}^n \times R^{n'}.$$

In the following we denote a smooth map and its local representation by the same symbol. Clearly $\pi^* : U \times \mathbb{R}^{n'} \to U$ sends any (p,l) into p.

Moreover

$$T\pi^* : U \times \mathbb{R}^{n'} \times \mathbb{R}^n \times \mathbb{R}^{n'} \to U \times \mathbb{R}^n$$

sends any quadrupel (p,l_1,w,l_2) with $p \in U$, $l_1,l_2 \in \mathbb{R}^{n'}$ and $w \in \mathbb{R}^n$ into (p,w). Thus

$$\theta : U \times \mathbb{R}^{n'} \times \mathbb{R}^n \times \mathbb{R}^{n'} \to \mathbb{R}$$

is given by

$$\theta(p,l_1,w,l_2) = -l_1(w).$$

Now the <u>canonical 2–form</u> ω on T^*M is given by

$$\omega := \partial\theta .$$

Clearly ω is smooth. Nevertheless let us represent ω locally in a chart U. To this end let

$$A : U \times \mathbb{R}^{n'} \to U \times \mathbb{R}^{n'} \times \mathbb{R}^n \times \mathbb{R}^{n'}$$

be a section of $\pi_{T^*U} : T\,T^*U \to T^*U$. A can be viewed as a smooth map

$$A : U \times \mathbb{R}^{n'} \to \mathbb{R}^n \times \mathbb{R}^{n'} .$$

We split A into A = (W,L) where

$$W : U \times \mathbb{R}^{n'} \to \mathbb{R}^n$$

$$L : U \times \mathbb{R}^{n'} \to \mathbb{R}^{n'}$$

are smooth maps. Then

$$\theta(A)(p,l) = -l(W(p,l))$$

for all $p \in U$ and $l \in \mathbb{R}^{n'}$. Given two vector fields A_1 and A_2 on T^*U we have

$$\omega(A_1,A_2) = \partial\theta(A_1,A_2)$$
$$= d(\theta(A_2))(A_1) - d(\theta(A_1))(A_2) - \theta[A_1,A_2].$$

As in the case of A above we write $A_1 = (W_1,L_1)$ and $A_2 = (W_2,L_2)$. Then

$$d(\theta(A_2))(A_1)(p,l) = -l(dW_2(A_1)(p,l)) - L_1(p,l)(W_2(p,l)).$$

Moreover

$$\theta[A_1,A_2](p,l) = -l(dW_2(A_1)(p,l)) + l(dW_1(A_2))(p,l).$$

Thus

$$\omega(A_1,A_2) = L_2(W_1) - L_1(W_2).$$

In particular this formula implies that ω is non degenerate. (T^*M,ω) is called a <u>symplectic manifold</u>, meaning that ω is a non degenerate 2–form on T^*M with $\partial\omega = 0$.

Let next

$$E : TM \longrightarrow \mathbb{R}$$

be a smooth function. The <u>fibre derivative</u> of E is a strong bundle map

$$\mathbb{F}E : TM \longrightarrow T^*M$$

defined by

$$\mathbb{F}E(v_p)(w_p) = dE(v_p)(w_p)$$

for all $v_p, w_p \in T_pM$ and $p \in M$, where w_p on the right hand side is considered as an element in $T_{v_p}(T_pM) \subset T^2M$. We define the pull back $(\mathbb{F}E)^*\theta$ of the canonical 1–form as in section 3.1 by

$$(\mathbb{F}E)^*\theta(v) = \theta \circ T(\mathbb{F}E)(v)$$

for all $v \in TM$. According to lemma 3.1.4 we have

$$\partial(\mathbb{F}L)^*\theta = (\mathbb{F}L)^*\omega\,.$$

Now let g be a Riemannian metric on M. Define

$$E_g : TM \longrightarrow \mathbb{R}$$

by

$$E_g(v_p) = \frac{1}{2}\, g(v_p,v_p)$$

for all $v_p \in T_pM$. Then

$$\mathbb{F}E_g(v_p)(w_p) = g(v_p,w_p)$$

and thus

$$\mathbb{F}E_g = g^{\#}$$

where

$$g^{\#} : TM \longrightarrow T^*M$$

is given by

$$g^{\#}(v_p)(w_p) = g(v_p,w_p)$$

for all $v_p, w_p \in T_pM$ and $p \in M$.

Now we define the <u>spray of the Riemannian metric</u> g as follows: Denote the pullback $(\mathbb{F}E_g)^*\omega$ of the canonical 2–form by ω_g. Since $g^{\#}$ is a bundle isomorphism, ω_g is non degenerate. Thus there is a unique vector field

$$S_g : TM \longrightarrow T^2M$$

such that

$$\omega_g(S_g,Z) = dE_g(Z)$$

for all $Z \in \Gamma(T^2M)$.

<u>Lemma 3.4.1</u> For any Riemannian metric g the vector field S_g is a spray on M. Moreover $dE_g(S_g) = 0$.

<u>Proof</u>: Since ω_g is skew symmetric we have $dE_g(S_g) = 0$. To show that S_g is a spray we proceed locally. Clearly g written on a chart U of M takes the form

$$g : U \longrightarrow S^2(\mathbb{R}^n,\mathbb{R})$$

where $S^2(\mathbb{R}^n,\mathbb{R})$ denotes the space of all $\mathbb{R}$–valued symmetric bilinear maps on $\mathbb{R}^n \times \mathbb{R}^n$. Moreover E_g represented on U is the smooth map

$$E_g : U \times \mathbb{R}^n \longrightarrow \mathbb{R}$$

given by

$$E_g(p,v) = \tfrac{1}{2}\, g(p)(v,v)$$

for all $v \in \mathbb{R}^n$ and $p \in U$. Hence

$$dE_g : U \times \mathbb{R}^n \times \mathbb{R}^n \times \mathbb{R}^n \longrightarrow \mathbb{R}$$

is given by

$$dE_g(p,v)(v_1,v_2) = \tfrac{1}{2}\, Dg(p)(v_1)(v,v) + g(p)(v,v_2)$$

with Dg(p) being the Fréchet derivative of g at p. Moreover, $\mathbb{F}E_g = g^{\#}$ takes

$(p,v) \in U \times \mathbb{R}^n$ into $(p,g(p)(v,\cdot)) \in U \times \mathbb{R}^{n'}$. Therefore

$$Tg^{\#} : U \times \mathbb{R}^n \times \mathbb{R}^n \times \mathbb{R}^n \longrightarrow U \times \mathbb{R}^{n'} \times \mathbb{R}^n \times \mathbb{R}^{n'}$$

sends any quadruple (p,v,v_1,v_2) into

$$(p,g(p)(v,\cdot),\ v_1,Dg(p)(v_1)(v,\cdot) + g(p)(v_2,\cdot)).$$

Now the pull back ω_g of the canonical 2–form ω via $g^{\#}$ is locally given by

$$\begin{aligned}
&\omega_g(p,v)((v_1,v_2),(w_1,w_2)) \\
&= \omega(p,g(p)(v,\cdot))(Tg^{\#}(p,v)(v_1,v_2),Tg^{\#}(p,v)(w_1,w_2)) \\
&= \omega(p,g(p)(v,\cdot))((v_1,Dg(p)(v_1)(v,\cdot) + g(p)(v_2,\cdot)),(w_1,Dg(p)(w_1)(v,\cdot) + g(p)(w_2,\cdot))) \\
&= Dg(p)(w_1)(v,v_1) - Dg(p)(v_1)(v,w_1) + g(p)(w_2,v_1) - g(p)(v_2,w_1).
\end{aligned}$$

On the other hand S_g locally looks like

$$S_g : U \times \mathbb{R}^n \longrightarrow U \times \mathbb{R}^n \times \mathbb{R}^n \times \mathbb{R}^n$$

sending any pair (p,v) into

$$S_g(p,v) = (p,v,v_1,v_2)$$

where v_1 and v_2 depend smoothly on (p,v). If $Z \in \Gamma TU$ is given by

$$Z(p,v) = (p,v,w_1,w_2)$$

then the equation $\omega(S_g,Z) = dE_g(Z)$ at the point (p,v) reads as

$$\begin{aligned}
&Dg(p)(w_1)(v,v_1) - Dg(p)(v_1)(v,w_1) + g(p)(w_2,v_1) - g(p)(v_2,w_1) \\
&= \tfrac{1}{2}\, Dg(p)(w_1)(v,v) + g(p)(v,w_2)\ .
\end{aligned}$$

This is valid for all $p \in U$ and $v,w_1,w_2 \in \mathbb{R}^n$. $w_1 = 0$ yields

$$g(p)(v,w_2) = g(p)(w_2,v_1)$$

for all $w_2 \in \mathbb{R}^n$ and thus $v = v_1$, which shows that S_g is a vector field of second order.

Setting $v_2 = \phi(p)(v)$ as in section 3.3 the above equation becomes

$$g(p)(\phi(p)(v),w_1) = \tfrac{1}{2}\, Dg(p)(w_1)(v,v) - Dg(p)(v)(v,w_1)$$

for all $p \in U$ and $v,w_1 \in \mathbb{R}^n$. This implies that

$$\phi(p) : \mathbb{R}^n \longrightarrow \mathbb{R}^n$$

is homogeneous of order two for all $p \in U$. Thus S_g is quadratically homogeneous and hence a spray.

From here on we proceed as in the previous section to construct the symmetric connection C with spray S_g and its covariant derivative ∇. Both C and ∇ are called the <u>connection of Levi–Cività</u> associated with g.

Given any connection $\tilde{\nabla}$ in TM, deduced from a Riemannian metric or not, we denote its torsion tensor by $T_{\tilde{\nabla}}$. Moreover, define $\tilde{\nabla}_X g$ by

$$\tilde{\nabla}_X g(Y,Z) := X(g(Y,Z)) - g(\tilde{\nabla}_X Y,Z) - g(Y,\tilde{\nabla}_X Y)$$

for all $X,Y,Z \in \Gamma TM$.

<u>Proposition 3.4.2</u> Let g be a Riemannian metric on a smooth manifold M and ∇ the associated Levi–Cività connection. Then

$$T_\nabla = 0 \text{ and } \nabla_X g = 0$$

for all $X \in \Gamma TM$. If $\tilde{\nabla}$ is any connection in TM satisfying $T_{\tilde{\nabla}} = 0$ and $\tilde{\nabla}_X g = 0$ for all $X \in \Gamma TM$, then

$$\tilde{\nabla} = \nabla .$$

<u>Proof</u>: Given any connection $\tilde{\nabla}$ satisfying $T_{\tilde{\nabla}} = 0$ and $\tilde{\nabla} g = 0$ we find as in the proof of proposition 3.2.2

$$\begin{aligned} g(\tilde{\nabla}_X Y,Z) = \tfrac{1}{2}(&X(g(Y,Z)) + Y(g(Z,X)) - Z(g(Y,X)) \\ &+ g(Z,[X,Y]) + g(Y,[Z,X]) - g(X,[Y,Z])) \end{aligned} \qquad (*)$$

for all $X,Y,Z \in \Gamma TM$. Thus $\tilde{\nabla}$ is unique. To show the existence of such a connection we fix $X,Y \in \Gamma TM$ and define a map

$$A : \Gamma TM \longrightarrow C^\infty(M,\mathbb{R})$$

by setting $A(Z)$ equal to the right hand side of (*) for each $Z \in \Gamma TM$. Then let $\tilde{\nabla}_X Y$ be given by

$$g(\tilde{\nabla}_X Y, Z) = A(Z)$$

for all $Z \in \Gamma TM$. It is easily verified that $\tilde{\nabla}$ is a covariant derivative with $T_{\tilde{\nabla}} = 0$ and $\tilde{\nabla} g = 0$. Inserting the spray $\tilde{S}$ of $\tilde{\nabla}$ into ω yields

$$\omega(\tilde{S}(v_p), Z(v_p)) = dE_g(Z(v_p))$$

for all $v_p \in TM$ and all $Z(v_p) \in T_{v_p}(TM)$. Thus $\tilde{S} = S_g$. Since $\tilde{\nabla}$ as well as the Levi–Civita connection ∇ is symmetric, we conclude $\nabla = \tilde{\nabla}$ by lemma 3.3.2.

3.5 Curvature tensors and the Bianchi identity

Given a connection on a smooth vector bundle we defined in section 2.10 its associated covariant derivative and the corresponding curvature. Using these definitions, each connection $\tilde{\nabla}$ on TM yields

$$R(X,Y)Z = \tilde{\nabla}_X \tilde{\nabla}_Y Z - \tilde{\nabla}_Y \tilde{\nabla}_X Z - \tilde{\nabla}_{[X,Y]} Z$$

for all $X,Y,Z \in \Gamma TM$.

Given any $C^\infty(M,\mathbb{R})$–trilinear map

$$A : \Gamma TM \times \Gamma TM \times \Gamma TM \longrightarrow \Gamma \mathbb{E}$$

where $\mathbb{E}$ is a smooth vector bundle over M, we write

$$\mathfrak{S}(A(X,Y,Z)) = A(X,Y,Z) + A(Z,X,Y) + A(Y,Z,X)$$

for all $X,Y,Z \in \Gamma TM$.

Using $\tilde{\nabla}_X(R)$ and $\tilde{\nabla}_X(T)$, two notions stemming from the general definition at the end of section 2.10, we immediately find

$$\mathfrak{S}(R(X,Y)Z) = \mathfrak{S}(T(T(X,Y),Z)) + \mathfrak{S}(\tilde{\nabla}_X(T)(Y,Z)).$$

Hence if $\tilde{\nabla}$ is torsion free

$$\mathfrak{S}(R(X,Y)Z) = 0$$

for all $X,Y,Z \in \Gamma TM$.

A straightforward calculation shows that for all $X,Y,Z \in \Gamma TM$

$$\mathfrak{S}(\tilde{\nabla}_X(R)(Y,Z)) + \mathfrak{S}(R(T(X,Y),Z)) = 0.$$

Here $R(Y,Z) : \Gamma TM \longrightarrow \Gamma TM$ is regarded as an TM–valued one–form. Thus if $\tilde{\nabla}$ is torsion free, we obtain the second <u>Bianchi identity</u>

$$\mathfrak{S}(\tilde{\nabla}_X(R)(Y,Z)) = 0.$$

In order to understand the curvature operator from a more formal standpoint, we extend the notion of the boundary operator ∂ introduced in 3.1 to <u>bundle valued forms</u> in the following way:

Let $\Psi \in \Gamma L(\Lambda^k TM,\mathbb{E})$, where $\mathbb{E}$ is a smooth vector bundle over M. As an example of such a form consider in case $k = 2$ and $\mathbb{E} = L(TM,TM)$ the form R given by

$$R(X,Y) = \tilde{\nabla}_X \circ \tilde{\nabla}_Y - \tilde{\nabla}_Y \circ \tilde{\nabla}_X - \tilde{\nabla}_{[X,Y]}$$

for all $X,Y \in \Gamma TM$, where $\tilde{\nabla}$ is a covariant derivative on TM. Now let ∇ be a covariant derivative on $\mathbb{E}$. Then for any non negative integer k we define

$$\partial_\nabla : \Gamma L(\Lambda^k TM,\mathbb{E}) \longrightarrow \Gamma L(\Lambda^{k+1} TM,\mathbb{E})$$

by

$$\begin{aligned}\partial_\nabla \Psi(X_0,\ldots,X_k) := &\sum_{r=0}^{k} (-1)^r \nabla_{X_r}(\Psi(X_0,\ldots,\hat{X}_r,\ldots,X_k)) \\ &+ \sum_{0\leq i<j\leq k}(-1)^{i+j}\, \Psi([X_i,X_j],X_0,\ldots,\hat{X}_i,\ldots,\hat{X}_j,\ldots,X_k)\end{aligned}$$

for all $X_0,\ldots,X_k \in \Gamma TM$. ∂_∇ is called the <u>exterior covariant derivative</u>.

If $\Psi \in \Gamma L(TM,\mathbb{E})$ then we have for all $X,Y \in \Gamma TM$

$$\partial_\nabla \Psi(X,Y) = \nabla_X \Psi(Y) - \nabla_Y \Psi(X) - \Psi([X,Y]),$$

a formula which will play an important role in the next section. Observe that for each covariant derivative ∇ on TM we have

$$\partial_{\nabla}(\mathrm{id}_{TM}) = T_{\nabla}$$

where $\mathrm{id}_{TM} : TM \longrightarrow TM$ is the identity map and T_{∇} is the torsion tensor of ∇.

Clearly if $\omega \in \Gamma\Lambda^k T^* M$ and $\Psi \in \Gamma L(\Lambda^r TM,\mathbb{E})$ then

$$\partial_{\nabla}(\omega \wedge \Psi) = \partial\omega \wedge \Psi + (-1)^k \omega \wedge \partial_{\nabla}\Psi .$$

In general $\partial_{\nabla} \circ \partial_{\nabla} = \partial_{\nabla}^2 \neq 0$. In fact, let $\mathbb{E} = TM$ and $Z : M \longrightarrow TM$ be any smooth vector field. Then

$$\partial_{\nabla}(Z)(X) = \nabla_X Z$$

and

$$\partial_{\nabla}^2(Z)(X,Y) = R(X,Y)Z$$

which is called the Ricci identity. If ∇Z is regarded as a TM–valued one form on M and ∇ is torsion free we have

$$R(X,Y)Z = \nabla_X(\nabla Z)(Y) - \nabla_Y(\nabla Z)(X) = \partial_{\nabla}(\nabla Z)(X,Y).$$

Returning to the general case, observe that if $R_{\mathbb{E}}$ denotes the curvature tensor of ∇ in $\mathbb{E}$, then for any $\mathbb{E}$–valued k–form Ψ on M

$$\partial_{\nabla}^2\Psi(X_0,\dots,X_{k+1}) =$$
$$= \sum_{\sigma} \mathrm{sign}(\sigma) R_{\mathbb{E}}(X_{\sigma(0)},X_{\sigma(1)})\Psi(X_{\sigma(2)},\dots,X_{\sigma(k+1)})$$

where the summation being extended over all permutations σ of $0,\dots,k+1$ with $\sigma(0) < \sigma(1)$ and $\sigma(2) < \cdots < \sigma(k+1)$. In short

$$\partial_{\nabla}^2 \Psi = R_{\mathbb{E}}(\Psi) .$$

If the connection ∇^L on L(TM,TM) induced by ∇ (sec. 2.10) is denoted by ∇ again the second Bianchi identity for the L(TM,TM)–valued two form R on M is written as

$$\partial_{\nabla}^2 R = 0.$$

Next we introduce the Ricci tensor for a connection ∇ on TM. Consider for given $Y,Z \in \Gamma TM$

$$\rho(Y,Z) : \Gamma TM \longrightarrow \Gamma TM$$

sending $X \in \Gamma TM$ into $R(X,Y)Z$. Then we define the <u>Ricci tensor</u>

$$\mathrm{Ric} : \Gamma TM \times \Gamma TM \longrightarrow C^{\infty}(M,\mathbb{R})$$

by

$$\mathrm{Ric}(Y,Z) = \mathrm{tr}\, \rho(Y,Z)$$

for all $Y,Z \in \Gamma TM$.

The rest of this section is devoted to the Levi–Cività connection ∇ of a Riemannian metric g on M. We introduce the <u>Riemannian curvature tensor</u> $R^{\bullet}$ by

$$R^{\bullet}(X,Y;Z,U) := g(R(X,Y)Z,U)$$

for all $X,Y,Z,U \in \Gamma TM$. In any book on differential geometry as e.g. in [Kl] elementary proofs of the following properties of the Riemannian curvature tensor are available:

$$R^{\bullet}(X,Y;Z,U) = -R^{\bullet}(Y,X;Z,U)$$
$$R^{\bullet}(X,Y;Z,U) = -R^{\bullet}(X,Y;U,Z)$$
$$R^{\bullet}(X,Y;Z,U) = R^{\bullet}(Z,U;X,Y).$$

We immediately deduce

<u>Lemma 3.5.1</u> The Ricci tensor of the Levi–Cività connection of any Riemannian metric is symmetric.

<u>Proof</u>: Let $e_1,\dots,e_n \in T_pM$ be an orthonormal basis. Then for any $v,w \in T_pM$

$$\mathrm{tr}\, \rho(v,w) = \sum_{i=1}^{n} R^{\bullet}(e_i,v;w,e_i),$$

and the symmetry follows from the properties of the Riemannian curvature tensor stated above.

Finally we turn to the scalar curvature. The Ricci tensor can be written uniquely via a smooth strong bundle map

$$r : TM \longrightarrow TM$$

such that

$$g(rX,Y) = Ric(X,Y)$$

for all $X,Y \in \Gamma TM$. Then the scalar curvature $\lambda \in C^{\infty}(M,\mathbb{R})$ is defined by

$$\lambda := \mathrm{tr}\ r.$$

A more instructive formula using the Weingarten map will be derived in the next section.

3.6 Embeddings, the Weingarten map and the second fundamental form, the equations of Gauss and Codazzi, the mean and the Gaussian curvature

In this section we consider two smooth manifolds M and N as well as a smooth embedding

$$j : M \longrightarrow N.$$

In addition we suppose that N carries a smooth Riemannian metric $\tilde{g}$. The connection of Levi–Cività of $\tilde{g}$ is denoted by $\tilde{\nabla}$. As in section 3.2 we define the normal bundle ν over M. The fibre ν_p of ν over $p \in M$ is given by

$$\nu_p := \{v \in T_{j(p)}N \mid \tilde{g}(j(p))(v,w) = 0 \ \forall \ w \in Tj(T_pM)\}.$$

It is easy to see that ν is a vector bundle over M trivializing over the same charts as TM does, namely over any chart U of M. Clearly $\tilde{g}$ induces a Riemannian metric g on M and a Riemannian structure g_ν on ν which are respectively given as follows. We set

$$g(X,Y) := j^{*}\tilde{g}(X,Y) = \tilde{g}(TjX,TjY)$$

for all $X,Y \in \Gamma TM$ and

$$g_\nu(s_1,s_2) = \tilde{g}(s_1,s_2)$$

for all $s_1,s_2 \in \Gamma\nu$. The connection $\tilde{\nabla}$ induces a connection ∇ on M determined by

$$Tj\,\nabla_X Y = (\tilde{\nabla}_X TjY)^T$$

for all $X,Y \in \Gamma TM$, where T means the pointwise formed component of $(\tilde{\nabla}_X TjY)$ in $Tj(TM)$ with respect to the decomposition

$$T_{j(p)}N = Tj(T_pM) \oplus \nu_p$$

for all $p \in M$. Here $\tilde{\nabla}_X TjY$ is defined as in section 2.10. The connection $\tilde{\nabla}$ also defines a connection ∇^ν in ν by setting

$$\nabla^\nu_X\, s = (\tilde{\nabla}_X s)^\nu$$

for each $X \in \Gamma TM$ and $s \in \Gamma\nu$, where ν denotes the pointwise formed component in ν. The following lemma is easy to prove:

<u>Lemma 3.6.1</u> The connection ∇ on TM induced by $\tilde{\nabla}$ is the Levi–Cività connection of g. The connection ∇^ν on ν satisfies

$$\nabla^\nu_X(g_\nu)(s_1,s_2) = 0$$

for all $X \in \Gamma TM$ and all $s_1,s_2 \in \Gamma\nu$.

Here $\nabla^\nu_X(g_\nu)$ is defined analogously as in section 3.4 by

$$\nabla^\nu_X(g_\nu)(s_1,s_2) = X(g_\nu(s_1,s_2)) - g_\nu(\nabla^\nu_X s_1,s_2) - g_\nu(s_1,\nabla^\nu_X s_2).$$

Let us examine the equation of a geodesic curve $\sigma : I \longrightarrow M$ in terms of $\tilde{\nabla}$. Since σ is a geodesic we have

$$\nabla_{\dot\sigma}\dot\sigma = 0.$$

Hence we obtain

Lemma 3.6.2 A smooth curve $\sigma : I \longrightarrow M$ defined on an open interval $I \subset \mathbb{R}$ is a geodesic with respect to ∇, iff

$$(\overline{\nabla}_{Tj\dot{\sigma}} Tj\dot{\sigma})(t) \in \nu_{\sigma(t)}$$

for all $t \in I$.

Next we turn to the codimension one case $k = m + 1$, with $k = \dim N$ and $m = \dim M$. For simplicity we assume M and N to be orientable, i.e. we assume the existence of nowhere vanishing smooth sections in $\Lambda^m T^* M$ and $\Lambda^k T^* N$. Since M is of codimension 1

$$\dim \nu_p = 1$$

for all $p \in M$. Thus ν is a line bundle. The orientability guarantees a smooth section $n : M \longrightarrow \nu$ such that

$$g_\nu(n,n) = 1.$$

This equation is essential. In fact it yields immediately

$$g_\nu(\nabla^\nu_X n,n) = 0$$

for all $X \in \Gamma TM$. Therefore each $X \in \Gamma TM$ determines a unique vector field $WX \in \Gamma TM$ such that

$$Tj\, WX = \overline{\nabla}_X n.$$

Note that $(WX)(p)$ depends only on $X(p)$. Thus we have a strong bundle map

$$W : TM \longrightarrow TM,$$

called the Weingarten map of the unit normal vector field n. The following lemma shows that $g(WX,Y)$ plays a special role in determining $\nabla_X Y$.

Lemma 3.6.3 For all $X,Y \in \Gamma TM$

$$\overline{\nabla}_X TjY = Tj\nabla_X Y - g(WX,Y)\cdot n.$$

Proof: We have

$$\begin{aligned} Tj\,\nabla_X Y &= \tilde{\nabla}_X TjY - \tilde{g}(\tilde{\nabla}_X TjY, n)\cdot n \\ &= \tilde{\nabla}_X TjY + \tilde{g}(TjY, \tilde{\nabla}_X n)\cdot n \\ &= \tilde{\nabla}_X TjY + \tilde{g}(TjY, TjWX)\cdot n \\ &= \tilde{\nabla}_X TjY + g(Y, WX)\cdot n. \end{aligned}$$

We introduce the two tensor h on M by

$$h(X,Y) = g(WX,Y)$$

for all $X,Y \in \Gamma TM$. h is called the <u>second fundamental form</u> of n. One of the fundamental properties of the Weingarten map is formulated in

<u>Proposition 3.6.4</u> W is self–adjoint meaning that

$$g(WX,Y) = g(WY,X)$$

for all choices of $X,Y \in \Gamma TM$. Thus h is symmetric.

Proof: Since $\tilde{\nabla}$ and ∇ are torsion free, we have

$$\tilde{\nabla}_X TjY - \tilde{\nabla}_Y TjX = Tj[X,Y] = Tj\nabla_X Y - Tj\nabla_Y X$$

and thus

$$g(WX,Y)\cdot n = g(WY,X)\cdot n$$

for all $X,Y \in \Gamma TM$.

<u>Corollary 3.6.5</u> In case of $\dim N = 1+\dim M$ a smooth curve $\sigma : I \longrightarrow M$ is a geodesic iff

$$(\tilde{\nabla}_{Tj\dot{\sigma}}\, Tj\dot{\sigma})(t) = -h(\dot{\sigma}(t),\dot{\sigma}(t))\cdot n$$

for all $t \in I$. Moreover $j \circ \sigma$ is a geodesic in N iff σ is a geodesic in M and $h(\dot{\sigma}(t),\dot{\sigma}(t)) = 0$ for all $t \in I$.

Our next goal is to relate the curvature tensors of ∇ and $\tilde{\nabla}$. We still assume $\dim N = \dim M+1$ and that both M and N are orientable. The Weingarten map W and the second fundamental form h are given relative to a fixed unit normal vector field n along j. Let R and $\tilde{R}$ denote the curvature of ∇ and $\tilde{\nabla}$ respectively. Then we have

Proposition 3.6.6 For all $X,Y,Z \in \Gamma TM$ the following equation holds:

$$\tilde{R}(TjX,TjY)TjZ =$$
$$= Tj(R(X,Y)Z + g(Z,WX)WY - g(Z,WY)WX)$$
$$+ (g(Z,\nabla_Y(W)X) - g(Z,\nabla_X(W)Y))\cdot n.$$

Proof: Using lemma 3.6.3 we compute

$$\begin{aligned}\tilde{\nabla}_X\tilde{\nabla}_Y TjZ &= \tilde{\nabla}_X Tj\nabla_Y Z - g(\nabla_X Z,WY)\cdot n - g(Z,\nabla_X(W)Y)\cdot n \\ &\quad - g(Z,W\nabla_X Y)\cdot n - g(Z,WY)\cdot TjWX \\ &= Tj\nabla_X\nabla_Y Z - g(Z,WY)\cdot TjWX - g(\nabla_Y Z,WX)\cdot n \\ &\quad - g(\nabla_X Z,WY)\cdot n - g(Z,\nabla_X(W)Y)\cdot n - g(Z,W\nabla_X Y)\cdot n.\end{aligned}$$

Hence

$$\begin{aligned}\tilde{R}(TjX,TjY)TjZ &= Tj\,R(X,Y)Z - g(Z,WY)\,TjWX + g(Z,WX)TjWY \\ &\quad - g(Z,\nabla_X(W)Y)\cdot n + g(Z,\nabla_T(W)X)\cdot n\,.\end{aligned}$$

Let $\tilde{R}^{\cdot}$ and $R^{\cdot}$ denote the Riemannian curvature tensors of $\tilde{g}$ and g respectively. Via the previous proposition we obtain the equations of Gauss and Codazzi:

Corollary 3.6.7 Let $X,Y,Z,U \in \Gamma TM$. Then the Gauss equation reads as

$$\tilde{R}^{\cdot}(TjX,TjY;TjZ,TjU) =$$
$$= R^{\cdot}(X,Y;Z,U) + h(Z,X)\cdot h(Y,U) - h(Z,Y)\cdot h(X,U)$$

and the equation of Codazzi as

$$\tilde{g}(\tilde{R}(TjX,TjY)TjZ,n) = g(Z,\nabla_Y(W)X) - g(Z,\nabla_X(W)Y)\,.$$

In particular, if $\tilde{R} = 0$ then

$$\nabla_Y(W)X = \nabla_X(W)Y$$

for all choices of $X,Y \in \Gamma TM$.

Next we will use the equation of Gauss to relate the scalar curvature of g to the second trace coefficient (the second coefficient in the characteristic polynomial) of the Weingarten map for a special type of embeddings. The situation is as follows: Let $N = \mathbb{R}^k$ and $\tilde{g} = \langle,\rangle$ a fixed scalar product, thus $\tilde{R} = 0$. Hence j embeds into a flat Euclidean space. As above we assume M to be orientable and $k = \dim M + 1$. The Weingarten map W is defined with respect to a fixed unit normal vector field n along j. The Ricci tensor at $p \in M$ is given by

$$\mathrm{Ric}(Y(p),Z(p)) = \sum_{i=1}^{k-1} R^{\cdot}(e_i,Y(p);Z(p),e_i).$$

Here $e_1,\dots,e_{k-1}$ is a basis of T_pM orthonormal with respect to $g = j^*\langle,\rangle$. Thus by the equation of Gauss we have

$$\mathrm{Ric}(Y(p),Z(p)) = \sum_{i=1}^{k-1} - g(W(p)e_i,Z(p))\cdot g(W(p)Y(p),e_i) + g(W(p)Z(p),Y(p))\cdot \mathrm{tr}\, W(p).$$

The Weingarten map $W(p) : T_pM \longrightarrow T_pM$ is selfadjoint. Thus $e_1,\dots,e_{k-1}$ can assumed to be an eigenbasis of W(p). Hence we have the following proposition:

<u>Proposition 3.6.8</u> Let $j : M \longrightarrow \mathbb{R}^k$ be a smooth embedding of an orientable manifold M into an Euclidean space $\mathbb{R}^k$ with $k-1 = \dim M$. Let moreover n be a unit normal vector field along j. Then the scalar curvature λ of M and the second trace coefficient $\alpha(W)$ of the Weingarten map W, given by n, are related by the formula

$$\lambda = -\mathrm{tr}\, W^2 + (\mathrm{tr}\, W)^2 = (-1)^{k-1} 2\alpha(W).$$

We conclude this section by introducing the <u>mean curvature</u> $H = H(n)$ of $j(M) \subset N$ by

$$H = \frac{1}{\dim M} \operatorname{tr} W$$

and the <u>Gaussian curvature</u> $\kappa = \kappa(n)$ by

$$\kappa = \det W .$$

By the theorem of Cayley–Hamilton the Weingarten map W of any embedding of an orientable three dimensional manifold M into a flat Euclidean space $\mathbb{R}^4$ satisfies

$$W^3 - \dim M \cdot H \cdot W^2 + \tfrac{\lambda}{2} W - \kappa \cdot \mathrm{id}_{TM} = 0 .$$

In case of dim M = 2 we have

$$W^2 - \dim M \cdot H \cdot W + \tfrac{\lambda}{2} \mathrm{id}_{TM} = 0 .$$

We leave it to the reader to show that the formulas in this section make all sense in case of a pseudo Riemannian metric if the signs are appropriately changed.

3.7 Geodesic spray of a right resp. left invariant metric on a Lie group

Let G be a Lie group with $\mathcal{G}$ as its Lie algebra. Recall from Chapter 1 that every $g \in G$ defines a left resp. right translation

$$L_g : G \longrightarrow G \quad \text{resp.} \quad R_g : G \longrightarrow G$$
$$g' \longmapsto g'g \qquad\qquad g' \longmapsto gg'$$

Assume next that $<,>$ is a <u>right invariant Riemannian metric</u> on G, meaning that $<,>$ satisfies

$$<v,w> = <TR_g(e)v, TR_g(e)w>$$

for every $v,w \in \mathcal{G}$. In analogy to the above introduced right invariance we define <u>left invariance</u> for a Riemannian metric via left translations.

Clearly <,>, being a Riemannian metric, determines a covariant derivative ∇ of the Levi–Cività type. By (*) from section 3.4 it is given by

$$<\nabla_X Y,Z> = \tfrac{1}{2}(X<Y,Z> + Y<Z,X> - Z<X,Y>)$$
$$+ \tfrac{1}{2}(<Z,[X,Y]> + <Y,[Z,X]> - <X,[Y,Z]>)$$

for all X,Y,Z $\in \Gamma TG$. A vector field X on G is called <u>right invariant</u>, if

$$X(g) = TR_g(e)\, X(e)$$

for all g $\in$ G. In case X_r, Y_r and Z_r are right invariant vector fields (r stands for right) the above equation for the Levi–Cività connection reduces, due to the right invariance of <,>, to

$$<\nabla_{X_r} Y_r, Z_r> = \tfrac{1}{2}(<Z_r,[X_r,Y_r]> - <Y_r,[X_r,Z_r]> - <X_r,[Y_r,Z_r]>).$$

This formula yields as we will see below an explicit expression for $\nabla_{X_r} Y_r$. Moreover since both $\nabla_{X_r} Y_r$ and the geodesic spray of <,> are determined by $\mathcal{G}$, the Lie algebra of G determined by left invariant vector fields, we pause here to relate the vector space of all left invariant vector fields with the vector space of all right invariant vector fields on G.

To this end we introduce $j : G \longrightarrow G$ sending any g into g^{-1}. Moreover let

$$\mu : G \times G \longrightarrow G$$

denote the multiplication in G. Thus we have

$$\mu(g,j(g)) = e$$

saying that

$$T\mu(g,g^{-1})(v_g,Tjv_g) = 0$$

for all $v_g \in T_gG$. Since

$$\mu(g_1,g_2) = R_{g_2}(g_1) = L_{g_1}(g_2)$$

we have

$$T\mu(g_1,g_2)(v_1,v_2) = TR_{g_2}(g_1)v_1 + TL_{g_1}(g_2)v_2$$

for $v_1 \in T_{g_1}G$ and $v_2 \in T_{g_2}G$. Hence

$$0 = T\mu(g,g^{-1})(v_g,Tjv_g)$$
$$= TR_{g^{-1}}(g)v_g + TL_g(g^{-1})Tj(g)v_g$$

and consequently

$$Tjv_g = -(TL_{g^{-1}} \circ TR_{g^{-1}})v_g$$

for all $g \in G$ and $v_g \in T_gG$.

Let $\xi \in \mathcal{G}$ and X and X_r the left and the right invariant vector fields generated by ξ respectively, i.e.

$$X(g) = TL_g(e)\xi \text{ and } X_r(g) = TR_g(e)\xi.$$

Then for each $g \in G$

$$Tj(g)X_r(g) = -TL_{g^{-1}}(e)\xi = -X(g^{-1}),$$

thus

$$Tj \circ X_r \circ j = -X.$$

This shows that X_r and $-X$ are j–related. If we let $\Gamma_l TG$ and $\Gamma_r TG$ be the vector spaces of all left and right invariant vector fields on G respectively, we obtain an isomorphism

$$j_* : \Gamma_r TG \longrightarrow \Gamma_l TG$$

given by

$$j_*(X_r) = Tj \circ X_r \circ j \text{ for all } X_r \in \Gamma_r TG.$$

By lemma 1.1.3 we have

$$j_*[X_r,Y_r] = [j_*X_r,j_*Y_r]$$

for any two right invariant vector fields X_r and Y_r. Given $\xi,\eta \in \mathcal{G}$ we introduce $[\ ,\ ]_r$ by

$$[\xi,\eta]_r = [X_r(\xi),X_r(\eta)](e)$$

where $X_r(\zeta)(g) = TR_g(e)\zeta$ for any $\zeta \in \mathcal{G}$. Clearly $\mathcal{G}$ together with $[\ ,\ ]_r$ is a Lie algebra as well. We denote it by $\mathcal{G}_r$. We have

$$[\xi,\eta]_r = -[\xi,\eta]$$

since $Tj(e) = -\mathrm{id}_{\mathcal{G}}$. Hence

$$Tj(e) : \mathcal{G}_r \longrightarrow \mathcal{G}$$

is even a Lie algebra isomorphism. We collect this material in

Proposition 3.7.1 The Lie algebras $\mathcal{G}_r$ and $\mathcal{G}$ defined by the right resp. the left invariant vector fields on G are isomorphic via $Tj(e)$, where $j : G \longrightarrow G$ maps every g into its inverse g^{-1}.

Now let us return to $\nabla_{X_r} Y_r$. Assume that X_r, Y_r and Z_r are given by

$$X_r(g) = TR_g(e)\xi,\ Y_r(g) = TR_g(e)\eta \text{ and } Z_r(g) = TR_g(e)\zeta$$

with $\xi,\eta,\zeta \in \mathcal{G}$. Since the Lie bracket of two right invariant vector fields is again right invariant, we have

$$<\nabla_{X_r} Y_r(g), Z_r(g)> = \tfrac{1}{2}\, TR_g(e)\, (-<[\xi,\eta],\zeta> + <\eta,[\xi,\zeta]> + <\xi,[\eta,\zeta]>).$$

For any $\lambda \in \mathcal{G}$ let ad_λ^* be the adjoint of ad_λ formed with respect to $<,>$. Then

$$\nabla_{X_r} Y_r(g) = -\tfrac{1}{2}\, TR_g(e)[\xi,\eta] + \tfrac{1}{2}\, T\, R_g(e)(\mathrm{ad}_\xi^*\eta + \mathrm{ad}_\eta^*\xi)$$

showing in particular that $\nabla_{X_r} Y_r$ is right invariant.

In case the metric is left invariant the terms on the right hand side switch signs. Thus we state

Proposition 3.7.2 Let $<,>$ be a right resp. left invariant Riemannian metric on a Lie group G and ξ,η be any two vectors in $\mathcal{G}$. Moreover let $X_r(\xi)$, $Y_r(\eta)$ be the right invariant vector fields generated by ξ,η. The corresponding left invariant ones are denoted by $X(\xi)$ and $Y(\eta)$ respectively. Then the Levi–Cività connection ∇ is given by

$$\nabla_{X_r(\xi)} Y_r(\eta)(g) = -\tfrac{1}{2}\, TR_g(e)[\xi,\eta] + \tfrac{1}{2}\, TR_g(e)(\mathrm{ad}_\xi^*\eta + \mathrm{ad}_\eta^*\xi)$$

in case <,> is right invariant and

$$\nabla_{X(\xi)}Y(\eta)(g) = \frac{1}{2}\,TL_g(e)[\xi,\eta] - \frac{1}{2}\,TL_g(e)(ad^*_\xi\eta + ad^*_\eta\xi)$$

if <,> is left invariant.

Next we determine the spray of say a right invariant metric. Let X,Y denote the right invariant vector fields generated by $\xi,\eta \in \mathcal{G}$ respectively. As introduced in section 3.3 the connection C associated with the covariant derivative ∇ is given by

$$C(X(g),Y(g)) = TY(g)X(g) - \nabla_X Y(g)$$

for all $g \in G$, where $\nabla_X Y(g)$ is considered as an element in $T_{Y(g)}T_gG \subset T^2G$. Thus we have

$$C(X(g),Y(g)) = TY(g)X(g) + \frac{1}{2}\,TR_g(e)([\xi,\eta] - ad^*_\xi\eta - ad^*_\eta\xi).$$

Hence if S denotes the spray of C

$$S(X(g)) = TX(g)X(g) - ad^*_\xi\xi.$$

Let us identify TG with $G \times \mathcal{G}$ by means of the trivialization given by right translation

$$G \times \mathcal{G} \longrightarrow TG$$

$$(g,\lambda) \longrightarrow TR_g(e)\lambda.$$

Moreover we have the obvious isomorphisms determined by the cartesian product and the above identification

$$T^2G \cong T(G\times\mathcal{G}) \cong TG \times T\mathcal{G} \cong G \times \mathcal{G} \times \mathcal{G} \times \mathcal{G}.$$

Now the right invariant vector fields generated by ξ and η read as

$$X(g) = (g,\xi) \text{ and } Y(g) = (g,\eta)$$

respectively. Thus we have

$$TY(g)X(g) = (g,\xi,\eta,0)$$

for all $g \in G$. Now observe that the canonical isomorphism

$$T_gG \longrightarrow T_{Y(g)}T_gG \subset T^2G$$

assigns to any $(g,\lambda) \in G \times \mathcal{G}$ the quadruple $(g,0,\eta,\lambda)$. Thus

$$\nabla_X Y(g) = (g,0,\eta,\tfrac{1}{2}([\xi,\eta] - ad^*_\xi\eta - ad^*_\eta\xi)).$$

Consequently the connection C and its spray S are given by

$$C((g,\xi),(g,\eta)) = (g,\xi,\eta,\tfrac{1}{2}([\xi,\eta] - \mathrm{ad}_{\xi}^{*}\eta - \mathrm{ad}_{\eta}^{*}\xi))$$

and

$$S(g,\xi) = (g,\xi,\xi, -\mathrm{ad}_{\xi}^{*}\xi)$$

for all $g \in G$ and all $\xi,\eta \in \mathcal{G}$.

If <,> is left invariant we choose the trivialization of TG by means of left translations. Then the fourth component of the left hand sides of the last two equations switches sign. Therefore we state

<u>Theorem 3.7.3</u> Let <,> be an either right or left invariant Riemannian metric on a Lie group G and $\xi \in \mathcal{G}$ any element. Let moreover $TG \cong G \times \mathcal{G}$ be established by either right or left translations accordingly to the left or right invariance of <,>. Then the map $u : G \times \mathcal{G} \longrightarrow \mathcal{G}$ of the geodesic spray

$$S : G \times \mathcal{G} \longrightarrow G \times \mathcal{G} \times \mathcal{G} \times \mathcal{G}$$

which itself has the form

$$S(g,\xi) = (g,\xi,\xi,u(g,\xi))$$

on each $(g,\xi) \in G \times \mathcal{G}$ satisfies either

$$u(g,\xi) = -\mathrm{ad}_{\xi}^{*}\xi \qquad \forall (g,\xi) \in G \times \mathcal{G}$$

in case <,> is right invariant or

$$u(g,\xi) = \mathrm{ad}_{\xi}^{*}\xi \qquad \forall (g,\xi) \in G \times \mathcal{G}$$

if <,> is left invariant.

<u>Remark</u>: If we identify $T^{*}G$ with $G \times \mathcal{G}'$ by sending $1_g \in T_g^{*}G$ into $(g,1_g \circ TR_g(e))$, we have the obvious isomorphisms

$$T(T^{*}G) \cong T(G \times \mathcal{G}') \cong TG \times T\mathcal{G}' \cong G \times \mathcal{G} \times \mathcal{G}' \times \mathcal{G}'.$$

In this trivialization the canonical one form on $T^{*}G$ reads as

$$\theta(g,\xi,\varphi,\Psi) = -\varphi(\xi)$$

for each $g \in G$, $\xi \in \mathcal{G}$ and $\varphi,\Psi \in \mathcal{G}'$. The canonical two form $\omega = \partial\theta$ is given by

$$\omega((g,\xi_1,\varphi,\Psi_1),(g,\xi_2,\varphi,\Psi_2)) = \Psi_2(\xi_1) - \Psi_2(\xi_2) - \varphi([\xi_1,\xi_2]).$$

Suppose that $<,>$ is a right invariant Riemannian metric on G and let A be the canonical isomorphism

$$G \times \mathcal{G} \longrightarrow G \times \mathcal{G}'$$

given by $A(g,\xi) = (g,<\xi,.>)$ for each pair $(g,\xi) \in G \times \mathcal{G}$. Then the pull back of ω via A takes the form

$$A^*\omega((g,\xi_1,\zeta,\eta_1),(g,\xi_2,\zeta,\eta_2))$$
$$= <\xi_1,\eta_2> - <\xi_2,\eta_1> - <\zeta,[\xi_1,\xi_2]>$$

for all $g \in G$ and all $\zeta,\xi_1,\xi_2,\eta_1,\eta_2 \in \mathcal{G}$.

References, Chapter 3

[D,IX] Dieudonné, J. Eléments d'Analyse 9, Gauthier–Villars, Paris, (1982).

[Hi] Hirsch, M.W. Differential Topology, Springer Verlag, New York, Heidelberg Berlin, (1976).

[Kl] Klingenberg, W. Riemannian Geometry, de Gruyter, Berlin, (1982).

[L] Lang, S. Differential Manifolds, Addison Wesley Publishing Company, Reading, Massachusetts, Menlo Park, California, (1972).

Chapter 4

PRINCIPAL BUNDLES AND CONNECTIONS

4.0 Preliminaries

Let M be a smooth manifold, as always assumed to be paracompact (e.g. second countable), and let G be a Lie group with Lie algebra $\mathfrak{g}$, the latter being identified with $T_e(G)$ or with the algebra of left–invariant vector fields on G; we also assume that the tangent bundle T(G) is trivialized by means of the (tangent maps of the) left translations in G.

A smooth left (right) action of G on the manifold M is a smooth map $l: G \times M \longrightarrow M$ ($r: M \times G \longrightarrow M$), usually written $(g,x) \longrightarrow gx$ (resp. $(x,g) \longrightarrow xg$) such that $ex = x$, $e \in G$ the identity, and $g(hx) = (gh)x$ (etc. for right actions). For each fixed $g \in G$, $x \longrightarrow gx$ is a diffeomorphism of M whose inverse is $x \longrightarrow g^{-1}x$; in this manner, the action induces an anti–homomorphism (homomorphism for right actions) of G into the group Diff(M) of diffeomorphisms of M. For the sake of definiteness, we deal here with left actions, the case of right actions being quite analogous:

For any fixed $x \in M$, the set $0_x = \{gx \mid g \in G\} = Gx$ is called the orbit through x (or: of x) under the given action, and the evidently closed subgroup $H_x \subset G$ defined by $H_x = \{g \in G \mid gx = x\}$ is the isotropy subgroup of x. Since H_x is closed, it is a Lie subgroup of G by ch. 1.3 and, in order to obtain more information on the orbits of G in M, we need the following general consideration:

Let $H \subset G$ be a closed subgroup, $\mathfrak{h} \subset \mathfrak{g}$ its Lie algebra (cf. ch. 1). Recall that the choice of a supplement of $\mathfrak{h}$ in $\mathfrak{g}$, i.e. of a splitting $\mathfrak{g} = \mathfrak{m} \oplus \mathfrak{h}$ defines a "canonical chart of the second kind" at $e \in G$: The map $\mu(m,h) = \exp(m)\exp(h)$ induces a diffeomorphism of some open neighbourhood $W \times V \subset \mathfrak{m} \times \mathfrak{h}$ onto an open neighbourhood U of $e \in G$. One may further assume that V,W are chosen so that $\mu((0) \times V) = U \cap H$, a canonical chart at $e \in H$. Let next $\pi : G \longrightarrow G/H$ be the projection onto the set of left H–cosets. Then $\hat{U} = \pi(U)$ is a neighbourhood of the "origin" $0 \in G/H$ (0 the neutral coset eH). It is not hard to see that π induces a homeomorphism of $\mu(W \times (0)) = \exp(W) \subset U$ onto $\hat{U}$ and thus, exp(W) yields a coordinate chart at 0. At any other point $y \in \mathfrak{h}/H$, a chart is obtained by using the left translations in G/H induced by the ones of G and we leave it to the reader to verify that one obtains, in this manner, a smooth atlas on G/H which then defines the desired left–invariant differentiable structure. Henceforth, it will be tacitly assumed that the "homogeneous space" G/H is equipped with this structure and further information on the very important class of manifolds thus obtained will follow in later sections.

As a first application, let us return to the orbits O_x of a left G–action on M: It is clear that for a fixed x , $g \longrightarrow gx$ is a differentiable map $G \longrightarrow M$ whose image is precisely O_x. Moreover, there now is the induced bijection ρ_x: $G/H_x \longrightarrow O_x$ which certainly remains continuous. In fact, as a map into M, ρ_x is an injective immersion with image O_x and therefore, the orbit O_x inherits a manifold structure with respect to which the inclusion $O_x \subset M$ is an immersion. At this point, caution is in order: The orbit O_x is not, in general, a submanifold of M! In fact, this will be the case if and only if the map ρ_x is a homeomorphism, i.e. iff. the quotient topology of G/H_x "coincides with" the relative topology of $O_x \subset M$, etc.

The obvious next question is that of the structure of the "orbit space" $G\backslash M$, quotient of M mod. the left G–action we consider here. The set $G\backslash M$ is, of course, given its quotient topology, i.e. the finest topology for which the natural map $\sigma : M \longrightarrow G\backslash M$ is continuous. It is easy to see that in our situation σ then also is an open map, so that the quotient topology is the unique topology for which σ is both continuous and open. $G\backslash M$ will not, in general, be a manifold – nor even a Hausdorff space. In fact, it is well–known from general topology that $G\backslash M$ is Hausdorff if and only if the graph of the equivalence relation $x \sim gx$ is closed in $M \times M$. A somewhat more workable condition in the current context is the following: Since M is completely regular, so is $G\backslash M$ and hence $G\backslash M$ is T_2 if and only if its points are closed. Clearly, if this is the case, then the orbits $O_x = \sigma^{-1}(\sigma(x))$ are closed in M. Conversely, the orbits are "G–saturated" by definition and therefore, if they are closed, so are their images, i.e. the points of $G\backslash M$. One concludes that $G\backslash M$ is T_2 if and only if the orbits of G are closed in M. Evidently, this condition always is satisfied if G is compact. In this case, by the way, the orbits even are closed submanifolds since each ρ_x has to be a homeomorphism (being a continuous bijection between two compact T_2 spaces).

However, even when $G\backslash M$ is Hausdorff, it need not be a manifold. Intuitively, the reason for this is the possible existence of different orbit types, i.e. of points with essentially different isotropies, so that there may be orbits of different dimensions; examples are quite easy to construct. Further information on the structure of $G\backslash M$ often can be obtained along the following lines: First of all, given the closed subgroup H of G, we denote by (H) its conjugacy class, i.e. the set of all subgroups of the form gHg^{-1}, $g \in G$. Given H,K, $(H) \leq (K)$ means that there is some $g \in G$ such that $H < gKg^{-1}$ and this defines a pre–order on the classes, i.e. a reflexive and transitive relation. Conjugacy classes enter here for the simple reason that if e.g. $y \in O_x$, then $(H_y) = (H_x)$ – the converse being false, in general. For each x, let now $M_{(H_x)}$ be the union of all orbits with $(H_y) = (H_x)$. Clearly, M is the disjoint union of these sets, all of which are G–saturated. Accordingly, the quotients $S_{(H_x)} = G\backslash M_{(H_x)}$ are well–defined and partition $G\backslash M$. Generally speaking, a "best possible scenario" now is the one where $G\backslash M$ is Hausdorff, each "stratum" $S_{(H_x)}$ is a manifold and for $(H_y) < (H_x)$, $S_{(H_y)}$ is contained in the topological boundary of $S_{(H_x)}$, etc. (a "stratification" of $G\backslash M$). We omit further details and return to more elementary considerations:

There is one general theorem concerning the quotient $G\backslash M$ which we mention here, once again without proof:

The orbit space $G\backslash M$ is a manifold if and only if the graph of the equivalence relation $x \sim gx$ is a closed submanifold on $M \times M$. As before, this condition may not be too easy to check in practice.

Aside from transitive actions, which correspond to homogeneous spaces G/H, cf. above, there are other special classes of importance; we mention the following: An action is called effective if $\bigcap_x H_x = (e)$, i.e. if $gx = x$ for all x implies $g = e$. In general, the above intersection is a closed normal subgroup N and, when desired, the original action of G may be replaced without essential loss of information by the induced action of G/N which will then be effective. Next, there is a considerably stronger, yet very important restriction on the action of G, namely that $gx = x$ for some x already should imply $g = e$. In this case, one says that the action is free: no $g \in G$ other than e possesses a fixpoint. Obviously, then, all orbits are of the same type, namely essentially G. It turns out that principal bundles fall in this class, with one additional condition: The action of G is said to be proper if for every compact set $K \subset M$, the set $\{g \in G \mid K \cap gK \neq \phi\}$ is compact in G. If $G = \Gamma$ is discrete, this reduces to the so–called properly discontinuous actions; an interesting example of this is the action of the fundamental group $\pi_1(M)$ as "decktransformations" in the universal covering $\tilde{M}$ of M. In this case, the quotient is M, a manifold. This result generalizes to the following theorem; in view of the conventions adopted in the later sections, we state it for right actions.

If G acts freely and properly on the right on M, then the quotient M/G admits a unique smooth structure for which the canonical map p: $M \longrightarrow M/G$ is a submersion.

In this situation, M/G is a "categorical quotient" in the sense that a map f: $M/G \longrightarrow N$ is smooth if and only if $f \circ p$: $M \longrightarrow N$ is. All details may be found in the literature; the results will be used below in sec. 4.1.

<u>Remark</u>: If with the above notations G is interpreted as a group of symmetries of M, then for $x \in M$, H_x is the corresponding symmetry group of the point x. H_x will, in general, be properly smaller than G: x admits "fewer symmetries than its orbit O_x" or all of M do.

Lastly, a terminological matter of some importance: Suppose that we are given two right actions $M \times G \longrightarrow M$, $N \times G \longrightarrow N$. A map $f: M \longrightarrow N$ then is called <u>equivariant</u> if $f(xg) = f(x)g$ for all $x \in M$ and $g \in G$. If G acts on N on the left, we use the corresponding right action, i.e. f will be equivariant if $f(xg) = g^{-1}f(x)$, etc. Such equivariant maps will later play an important role in connection with section spaces of certain fibre bundles, etc., cf. sec. 4.3; cf. also already in sec. 4.1 for "gauge transformations".

4.1 Principal bundles

We begin with the definition of a principal bundle over M by means of "transition maps", analogous to the corresponding description of e.g. vector bundles in ch. 2:

Definition 4.1.1 Let G and M be as before. A principal bundle over M with group G is a manifold P together with a smooth map $\pi : P \longrightarrow M$ such that the following conditions are satisfied:

(i) Each point $x \in M$ has an open neighbourhood U such that there is a diffeomorphism $\Psi_U: \pi^{-1}(U) \longrightarrow U \times G$ which is of the form

$$\Psi_U(p) = (\pi(p), \Psi_{U,\pi(p)}(p))$$

(bundle isomorphism!) where each $\Psi_{U,\pi(p)}$ is a diffeomorphism of the fibre at $\pi(p)$ onto G.

(ii) Given two open sets U,V and maps Ψ_U, Ψ_V as in (i) with $U \cap V \neq \phi$ the automorphism $\Psi_{UV} = \Psi_U \Psi_V^{-1}$ of $(U \cap V) \times G$ is of the form

$$\Psi_{UV}(x,g) = (x, g_{UV}(x)g)$$

where g_{UV} is a smooth map $U \times V \longrightarrow G$.

Thus, $\pi : P \longrightarrow M$ is fibre bundle, i.e. is locally trivial (condition (i)) with fibre G and structure group G operating on itself by left translations (condition (ii)). One easily sees that (ii) amounts to requiring that Ψ_{UV} should be equivariant for the action of G on $(U \cap V) \times G$ by right translations in the second factor, i.e. the action $(x,g)\cdot h = (x,gh)$: indeed, (ii) shows that $\Psi_{UV}(x,g) = \Psi_{UV}(x,e)g$, implying the equivariance. Conversely, if Ψ_{UV} is equivariant, set $\Psi_{UV}(x,e) = (x, g_{UV}(x))$; (ii) then follows immediately.

Evidently $g_{UV}(x) = g_{VU}(x)^{-1}$. Moreover, if U,V,W are three overlapping open sets as in (i) and if Ψ_U, Ψ_V and Ψ_W are trivializations of P over these sets, then it is immediate that over $U \cap V \cap W$, $g_{UW} = g_{UV} g_{VW}$.

Since every point of M has an open neighbourhood U over which there is a trivialization Ψ_U of P, we may choose an open covering (U_α) of M such that P is trivial over each U_α by means of a map Ψ_α in the sense of (i). We then write $\Psi_{\alpha\beta}$ for $\Psi_\alpha \Psi_\beta^{-1}$ (over $U_\alpha \cap U_\beta$). In particular, we obtain a family of smooth maps $g_{\alpha\beta} : U_\alpha \cap U_\beta \to G$ from (ii) with the property that $g_{\alpha\alpha} = 1$, $g_{\beta\alpha} = g_{\alpha\beta}^{-1}$, $g_{\alpha\gamma} = g_{\alpha\beta} g_{\beta\gamma}$ (on the respective common domains). Such a family of "transition maps" is also called a "1–cocycle" for the open covering (U_α) with values in (the sheaf of smooth maps into) G; it permits the reconstruction of P (up to isomorphism, cf. below) as follows:

In the disjoint union $\bigcup_\alpha (\{\alpha\} \times U_\alpha \times G)$, one defines an equivalence relation $\sim$ by

$$(\alpha,x,g) \sim (\beta,y,h) \text{ iff. } x = y \text{ and } h = g_{\beta\alpha}(x)g. \tag{4.1.2}$$

The quotient mod. this relation then is (isomorphic to) P, i.e. possesses a natural fibre–preserving diffeomorphism onto P, obtained by "patching" the maps Ψ_α^{-1}.

More generally, if we are given a 1–cocycle with respect to the open cover (U_α) as above, then the construction just outlined will yield a principal bundle over M with group G ("principal G–bundle" for short) and every principal G–bundle is obtained in this manner. For the details, we refer to [S], [H], [K,N], [G,H,V], [D,I].

We next describe an important first class of <u>examples</u>, namely the so–called <u>frame bundles</u> of vector bundles:

$\mathbb{E}$ is a vector bundle with fibre E, a real or complex vector space of finite dimension; L(E) is the ring of endomorphisms of E, GL(E) its group of units. Thus, the choice of a basis $(e_1,\ldots,e_k)$ of E yields an isomorphism of L(E) onto the ring of all $(k\times k)$–matrices, under which GL(E) maps to the non–singular such matrices. In addition to $\mathbb{E}$, let us also introduce the trivial bundle $E_M = M \times E$ with the same fibre and now consider the <u>bundle</u> $L(E_{\mu},\mathbb{E})$ of "linear maps" $E_M \longrightarrow \mathbb{E}$; the fibre at $x \in M$ of this bundle therefore is the space $L(E,E_x)$ of linear maps $E \longrightarrow E_x$, the latter the fibre at x of $\mathbb{E}$. We know from earlier (ch. 2) that as here $L(E_M,\mathbb{E})$ is a fibre bundle with fibre L(E); in fact, in the current case, this is very easy to see directly: Suppose that over the open set $U \subset M$, $\mathbb{E}$ is trivial under a map $\Psi_U\colon \mathbb{E}|U \longrightarrow U \times E$. Then $L(E_M,\mathbb{E})|U \cong U \times L(E)$ under the map $\beta \longrightarrow \alpha_{U,x}\,\beta$ for β in the fibre at x, where $\Psi_{U,x}$ is interpreted as a linear isomorphism $E_x \longrightarrow E$. If $\mathbb{E}$ is trivial over U and V and if Ψ_{UV} is given by the map $g_{UV}\colon U \cap V \longrightarrow GL(E)$, then it is clear that $\Psi_{U,x}\Psi_{V,x}^{-1}\beta = g_{UV}(x)\beta$. From this it now is evident that if β lies in GL(E), then so does its transform – and conversely. One concludes that the transition maps of $L(E_M,\mathbb{E})$ also yield – by restriction – transition maps for the set $\mathcal{F}(\mathbb{E}) = \bigcup_x Is(E,E_x)$ which make this into a fibre bundle over M with group and fibre GL(E), called the (linear) <u>frame bundle</u> of $\mathbb{E}$. The term "frame bundle" is motivated by the following remarks: Choose a basis $(e_1,\ldots,e_k)$ of E ("linear frame of E"). Then, if α is an isomorphism $E \cong E_x$, the images $\alpha e_1,\ldots,\alpha e_k$ clearly are a basis of E_x and, conversely, every isomorphism $E \cong E_x$ arises in this manner from a basis choice in E_x. Note also that a local section $s\colon U \longrightarrow \mathcal{F}(\mathbb{E})$ is a <u>local frame</u> for $\mathbb{E}$: at each $x \in U$, the $s(x)e_i$, $1 \leq i \leq k$, form a basis of the fibre E_x. In this sense, s defines a "moving frame" (repère mobile) of $\mathbb{E}$ over U. Historically, it was this idea of "moving frames", used systematically for the first time by E. Cartan, which was the original of the notion of a principal bundle (due to Ehresmann, cf. the notes in [S]).

The group GL(E) operates on $\mathcal{F}(\mathbb{E})$ from the right in a natural way: If $\beta : E \cong E_x$ and $\alpha \in GL(E)$, clearly $\beta\alpha: E \cong E_x$. It is obvious that this action is free and that its orbits are precisely the fibres of $\mathcal{F}(\mathbb{E})$. This phenomenon is quite general:

If $\pi : P \longrightarrow M$ is a principal G–bundle in the sense of definition 4.1.1, then there is a natural free right G–action $P \times G \longrightarrow P$ whose orbits are the fibres of P, obtained from the free right action of G on $U \times G$ used earlier. Indeed, pulling back these actions under the Ψ_U of definition 4.1.1 yields an action of G on all of P since the g_{UV} act on G from the left, hence commute with right multiplication. It can be shown that, in a sense, these last remarks characterize principal G–bundles, i.e. that we have the following definition, equivalent with definition 4.1.1 (cf. [B], [D,I]):

Definition 4.1.3 Let G be a Lie group. A principal G–bundle is a manifold P together with a right G–action $P \times G \longrightarrow P$ such that

(a) The action $r: P \times G \longrightarrow P$ is free;

(b) the quotient $M = P/G$ is a manifold;

(c) the natural map $\pi: P \longrightarrow M$ is a submersion.

Under these conditions, $M = P/G$ is the base of the bundle P and one also calls P a principal G–bundle over M. It can also be shown that (b) and (c) together may be replaced by the requirement that $P \times G \longrightarrow P$ be a proper action; in other words: a principal G–bundle is a free and proper right G–action.

Remarks:

(i) In all this, right actions may, of course, be replaced by left ones with the obvious changes; we chose the former partly because of the original examples given by frame bundles, partly because it seems to be quite standard by now to let transition maps operate from the left.

(ii) If Γ is a discrete group operating freely and properly discontinuously on P, then P/Γ is a manifold by an easy exercise mentioned earlier and $P \longrightarrow P/\Gamma$ is a local diffeomorphism ("P is étale over P/Γ"); since the conditions of definition 4.1.3 are satisfied, P is a principal Γ–bundle over P/Γ. If, in particular, P is connected and simply connected, then Γ is the fundamental group of P/Γ. Any connected manifold M can be obtained in this manner by letting P be the universal covering of M and $\Gamma = \pi_1(M)$, cf. the remarks in sec. 4.0.

(iii) Given definition 4.1.3, the following result now is quite clear: If $\pi : P \longrightarrow M$ is a fibre bundle which admits a free right G–action whose orbits are precisely the fibres of P, then P is a principal G–bundle.

Since π: $P \longrightarrow M$ is a submersion, it is of constant rank and therefore $\ker(T\pi) \subset T(P)$ is a subbundle (cf. ch.2), the vertical bundle V(P) of P (also called the bundle of vertical tangents or of tangents to the fibres of P). A tangent vector $u \in T_p(P)$ lies in $V_p(P)$ iff. it is tangent to the fibre through p (which is, in fact, a closed submanifold of P). One sees immediately that $\text{rank}(V(P)) = \dim(P) - \dim(M) = \dim(G)$. The module $\Gamma(V(P))$ of smooth sections of V(P) will also be denoted by $\mathfrak{X}^V(P)$; it is the module of vertical vector fields on P and a vector field X lies in $\mathfrak{X}^V(P)$ iff. $T\pi \cdot X = 0$. The principal right action of G leads to a special set of such vector fields, the so–called fundamental vector fields on P, in the following manner:

For h in the Lie $\mathfrak{g}$ of G, consider the 1–parameter subgroup exp(th), $t \in \mathbb{R}$, and its action on P by $(p,t) \longrightarrow p\exp(th)$. This is smooth in (p,t) and induces the following vector field:

$$A_h(p) = \frac{\partial}{\partial t} p\exp(th)|_{t=0} ; \tag{4.1.4}$$

clearly, $Z_h \in \mathfrak{X}^V(P)$. The map $h \longrightarrow Z_h$ is linear and may be generalized to a $C^\infty(P)$–linear map $f \longrightarrow Z_f$ of $C^\infty(P,\mathfrak{g})$ into $\mathfrak{X}^V(P)$ by setting

$$Z_f(p) = \frac{\partial}{\partial t} p\exp(tf(p))|_{t=0} . \tag{4.1.5}$$

One can show that this construction establishes an isomorphism of the $C^\infty(P)$–modules $C^\infty(P,\mathfrak{g})$ and $\mathfrak{X}^V(P)$. This statement is equivalent with the following:

<u>Proposition 4.1.6</u> The map $(p,h) \in p\times\mathfrak{g} \longrightarrow Z_h(p) \in V_p(P)$ yields an isomorphism trivial bundle $p\times\mathfrak{g}$ onto V(P): the vertical bundle of P is trivial.

Using Tr: $T(P)\times T(G) \longrightarrow T(P)$ together with the standard trivialization of T(G) to $G\times\mathfrak{g}$ by means of left translations, the map of the proposition may be described slightly differently, namely by means of the partial tangent map T_2r: indeed, it is easy to see that $Z_h(p) = T_2r(p,e)h$. Since G operates freely on P, it follows immediately that $h \longrightarrow T_2r(p,e)h$ is injective into $V_p(P)$, hence an isomorphism. These remarks essentially prove the proposition.

As a consequence, if $(h_1,...,h_n)$ is a basis of $\mathfrak{g}$, then any vertical vector field may be written uniquely in the form $\Sigma\, \lambda_i Z_{h_i}$ with $\lambda_i \in C^\infty(P)$. This in turn shows that for any differential form α on P, the values of α "along V(P)" are entirely determined by its values on fundamental vector fields. This observation will be used again below, sec. 4.4.

For $g \in G$, $p \longrightarrow pg$ is a diffeomorphism r_g of P whose tangent map $T(P) \longrightarrow T(P)$ we denote by R_g (or R_{g*}). By means of $g \longrightarrow R_g$, G now acts on T(P) and this action covers the principal action on P by construction: for each $p \in P$, R_g yields an isomorphism of $T_p(P)$ onto $T_{pg}(P)$. A vector field X on P is said to be invariant if

$$R_g X = X, \text{ i.e. } R_g X(p) = X(pg) \qquad (4.1.7)$$

for $p \in P$, $g \in G$; this simply says that X is an equivariant map $P \longrightarrow T(P)$. Similarly, a scalar differential form α is invariant if $r_g^* \alpha = \alpha$, $g \in G$. The following somewhat more general notion will soon become very important:

Suppose that $\lambda : G \longrightarrow GL(E)$ is a (smooth) representation of G in the finite–dimensional (real or complex) vector space E. Then the E–valued differential form β on P is said to be equivariant (with respect to λ) if

$$r_g^* \beta = \lambda(g^{-1})\beta, \ g \in G. \qquad (4.1.8)$$

The space of equivariant E–valued forms will be denoted by $A_G^{\cdot}(P,E)$ (the representation λ being fixed) whenever ambiguity is unlikely. This is a graded module over the graded algebra $A^{\cdot}(P)^G$ of G–invariant scalar forms, in particular a graded module over the ring $C^{\infty}(P)^G$ of invariant smooth functions on P, thus over $C^{\infty}(M)$ since $C^{\infty}(M) \cong C^{\infty}(P)^G$ in an obvious way.

The equivariant forms in the sense of (4.1.8) can also be described as the invariants of a suitable action of G: Firstly, the module $A^{\cdot}(P,E)$ of E–valued forms is naturally isomorphic to the tensor product $A^{\cdot}(P) \otimes E$, cf. ch. 2. Secondly, one defines a left G–action on $A^{\cdot}(P) \otimes E$ by means of

$$g \cdot \beta = r_g^* \alpha \otimes \lambda(g)e \qquad (4.1.9)$$

if $\beta = \alpha \otimes e$, and extension to all of $A^{\cdot}(P) \otimes E$. Then it is easy to see that β is equivariant iff. $g.\beta = \beta$. Accordingly, $A_G^{\cdot}(P,E) \cong (A^{\cdot}(P) \otimes E)^G$ as $C^{\infty}(M)$–modules. This description of equivariant forms is often used in practice to obtain very explicit descriptions of certain geometric objects such as e.g. invariant connections, Laplace– and Dirac–operators on homogeneous spaces; cf. the literature.

A differential form α on P is called <u>horizontal</u> if

$$i(X)\alpha = 0 \tag{4.1.10}$$

for $X \in \mathfrak{X}^v(P)$; the module of horizontal E–valued forms is denoted by $A_h^{\cdot}(P,E)$. In particular, $A_h^{\cdot}(P)$ is the module of horizontal scalar forms which, moreover, is a graded algebra over $C^\infty(P)$ with respect to the natural grading and exterior multiplication. Its intersection with $A^{\cdot}(P)^G$ is the $C^\infty(M)$–module $A_{h,G}(P) := A_b^{\cdot}(P)$ of invariant horizontal forms, also called "basic forms":

$$A_b^{\cdot}(P) = A^{\cdot}(P)^G \cap A_h^{\cdot}(P); \tag{4.1.11}$$

it can be shown quite easily that the pull–back map π^* induces an isomorphism

$$\pi^* : A^{\cdot}(M) \cong A_b^{\cdot}(P) \tag{4.1.12}$$

over $C^\infty(M)$, whence the terminology "basic form". This result will be generalized later (sec. 4.3); we also refer to the literature, e.g. [K,N], [G,H,V]. The usefulness of these notions will become clearer later on, e.g. in sec. 4.4. In the sequel, we shall generally use the induction $A_{h,G}^{\cdot}$ – in the interest of clarity, we hope.

For later use, we add the following remarks:

Since G operates on the right on P and, as was just pointed out, also on T(P), there now is a natural right action of G on $\mathfrak{X}(P)$, the space of vector fields on P, given by: $\hat{R}_g X := R_g \cdot X \cdot r_g^{-1}$; more explicitly, this means that

$$(\hat{R}_g X)(p) = R_g X(pg^{-1}) \tag{4.1.13}$$

for $p \in P$ and $g \in G$. it then is clear that X is invariant in the sense of (4.1.7) if and only if

$$\hat{R}_g X = X. \tag{4.1.14}$$

Moreover, one sees readily that $\mathfrak{X}^v(P)$ is a G–invariant subspace and thus, $\mathfrak{X}^v_I(P) = \mathfrak{X}^v(P)^G = \mathfrak{X}^v(P) \cap \mathfrak{X}_I(P)$ is well–defined. We next derive a different description of this $C^\infty(M)$–module which will be of use later on (cf. sec. 4.4):

Let $Z_f \in \mathfrak{X}^v(P)$ (i.e. $f \in C^\infty(P,\mathfrak{g})$). Then, for $p \in P$ and $g \in G$:

$$R_g Z_{f(p)}(p) = Z_{\mathrm{Ad}(g^{-1})f(p)}(pg)\,. \tag{4.1.15}$$

Indeed,

$$R_g Z_{f(p)}(p) = R_g(\frac{\partial}{\partial t} p\cdot \exp(tf(p))|_{t=0})$$

$$= \frac{\partial}{\partial t}\, pg \exp(tf(p))|_{t=0}$$

$$= \frac{\partial}{\partial t}\, pg\cdot g^{-1}\exp(tf(p))g|_{t=0}$$

$$= \frac{\partial}{\partial t}\, pg \exp(t\mathrm{Ad}(g^{-1})f(p))|_{t=0}$$

$$= Z_{\mathrm{Ad}(g^{-1})f(p)}(pg).$$

Note here that, in general, the right–hand side of (4.1.15) is <u>not</u> the value at pg of Z_f. However, suppose that Z_f is invariant: This amounts to the condition $R_g Z_{f(p)}(p) = Z_{f(pg)}(pg)$. Comparison with (4.1.15) now shows that we must have $f(pg) = \mathrm{Ad}(g^{-1})f(p)$, i.e. that $f \in C^\infty_G(P,\mathfrak{g})$. The converse is equally obvious and we conclude:

Proposition 4.1.16 The isomorphism $f \longrightarrow Z_f$ of $C^\infty(P,\mathfrak{g})$ onto $\mathfrak{X}^V(P)$ restricts to an isomorphism

$$C^\infty_G(P,\mathfrak{g}) \cong \mathfrak{X}^V_I(P) \tag{4.1.17}$$

(over $C^\infty(M)$).

Next, we introduce the notion of "homomorphism" or "bundle map" appropriate for principal bundles:

Suppose that $\pi:P \longrightarrow M$ and $\rho:Q \longrightarrow N$ are principal bundles with group G_1,G_2, respectively, that $\lambda: G_1 \longrightarrow G_2$ is a (smooth) group homomorphism and $f:M \longrightarrow N$ is a differentiable map. Under these assumptions,

Definition 4.1.18 A homomorphism or bundle map $\Psi:P \longrightarrow Q$ over f, with respect to λ, is a map Ψ such that $\rho\Psi = f\pi$ and that Ψ is G_1–equivariant.

The last condition is meaningful since G_1 acts (from the right) on Q via $\lambda : G_1 \longrightarrow G_2$. More explicitly, the definition means the following: Ψ maps each fibre P_x into the corresponding fibre $Q_{f(x)} = \rho^{-1}(f(x))$ and, moreover $\Psi(pg) = \Psi(p)\lambda(g)$ for $p \in P$, $g \in G_1$. Obviously, compositions of bundle maps are bundle maps, etc., so that principal bundles now form a well–defined category where, in particular, the notion of isomorphism has the usual meaning. In particular, when we speak of bundle maps $P \longrightarrow Q$ where $N = M$ without specifying $f:M \longrightarrow M$, we mean bundle maps over the identity of M; on occasion, for the sake of clarity, these will also be referred to as "strong bundle maps". Note also that, in principle, the general notion of a bundle map or homomorphism over f may be reduced to that of a strong bundle map by the following device: If $\rho : Q \longrightarrow N$ is a principal G_2–bundle and $f:M \longrightarrow M$ is a map, then the pullback $f^*Q = M\times_N Q = \{(x,q) \mid f(x) = \rho(q)\} \subset X\times Q$, the fibred product of M and

Q over N, is easily seen to be a fibre bundle over N under the restriction $\bar{\rho}$ to f^*Q in such a manner that its orbits are precisely the fibres of f^*Q is a principal G_2–bundle over M, the pull–back in the category of principal bundles. Now, if $\Psi : P \longrightarrow Q$ is a bundle map over f (with respect to λ), then Ψ yields a strong bundle map $P \longrightarrow f^*Q$ by the obvious construction and in this manner one obtains a natural bijection between the two sets of bundle maps, as is readily verified.

The following special cases are especially useful in practice:

(I) Suppose that $G_1 = G_2 = G$, $\lambda = \text{id.}$ and that we are dealing with bundle maps $P \longrightarrow P$ (over differentiable maps of M into itself). The set of all such bundle maps then is $C^\infty_G(P,P)$, the set of all equivariant smooth maps of P into itself. Among these, there are the invertible such maps which form the group (under composition) of equivariant diffeomorphisms of P, denoted by $\mathrm{Diff}_G(P)$ or Aut(P) since it is the automorphism group of the principal G–action on P. Given $\alpha \in \mathrm{Aut}(P)$, there is a unique diffeomorphism f of M such that $\pi\alpha = f\pi$ and one thus obtains a homomorphism of Aut(P) into Diff(M) (which need not be surjective!). The kernel of this homomorphism is a normal subgroup of Aut(P), denoted by $\mathcal{G}_P$ and known by now as the gauge group of P (or, more precisely, the gauge transformation group). More explicitly, then, $\alpha \in \mathcal{G}_P$ if and only if α is a diffeomorphism of P such that $\alpha(pg) = \alpha(p)g$ for $p \in P$, $g \in G$ and, in addition, $\pi\alpha = \pi$ (i.e. $\alpha(P_x) = P_x$ for all $x \in M$).

A perhaps more concrete and often very useful description of the gauge group is the following: If $\alpha \in \mathcal{G}_P$, then for each $p \in P$, $\alpha(p)$ lies in the same fibre as p, hence is of the form pf(p) for a unique $f(p) \in G$. The $f \in C^\infty(P,G)$ and the fact that α is equivariant is equivalent with the condition $f(pg) = g^{-1}f(p)g$, i.e. with the equivariance of f with respect to the action of G on itself by inner automorphisms. It is an easy matter to verify that $\alpha \longrightarrow f$ yields a canonical isomorphism of groups

$$\mathcal{G}_P \cong C^\infty_G(P,G) \tag{4.1.19}$$

where the right–hand term carries its group structure of pointwise multiplication.

Lastly, let us mention the following for later use: If $\alpha \in \mathrm{Aut}(P)$, then clearly T_α is a bundle map $T(P) \longrightarrow T(P)$ over α and this map <u>preserves the vertical bundle</u> $V(P)$: $T_\alpha(V(P)) = V(P)$. To show this, it suffices to prove that if Z_h, $h \in \mathcal{G}$, is a fundamental vector field, then $T_\alpha \cdot Z_h = Z_h$. This assertion is immediate from the equivariance of α since this means that $\alpha(p\exp(th)) = \alpha(p)\exp(th)$; differentiation with respect to t now yield the claim.

(II) Let now $G_1 = H$, $G_2 = G$ and suppose that f = id., i.e. that we are dealing with strong bundle maps again. Such a map then often is called a <u>reduction</u> of the group of Q "to H" (with respect to λ). The terminology is motivated by the special case where H is either a closed subgroup or an immersed subgroup of G and λ is the inclusion. It is important to note, however, that we do not assume λ to be injective (nor even locally injective) since there are very important practical situations where this condition fails to be satisfied; some such cases will be dealt with later. Given λ, the question of the existence of a "reduction" of a principal G–bundle to a principal H–bundle is, in general, quite intractable. In somewhat more special cases, however, one has a well–known standard result:

Let $H \subset G$ be a <u>closed</u> subgroup (λ = inclusion) and suppose that P is a principal G–bundle. The principal right G–action then restricts to a free right action of H on P and P/H again is a manifold such that $P \longrightarrow P/H$ is a principal H–bundle. Moreover, P/H is a bundle over M with fibre the homogeneous space G/H. The result alluded to now is that:

Proposition 4.1.20 The bundle P admits a reduction to a principal H–bundle if and only if the bundle $P/H \longrightarrow M$ has a global section.

An outline of the proof follows: Let $\mu : P \longrightarrow P/H$ be the natural projection; then $\mu(pg) = \mu(p)$ iff. $g \in H$. Suppose now that Q is a principal H–bundle which is a reduction of P. Then Q is embedded into P as a closed submanifold and $\mu|Q$ maps each fibre of Q to one point in P/H. Write s(x) for the point $\mu(\pi^{-1}(x) \cap Q)$. Then s is a section of P/H.

Conversely, if $s \in \Gamma(P/H)$ is given, set $Q = \mu^{-1}(s(M)) \subset P$. Now s(M) is a closed submanifold of P/H and μ is transversal to s(M); hence Q is a closed submanifold of P (the transversality of μ is immediate from the fact that it is a submersion!). Moreover, $\pi(Q) = M$, i.e. Q meets every fibre of P. Also, if $p \in Q$ and $g \in H$, then $pg \in Q$ because of $\mu(pg) = \mu(p)$ – and conversely, as pointed above. Thus, H operates freely on Q and $Q/H = M$. One shows that Q is a fibre bundle over M and then the desired result follows by earlier remarks. For more detail, we refer to the literature, e.g. [B], [G,H,V], [D,I], [K,N] and the classic reference [S], sec. 12.

An immediate consequence, e.g., is the following: Suppose that $K \subset G$ is a maximal compact subgroup. Then a principal G–bundle P always admits a reduction to a principal K–bundle. Indeed, G/K is diffeomorphic to some euclidean space, hence (smoothly) contractible. The bundle P/K thus has a contractible fibre and therefore admits a section, c.f. e.g. [S], loc. cit.

In contrast to the vector bundles discussed earlier, principal bundles will not, in general, admit global sections. This results from the following considerations:

Suppose that $U \subset M$ is open and that $s: U \longrightarrow P$ is a section of the principal G–bundle P, i.e. is a smooth map with $\pi s = \text{id}$. Then $P|U$ is <u>trivial</u>. The map $(x,g) \longrightarrow s(x)g$ is a bundle isomorphism of $U \times G$ onto $P|U$, as is easily seen. Conversely, if $U \times G \cong P|U$ (over the identity), then the trivial section $x \longrightarrow (x,e)$ of $U \times G$ is transformed into a section of $P|U$. Thus: $P|U$ is trivial iff. there is a section $s: U \longrightarrow P$. This holds, of course, also when $U = M$; a principal bundle thus admits a global section iff. it is trivial.

Lastly, let s be a section of $P|U$. Then any other section t can be written in the form $t = sf$ with $f \in C^{\infty}(U,G)$ in a unique way; it is not hard to see that f has to be smooth. The converse being obvious, one now has a bijection of $\Gamma(U,P)$ onto $C^{\infty}(U,G)$, determined by the choice of s. In this sense, any two sections of $P|U$ "differ" by an element of $C^{\infty}(U,G)$. This last set, a group under pointwise multiplication, is none other than the gauge group of the trivial bundle $U \times G$.

4.2 <u>Examples</u>

I) In 4.1 we introduced the (linear) frame bundle of a vector bundle as an example of a principle bundle. We now specialize this to the <u>frame bundle of a manifold:</u>

Given the manifold M, its (linear) frame bundle $\mathcal{F}_M$ is defined to be the frame bundle $\mathcal{F}(T(M))$ of the tangent bundle $T(M)$ of M.

Let M be modelled on the (finite–dimensional) vector space E; then $\mathcal{F}_M$ is a principal GL(E)–bundle over M. Suppose now that G is a Lie group and $\lambda: G \longrightarrow GL(E)$

a representation of G in E. Then a (G,λ)–structure on M is a reduction of $\mathcal{F}_M$ to a principal G–bundle (with respect to λ). If G is a closed subgroup of GL(E) and λ is the inclusion, we simply speak of a G–structure on M (cf. e.g. [St]). Classical examples are the following:

(i) E is Euclidean and $G = O(E)$ (the orthogonal group, $\cong O(n)$ if n is the dimension of E). A G–structure on M then is a Riemannian structure. Such structures always exist since O(E) is maximal compact in GL(E).

(ii) E is oriented and $GL^+(E) \subset GL(E)$ is the group of orientation–preserving automorphisms of E(i.e. those with det > 0). For $G = GL^+(E)$, a G–structure simply is an orientation of M. If M is orientable, one may also reduce GL(E) to $SL(E) \subset GL^+(E)$ since $GL^+(E)/SL(E) = \mathbb{R}^+$, a contractible space.

Combining this with (i), one defines an oriented Riemannian structure on M as a reduction to SO(E); since SO(E) is maximal compact in SL(E), a reduction to SO(E) SO(E) is maximal compact in SL(E), a reduction to SO(E) is always possible if M is orientable, i.e. if $\mathcal{F}_M$ admits a reduction to SL(E). The obstruction to a reduction to SL(E) is topological in nature: it is the first Stiefel–Whitney class of M in $H^1(M;\mathbb{Z}_2)$; cf. the appropriate literature.

(iii) Suppose next that E carries a non–generate quadratic form of type (p,q) with $p > 0$, $q > 0$ and let $O_{p,q}(E)$ be its orthogonal group ($\cong O(p,q)$). Also choose a splitting $E = E^+ \oplus E^-$, orthogonal for the given form, such that on E^+, the form is positive–definite, on E^- negative–definite. Then $O(E^+) \times O(E^-)$ is a maximal compact subgroup of $O_{p,q}(E)$. A reduction of $\mathcal{F}_M$ to a principal $O_{p,q}(E)$–bundle is a pseudo–Riemannian structure of type (p,q) on M. If such a structure exists, we now see that $\mathcal{F}_M$ can be further reduced to a principal $O(E^+) \times O(E^-)$ – bundle. It follows that this implies that T(M) splits into an orthogonal sum $T(M) = T^+(M) \oplus T^-(M)$ such that the "metric" of M is positive–definite on $T^+(M)$, negative–definite on $T^-(M)$. One notes here that a reduction to $O_{p,q}(E)$ may not exist – contrarily to what happened in the Riemannian case.

(iv) E is even–dimensional and carries a symplectic form, G = Sp(E) is its (linear) symplectic group. A G–structure on M is a symplectic structure and M with a choice of such a structure is a symplectic manifold. This definition is equivalent with the perhaps more familiar one which defines a symplectic manifold as a pair (M,ω) of a manifold M together with an everywhere ;non–singular 2–form ω (the symplectic form of M). Such manifolds are well–known to play a crucial role in a number of areas of Mathematical Physics and Mathematics (classical Mechanics in the Hamiltonian formalism, non–classical analogues of this, including geometric quantization; Hamiltonian G–spaces, co–adjoint orbits, etc.).

(v) E is a real even dimensional space and carries a complex structure whose automorphism group is $G = GL_{\mathbb{C}}(E)$. A G–structure on M then is known as an almost–complex structure on M.

(vi) There is a class of examples of great practical importance where G is not a subgroup of the structure group and λ is not injective: the so–called <u>spin structures</u>, roughly described as follows:

Suppose that M is an oriented Riemannian manifold and let P be its tangent principal SO(E)–bundle, i.e. the bundle of oriented orthonormal frames of M.

Assuming $\dim(E) \geq 3$, the universal covering group of SO(E) is Spin(E) and there is a short exact sequence $1 \longrightarrow \mathbb{Z}_2 \longrightarrow \text{Spin}(E) \xrightarrow{\lambda} SO(E) \longrightarrow 1$ where λ is the covering map; $\mathbb{Z}_2$ is embedded as $\{\pm 1\}$.

A spin structure on M is a principal Spin(E)–bundle Q which is a "reduction" of P, i.e. which admits a bundle map $\varphi: Q \longrightarrow P$ with respect to λ. It then follows that Q is 2–to–1 covering of P. Such a structure need not exist: A necessary and sufficient condition is the vanishing of the second Stiefel–Whitney class $w_2(M) \in H^2(M,\mathbb{Z}_2)$—where the appearance of the $\mathbb{Z}_2$–cohomology is no great surprise because of the short exact sequence just mentioned. Details are left to the literature except for one remark: It is on spin manifolds that the notion of a Dirac operator makes sense.

At this point, we briefly mention an important special feature of the frame bundle $\mathcal{F}_M$ of a manifold M: Suppose that M is modelled on the (finite–dimensional) real vector space E. Then there exists a canonically defined horizontal equivariant E–valued 1–form θ on $\mathcal{F}_M$ (the "patching form"):

Let $\pi : \mathcal{F}_M \longrightarrow M$ be the projection. A point $u \in \mathcal{F}_M$ with $\pi(u) = x$ is an isomorphism $E \cong T_x(M)$. On the other hand, it $t \in T_u(\mathcal{F}_M)$, then $T\pi \cdot t$ lies in $T_x(M)$ and thus the following is well–defined:

$$\theta(t) = u^{-1}(T\pi \cdot t);$$

this yields a linear surjection $T_u(\mathcal{F}_M) \longrightarrow E$ for each u, easily seen to be an E–valued 1–form on $\mathcal{F}_M$ of class C^∞. By its very construction, θ is horizontal. Moreover, since $(ug)^{-1} = g^{-1}u^{-1}$ and $T\pi(R_g t) = T\pi \cdot t$ for $g \in GL(E)$, θ is equivariant.

In sec. 4.3, the form θ will be used to construct an important 2–form (on M) associated with a linear connection on M, namely its torsion.

II) Let G be a Lie group, $H \subset G$ a closed subgroup. The homogeneous space G/H is a manifold in a natural way and G operates smoothly and transitively on G/H by left translations since in G, these commute with the right operation of H and so descend to the quotient. The projection $\pi: G \longrightarrow G/H$ is a submersion with "enough" local sections, so that $\pi: G \longrightarrow G/H$ becomes a principal H–bundle. This basic fact is of great importance and will be used frequently in the sequel. By a common abuse of notations, left translation by $g \in G$ in G/H simply is denoted by $x \longrightarrow gx$. Since $x = kH$ for some $k \in G$, gx thus is the coset gkh; we also denote the "neutral coset" $H = eH$ by O ("origin of G/H") and note for later use that $hO = O$ for $h \in H$.

Somewhat more generally, suppose that $K \subset K$ are closed subgroups of Gs; then $\pi: G \longrightarrow G/H$ and $\rho: G \longrightarrow G/K$ are principal bundles and it now is easy to see that $\sigma : G/K \longrightarrow G/H$, induced by $K \subset H$, is a fibre bundle with fibre the homogeneous space H/K. We shall see later that this bundle belongs to a special class, called "bundles associated with $\pi: G \longrightarrow G/H$" (or, stretching the terminology apparently introduced by Bott for vector bundles associated with $G \longrightarrow G/H$, "homogeneous bundles").

At this point, let us mention some examples which should illustrate the importance of the two constructions just introduced and, in particular, of certain homogeneous spaces of classical groups:

(i) Equip $\mathbb{R}^n$ with the usual Euclidean structure $(x,y) = \Sigma\, \xi_i \eta_i$ for $x = (\xi_i,\dots,\xi_n)$ and $y = (\eta_1,\dots,\eta_n)$. The group of linear isometries of $\mathbb{R}^n$ then is O(n), the group of orthogonal $(n \times n)$–matrices which operates in the standard manner on $\mathbb{R}^n$. The orbits of this action are the spheres centered at 0 of radius $r \geq 0$. Let, in particular, $S = S^{n-1}$ be the unit sphere. O(n) operates transitively on S^{n-1} and, if we choose $e \in S^{n-1}$ to be $e = (1,0,\dots,0)$, the isotropy of may be identified with the group O(n–1), embedded into

O(n) as the group of all matrices of the form

$$\begin{bmatrix} 1 & 0 \\ 0 & A \end{bmatrix}, \ A \in O(n-1).$$

Since S^{n-1} is a closed (even compact!) submanifold of $\mathbb{R}^n$, the natural bijection $O(n)/O(n-1) \longrightarrow S^{n-1}$ is a diffeomorphism.

The same holds for the action of the group SO(n). In particular, both O(n) and SO(n) are principal bundles over S^{n-1}.

(ii) The remarks of (i) may be generalized in the following manner: Consider the set $S_{n,k}$ of all orthonormal sequences $(e_1,...,e_k)$ of vectors of $\mathbb{R}^n$ ("orthonormal k–frames" in $\mathbb{R}^n$; $k \leq n$). One verifies easily that O(n) operates transitively on $S_{n,k}$. Let $e_i = (0,...,1,...,0)$, the standard basis of $\mathbb{R}^n$. If $\varphi \in O(n)$ leaves $e_1,...,e_k$ pointwise fixed, then it also maps the orthogonal complement $[(e_1,...,e_k)]^\perp$ onto itself. With this, one identifies the isotropy of $(e_1,...,e_k)$ with the matrices of the form

$$\begin{bmatrix} I_k & 0 \\ 0 & A \end{bmatrix}, \ A \in O(n-k),$$

a subgroup isomorphic to O(n–k). The natural bijection of $O(n)/O(n-k)$ onto $S_{n,k}$ is now used to make the latter into a compact manifold, the Stiefel manifold of orthonormal k–frames in $\mathbb{R}^n$.

Accordingly, O(n) now becomes a principal O(n–k)–bundle over $S_{n,k}$. Note that for $k = 1$, $S_{n,1}$ is the unit sphere while $S_{n,n} = O(n)$

(iii) Similar constructions apply to $\mathbb{C}^n$ and $\mathbb{H}^n$, replacing the orthogonal groups of (ii) by U(n), U(n,$\mathbb{H}$), etc. or, for $1 \leq k < n$, by SU(n), etc. We refer for more detail to the literature, e.g. [G,H,V], [K,N], [S].

(iv) Let next $G_{n,k}(\mathbb{R})$ be the set of all k–dimensional subspaces of $\mathbb{R}^n$. Once again, O(n) operates transitively on this set. Embed $\mathbb{R}^k$ as the subspace spanned by the first k standard basis vectors. The isotropy of $\mathbb{R}^k$ then consists of all orthogonal maps which map $\mathbb{R}^k$ and hence $(\mathbb{R}^k)^\perp$ onto itself. This group consists of all orthogonal matrices of the form

$$\begin{bmatrix} A & 0 \\ 0 & B \end{bmatrix}$$

with $A \in O(k)$, $B \in O(n-1)$ and is isomorphic to $O(k)\times O(n-k)$.

We transfer the differentiable structure of $O(n)/O(k)\times O(n-k)$ to $G_{n,k}(\mathbb{R})$ which then becomes a compact manifold, called the <u>Grassmannian manifold</u> of k–dimensional subspaces of $\mathbb{R}^n$. Its differentiable structure is determined by the requirement that the natural action of O(n) should be differentiable. For $k = 1$, one obtains the real projective space $P_{n-1}(\mathbb{R})$. Moreover, as above, for $1 \leq k < n$, $G_{n,k}(\mathbb{R})$ may also be identified with $SO(n)/S(O(k)\times O(n-k))$ where the denominator denotes the group of matrices $\begin{bmatrix} A & 0 \\ 0 & B \end{bmatrix}$ with $A \in O(k)$, $B \in O(n-k)$ such that $\det(A)\det(B) = 1$.

Thus, O(n) is a principal bundle over $G_{n,k}(\mathbb{R})$. However, this Grassmannian also provides an example of a fibration of the type $G/K \longrightarrow G/H$ mentioned above, in the following manner: In (ii), the Stiefel manifolds were identified with the homogeneous spaces $O(n)/O(n-k)$ where $O(n-k)$ is embedded as $I_k\times O(n-k)$. Clearly, then, $I_k\times O(n-k)$ is a closed subgroup of $O(k) \times O(n-k)$; accordingly, $S_{n,k}$ is a fibre bundle over $G_{n,k}(\mathbb{R})$ with fibre the homogeneous space $O(k)\times O(n-k)/I_k\times O(n-k) = O(k)$ where this last map is an isomorphism of Lie groups. One easily sees that O(k) operates freely from the right on $S_{n,k}$ in such a way that its orbits are precisely the fibres of the bundle $S_{n,k} \longrightarrow G_{n,k}(\mathbb{R})$ which therefore becomes a principal O(k)–bundle over $G_{n,k}(\mathbb{R})$, called the <u>universal</u> principal O(k)–bundle over $G_{n,k}(\mathbb{R})$.

Analogous remarks apply in the complex (and the quaternionic) case. For details, we refer to the literature, e.g. [G,H,V], ch. V, [D,I], ch.XVI, [K,N], [S],

(v) The case k = 1, i.e. projective spaces, may merit special attention: In the real case, $O(1) = \{\pm 1\} \cong \mathbb{Z}_2$. Furthermore, the Stiefel manifold $S_{n,1}$ simply is the unit sphere S^{n-1} in $\mathbb{R}^n$ on which $\mathbb{Z}_2$ now acts by its usual antipodal action: $x \longrightarrow x$ and $x \longrightarrow -x$. Thus, we recover the realization of $P_{n-1}(\mathbb{R})$ as the quotient of S^{n-1} modulo the antipodal $\mathbb{Z}_2$–action and we now conclude that S^{n-1} is a principal $\mathbb{Z}_2$–bundle over $P_{n-1}(\mathbb{R})$. Since for $n > 2$, S^{n-1} is connected and simply connected, it is the universal covering manifold of $P_{n-1}(\mathbb{R})$ whose fundamental group thus is $\mathbb{Z}_2$.

In the complex case, $P_{n-1}(\mathbb{C})$ is the quotient $U(n)/U(1)\times U(n-1)$ or $SU(n)/S(U1)\times U(n-1))$. Again $U(n)/U(n-1)$ is the sphere $S^{2n-1} \subset \mathbb{C}^n$ and $P_{n-1}(\mathbb{C})$ now is the quotient of S^{2n-1} modulo the obvious action of

$$U(1) = S^1 = \{z \in \mathbb{C} \mid |z| = 1\}.$$

Thus, S^{2n-1} becomes a principal S^1–bundle over $P_{n-1}(\mathbb{C})$; these are the Hopf fibrations of odd–dimensional spheres (classical case: $n = 2$ where $P_1(\mathbb{C}) = S^2$, so that S^3 now is a principal S^1–bundle over S^2).

These fibrations are closely related to another class of principal bundles over the $P_{n-1}(\mathbb{C})$, this time with non–compact group: Let $\mathbb{C}^*$ be the multiplicative group of non–zero complex numbers. This group acts freely on any $\dot{\mathbb{C}}^n = \mathbb{C}^n \backslash \{0\}$ such that the quotient is precisely $P_{n-1}(\mathbb{C})$. It is easy to see that $\dot{\mathbb{C}}^n \longrightarrow P_{n-1}(\mathbb{C})$ is a principal $\mathbb{C}^*$–bundle.

In the quaternionic case one interprets S^3 as the group of unit quaternions ($\cong SU(2)$!) and uses the corresponding action of S^3 on the sphere $S^{4n-1} \subset \mathbb{H}^n$ to obtain $P_{n-1}(\mathbb{H})$. Thus, spheres of dimension $\equiv 3$ (mod.4) are principal S^3–bundles over quaternionic projective spaces.

4.3 Associated bundles

Let $\pi : P \longrightarrow M$ be a principal bundle with group G. Then P can be used to construct a large class of fibre bundles in the following manner:

Suppose that G acts from the left on the manifold F (by automorphisms of F, depending on the given structure) and write this action as $(g,y) \longrightarrow gy$. Then there is a natural right G–action on $P \times F$, namely

$$(p,y)g = (gp,g^{-1}y). \tag{4.3.1}$$

We denote the quotient $(P \times F)/G$ by $P \times_G F$. One shows that this is a manifold and, in fact, a fibre bundle over M with fibre F, the projection being induced by π. This bundle $\mathbb{F}$ is said to be associated with P under the action $G \times F \longrightarrow F$.

It turns out that any fibre bundle over M with structure group G is associated with some principal G–bundle; the argument follows the standard "black box" prescription, cf. [G,H,V], [H], [K,N], [S], and proceeds roughly as follows: Suppose that (U_α) is an open cover of M such that each $\mathbb{F} | U_\alpha$ is trivial (i.e. $\cong U_\alpha \times F$) and let $(g_{\alpha\beta})$ be the 1–cocycle defined by these local trivializations; by assumption, $g_{\alpha\beta} : U_\alpha \cap U_\beta \longrightarrow G$ since G is the structure group of $\mathbb{F}$. Next, we let G act on itself by left translations and form the quotient of the disjoint union $\bigcup_\alpha (\{\alpha\} \times U_\alpha \times G)$mod. the equivalence relation

$$(\alpha,x,g) \sim (\beta,y,h) \text{ iff } x = y \text{ and } h = g_{\beta\alpha}(x)g, \tag{4.3.2}$$

cf. (4.1.2). One shows that the quotient P may be interpreted as a principal G–bundle over M and that $\mathbb{F}$ is associated with this bundle under the action $G \times F \longrightarrow F$. The isomorphism class of P is uniquely determined by that of $\mathbb{F}$.

We conclude that a fibre bundle over M is associated with some principal bundle precisely when the bundle has a finite–dimensional structure group; this need not be the case, cf. e.g. [G,H,V], for a counterexample.

Remark: If we were to admit "infinite–dimensional differentiable groups", then every fibre bundle would be associated with some principal bundle; the group in question could be e.g. the diffeomorphism group Diff(F) of the fibre F. In practice, this point–of–view appears to be without great merit – except in certain special situations where infinite–dimensional smooth fibrations arise naturally.

We mention here some rather immediate examples of associated bundles: First of all, let G act on itself by inner automorphisms and form the bundle $\mathcal{G} = P \times_G G$. Since G operates by Lie group automorphisms, this actually is a bundle of groups, canonically associated with P. Secondly, a "linearized" version is obtained by letting G act on its Lie algebra $\mathfrak{g}$ by means of the adjoint representation Ad. This leads to the bundle $P(\mathfrak{g}) = P \times_G \mathfrak{g}$ of Lie algebras, called the adjoint bundle of P.

In particular, the adjoint bundle provides a first example of a vector bundle associated with P. The general construction of such bundles is quite clear: One starts with representation $\lambda : G \longrightarrow GL(E)$ of G in a real or complex vector space E (of finite dimension) and then one forms the bundle $\mathbb{E} = P \times_G E$. Since G acts linearly on E, this will, indeed, be a vector bundle. Recall now from earlier that there is a natural principal GL(E)–bundle associated with $\mathbb{E}$, namely its frame bundle $\mathcal{F}(\mathbb{E})$. If $\mathbb{E}$ is constructed in the form $P\times_G E$, we can obtain $\mathcal{F}(\mathbb{E})$ in a similar manner: One lets G act on GL(E) by left translations via λ, i.e. by $(g,\alpha) \longrightarrow \lambda(g)\alpha$. Then one forms the associated bundle $P\times_G GL(E)$ and observes that this will be a principal GL(E)–bundle since GL(E) acts freely from the right in such a way that its orbits are the fibres of the bundle. Moreover, it is not hard to see that this principal bundle is isomorphic to $\mathcal{F}(\mathbb{E})$. In fact, P is a reduction of $\mathcal{F}(\mathbb{E})$ by means of λ ("G–structure in $\mathbb{E}$").

In all generality, one observes that if P is a reduction of the principal H–bundle Q by means of the homomorphism $\lambda\colon G \longrightarrow H$, then $Q \cong P\times_G H$ where G acts on H by left translations via λ; the converse also holds.

Under the conditions of the assertion, Q may also be called an <u>extension</u> of P (by means of λ) to a principal H–bundle.

In particular, the notion of a G–structure on the manifold M may be re–formulated as follows: A G–structure is a principal G–bundle $P \longrightarrow M$ together with a representation $\lambda\colon G \longrightarrow GL(H)$, M being modelled on H, such that $P\times_G GL(H) = \mathcal{F}_M$ where G acts on GL(H) as above.

The construction of an associated bundle $\mathbb{F}$ leads to a very useful description of the space $\Gamma(\mathbb{F})$ of sections of $\mathbb{F}$ over M in terms of the data P and $G\times F \longrightarrow F$: By construction, there is a natural bundle map $\sigma\colon P\times F \longrightarrow \mathbb{F}$ over π, namely the canonical map $P\times F \longrightarrow (P\times F)/G$ which maps the pair (p,y) to its equivalence class [p,y]. As always, this may be viewed as a strong bundle map $P\times F \longrightarrow \pi^*\mathbb{F}$. It is easy to see that this is a fibre–preserving diffeomorphism which yields a trivialization of $\pi^*\mathbb{F}$ over P. Accordingly,

$$\Gamma(\pi^*\mathbb{F}) \cong C^\infty(P,F), \qquad (4.3.3)$$

a natural bijection preserving what additional structures $\mathbb{F}$ resp. F have, cf. below. On the other hand, $s \longrightarrow s \circ \pi = \pi^* s$ yields an injection

$$\pi^*: \Gamma(\mathbb{F}) \longrightarrow \Gamma(\pi^* \mathbb{F}); \tag{4.3.4}$$

the composition of these maps yields an injection $\Gamma(\mathbb{F}) \longrightarrow C^\infty(P,F)$ still denoted by π^*. Then

$$\mathrm{im}(\pi^*) = C^\infty_G(P,F), \tag{4.3.5}$$

the space of G–equivariant smooth maps $P \longrightarrow F$; we indicate some of the details of the construction:

First of all, recall that $\pi^* \mathbb{F} = \{(p,y) \mid \pi(p) = \rho(y)\}$ where ρ is the bundle projection $\mathbb{F} \longrightarrow M$. If $s \in \Gamma(F)$, define $\tilde{s} \in \Gamma(\pi^* \mathbb{F})$ by

$$\tilde{s}(p) = (p, s(\pi(p))).$$

Next, if $y \in \rho^{-1}(x)$, y is of the form $y = [p,e]$, $\pi(p) = x$, $e \in F$; in particular, then, $s(\pi(p)) = [p,f(p)]$ with $f(p) \in F$. With this, we set $\pi^* s = f$ (by abuse of notations!). Thus, $\pi^* s$ now denotes the map $P \longrightarrow F$ which corresponds to $\tilde{s}$ under the isomorphism $\Gamma(\pi^* \mathbb{F}) \cong C^\infty(P,F)$ induced by the trivialization $(p,[p,e]) \longrightarrow (p,e)$ of $\pi^* \mathbb{F}$. It is an easy matter to verify that f is equivariant – and conversely: any equivariant map $f: P \longrightarrow F$ defines a section of $\pi^* \mathbb{F}$ which is a lift of a section of $\mathbb{F}$. this establishes (4.3.4) and (4.3.5). In particular:

(i) $\pi^* : \Gamma(\mathscr{G}) \cong C^\infty_G(P,G)$, a group isomorphism. In other words: the gauge group $\mathcal{G}_P$ may be interpreted as $\Gamma(\mathscr{G})$.

(ii) $\pi^* : \Gamma(P(\mathfrak{g})) \cong C^\infty_G(P,\mathfrak{g})$, a Lie algebra and $C^\infty(M)$–module isomorphism. This provides another realization of the algebra of "infinitesimal gauge transformations".

(iii) More generally, if $\mathbb{E} = P\times_G E$ is an associated vector bundle (by means of $\lambda : G \longrightarrow GL(E)$), then $\pi^* : \Gamma(\mathbb{E}) \cong C^\infty_G(P,E)$, an isomorphism of $C^\infty(M)$–modules which will be very useful later on.

Next we describe an especially important class of bundles, primarily vector bundles, arising from the general construction mentioned, the "homogeneous" bundles on homogeneous spaces:

Let G be a Lie group, $H \subset G$ a closed subgroup. We know from earlier that $\pi:G \longrightarrow G/H$ then is a principal H–bundle (where G/H again denotes the space of left cosets mod.H). The bundles associated with this principal bundle sometimes are called homogeneous bundles, particularly in the case of vector bundles; systematic use of the term appears to go back to Bott ([Bo], in the holomorphic case). Such bundles are determined by a left action $H\times F \longrightarrow F$, so that homogeneous vector bundles are given by the (finite–dimensional real or complex) representations of H and are of great importance in representation theory (Mackey induction invariant differential operators, etc.).

On the other hand, one also may consider those bundle over G/H which admit "liftings" of the natural left action of G on G/H to bundle maps over these "left translations" ("G–bundles" in some of the literature; the term, however, has other meanings and will not be used here in this meaning): $\mathbb{F} \longrightarrow G/H$ is a fibre bundle and G acts from the left on $\mathbb{F}$ by fibre–preserving maps over $G\times G/H \longrightarrow G/H$ which preserve what structure the fibres of $\mathbb{F}$ have. Thus, there is an operation $(g,u) \longrightarrow gu$ on $\mathbb{F}$ such that if $u \in F_x$, then $gu_A \in F_{gx}$ and $g_x = g|F_x : F_x \cong F_{gx}$. One then shows that these bundles coincide with what we just called homogeneous bundles. An outline of the argument is the following: If $\mathbb{F} = G\times_H F$, then G still acts on itself by left translations which, of course, commute with the right translations by elements of H and so descend to the quotient $G\times_H F$: $g[k,u] = [gk,u]$. This yields the action covering $x \longrightarrow gx$, $x \in G/H$. For the converse, observe that the neutral coset $0 \in G/H$ remains fixed under all $h \in H$ and, in fact, H is its isotropy. Thus, if $\mathbb{F} \longrightarrow G/H$ admits a lift of the G–action on G/H, then H will act on the fibre F_0; one shows that $G\times_H F_o \cong \mathbb{F}$ (using that $G \longrightarrow G/H$ is a principal H–bundle).

The last remarks apply, in particular, to the tangent bundle T(G/H), hence to all standard tensor bundles over G/H, since it is clear that the diffeomorphism $x \longrightarrow gx$ induces an isomorphism L_g of T(G/H) (over $x \longrightarrow gx$). Now for $h \in H$, $g \in G$, $\pi(hgh^{-1}) = \pi(hg) = h\cdot\pi(g)$. Differentiation with respect to g at g = e yields

$$T_e\pi \circ \mathrm{Ad}(h) = L_h \circ T_e\pi .$$

Since 0 is fixed under H, L_h is the representation of H on $T_0(G/H)$. On the other hand, $T_e\pi$ maps $T_e(G) = \mathfrak{g}$ onto $T_0(G/H)$ with kernel $\mathfrak{h} \subset \mathfrak{g}$ and so induces an isomorphism $\mathfrak{g}/\mathfrak{h} \cong T_0(G/H)$. The identity must derived shows that this isomorphism is equivariant with respect to the action of H on $\mathfrak{g}/\mathfrak{h}$ induced by Ad, denoted by $\hat{\mathrm{Ad}}$, and the representation $h \longrightarrow L_h$ of H in $T_0(G/H)$; this is also stated by saying that the isomorphism is an intertwining operator for the two representations of H. Accordingly,

$$T(G/H) \cong G\times_H(\mathfrak{g}/\mathfrak{h}),$$

H acting on $\mathfrak{g}/\mathfrak{h}$ by means of $\hat{\mathrm{Ad}}$. Similarly, $T^*(G/H)$ then is obtained as the homogeneous bundle with fibre $(\mathfrak{g}/\mathfrak{h})^* = \mathfrak{h}^\circ \subset \mathfrak{g}^*$ with respect to the contragredient representation $h \longrightarrow {}^t\mathrm{Ad}(h^{-1}) := \mathrm{Ad}^*(h)$, etc.

The situation is further simplified in an important special case which will be useful later (cf. sec. 4.5, below): Suppose that G/H is reductive, i.e. that $\mathfrak{h} \subset \mathfrak{g}$ admits a summand $\mathfrak{m}$ such that $\mathrm{Ad}(\mathfrak{m}) \subset \mathfrak{m}$. In this case, $\mathfrak{g} = \mathfrak{h}\oplus\mathfrak{m}$ and both summands are stable under Ad(H). It is clear then that the natural isomorphism $\mathfrak{m} \cong \mathfrak{g}/\mathfrak{h}$ induced by $\mathfrak{g} \longrightarrow \mathfrak{g}/\mathfrak{h}$

is equivariant. Accordingly, if G/H is reductive, its tangent bundle T(G/H) is $\cong G\times_H \mathfrak{m}$ where H acts on $\mathfrak{m}$ by means of Ad. Similarly, $T^*(G/H) \cong G\times_H \mathfrak{m}^*$, etc. Note also that (iii) above leads to a particularly simple description of the vector fields on G/H: the module $\mathfrak{X}(G/H) = \Gamma(T(G/H))$ is isomorphic to $C_H^\infty(G,\mathfrak{m})$ analogous remarks apply to all standard tensors on G/H.

If $\mathbb{E} = G\times_H E$ is a homogeneous vector bundle defined eg. by the representation λ of H in E (λ: $H \longrightarrow GL(E)$), then G also operates from the left on $\Gamma(\mathbb{E})$: indeed, it operates on $\mathbb{E}$ and G/H and so we may define $(gs)(x) = gs(g^{-1}x)$ for $g \in G$ and $s \in \Gamma(\mathbb{E})$. Evidently, this yields a representation of G in the (infinite–dimensional!) vector space $\Gamma(\mathbb{E})$ or – by transport – on $C_H^\infty(G,E)$ where the representation coincides with the left regular action of G; these observations lead to the construction of representations of G <u>induced</u> by those of G (Mackey). Moreover, (iii) above also is useful for an explicit description of certain differential operators on G/H, cf. the literature.

Suppose now that $K \subset H$ is a closed subgroup, so that H/K is a manifold and, in fact, $H \longrightarrow H/K$ is a principal K–bundle. $G \longrightarrow G/K$ also is a principal K–bundle by general principles and we have the following commuting diagram of natural maps:

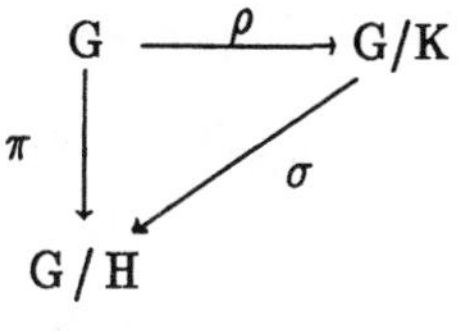

The fibration σ has fibre H/K and is a fibre bundle, cf. sec. 3.2, II). One then shows that, as fibre bundles over G/H,

$$G\times_H H/K \cong G/K \tag{4.3.6}$$

under the map $[g,eK] \longrightarrow gK$. Firstly, this map is well–defined since the representative $(gh,h^{-1}K)$ of [g,eK] maps to $gh\cdot h^{-1}K = gK$ for any $h \in H$. Secondly, the domain of the map is all of $G \times_H H/K$ since $(1,hK) \sim (1h,eK)$ and it is clear that the map is surjective and fibre–preserving (over the identity of G/H). Because of $K \subset H$, the map also is injective and the claim follows. In particular, thus, the bundle $G/K \longrightarrow G/H$ is associated with $G \longrightarrow G/H$.

In the situation of (4.3.6), suppose now that $\lambda\colon K \longrightarrow GL(E)$ is a given representation. This defines a homogeneous vector bundle $\mathbb{E} \longrightarrow G/K$ and also a homogeneous vector bundle $\mathbb{E}_0 \longrightarrow H/K$. From earlier observations we know that the standard action of H on H/K lifts to an action $H \times \mathbb{E}_0 \longrightarrow \mathbb{E}_0$ by bundle maps. This left action of H on $\mathbb{E}_0$ now may be used to construct the associated bundle $G \times_H \mathbb{E}_0$ over G/H; this bundle also is fibred over G/K and as such is $\cong \mathbb{E}$. In other words: the bundle $\mathbb{E} \longrightarrow G/K$ is locally trivial <u>over</u> G/H <u>with fibre</u> $\mathbb{E}_0$. If the representation λ extends to H, then $\mathbb{E}_0 \cong H/K \times E$ is trivial and $\mathbb{E} \longrightarrow G/K$ is the pull–back under σ of the homogeneous vector bundle $G \times_H E \longrightarrow G/H$. We leave the details to the reader.

For some simple examples of homogeneous vector bundles we return to the Grassmannian manifolds mentioned in 4.2 where we also introduced the universal k–frame bundle (Stiefel manifolds) over $G_{n,k}(\mathbb{R})$, $G_{n,k}(\mathbb{C})$ which are principal $0(k)$–, $U(k)$–bundles, resp. By means of the natural actions of these groups on $\mathbb{R}^k$ and $\mathbb{C}^k$, we obtain associated vector bundles over $G_{n,k}(\mathbb{R})$ and $G_{n,k}(\mathbb{C})$. One can show that these are subbundles of the trivial vector bundles $G_{n,k}(\mathbb{R}) \times \mathbb{R}^n$ and $G_{n,k}(\mathbb{C}) \times \mathbb{C}^n$ obtained by choosing, for each x, the subspace represented by x as the fibre E_x. One thus obtains the universal vector bundles over the Grassmannians $G_{n,k}$.

Let us return to the general case of bundles associated with a principal G–bundle $\pi\colon P \longrightarrow M$ and suppose that $\mathbb{F} = P \times_G F$ is determined by the left action $G \times F \longrightarrow F$. Assume next that Q is a principal H–bundle which is a reduction of P by means of the homomorphism $\lambda : H \longrightarrow G$; in this case,

$$P \cong Q \times_H G$$

where H operates on G by left translations via λ. The claim then is that

$$\mathbb{F} \cong Q \times_H F, \tag{4.3.7}$$

i.e. the group of the associated bundle $\mathbb{F}$ also is "reduced to H". We thus have the following formula:

$$P \times_G F \cong (Q \times_H G) \times_G F \cong Q \times_H F,$$

a kind of "transitivity condition": a bundle associated to one associated with Q is itself associated to Q. We omit the verification.

In particular, if the manifold M admits a G–structure via the homomorphism $\lambda : G \longrightarrow GL(H)$ (M being modelled on H), then the structure group of all standard tensor bundles over M also is reduced to G since they all are associated with the frame bundle $\mathcal{F}_M$.

Next, let $\mathbb{E}$ be a vector bundle associated with P under the representation $\lambda: G \longrightarrow GL(E)$ and let $A^\cdot(M,\mathbb{E})$ denote the $C^\infty(M)$–module of $\mathbb{E}$–valued differential forms on M (i.e. $A^\cdot(M,\mathbb{E}) = \Gamma(\Lambda^\cdot T^*(M) \otimes (\mathbb{E})$, cf. ch. 2). Using (4.1.8), etc. and (4.3.5) above, one then concludes that

$$\pi^* : A^\cdot(M,\mathbb{E}) \cong A^\cdot_{h,G}(P,\mathbb{E}),$$

the space of <u>horizontal equivariant</u> E–valued forms on P; this is an isomorphism of the respective $C^\infty(M)$–module structures which will prove very useful in some later developments (cf. sec^s 4.4, 4.5).

If both $\mathbb{E}$ and $\mathbb{F}$ are associated with P under $\lambda: G \longrightarrow GL(E)$ and $\mu: G \longrightarrow GL(F$ respectively, and if $\rho: E \longrightarrow F$ is an <u>equivariant</u> linear map, then φ evidently induces a $C^\infty(M)$–linear map $C^\infty_G(P,E) \longrightarrow C_G(P,F)$ or, descending to M again, $\Gamma(\mathbb{E}) \longrightarrow \Gamma(\mathbb{F})$ which necessarily is induced by a bundle map $\hat{\varphi}$ from $\mathbb{E}$ to $\mathbb{F}$; one can define $\hat{\varphi}$ as follows: $\hat{\varphi}([p,y]) = [p,\varphi y]$, as one readily verifies. In some sense, this construction yields the bundle maps $\mathbb{E} \longrightarrow \mathbb{F}$ which "are compatible with the G–structures" of $\mathbb{E}$, $\mathbb{F}$. Similar remarks apply, of course, when $\mathbb{E}$ and $\mathbb{F}$ are not necessarily vector bundles: it is the equivariance of φ which matters.

4.4 Connections

If $\pi: P \longrightarrow M$ is a principal G–bundle, we introduced the vertical bundle $V(P) = \ker(T\pi) \subset T(P)$ already in 4.1 where we also showed that V(P) is an invariant (and, in fact, trivial) subbundle of T(P). Since all manifolds are tacitly assumed paracompact, V(P) certainly admits supplementary subbundles in T(P); such a supplement, however, need not be invariant under G.

Definition 4.4.1 A connection on P is a choice of an invariant subbundle H(P) of T(P), the horizontal bundle of the connection, such that

$$T(P) = V(P) \oplus H(P).$$

A connection on P thus is an invariant splitting $T(P) = V(P) \oplus H(P)$. While the vertical bundle is defined naturally, the notion of "horizontal vectors" involves a choice; dually, horizontal forms are defined naturally, while "vertical forms", interpreted as differential forms on P, depend on the choice of a horizontal bundle. This applies, in particular, to the interpretation of $V^*(P)$ as a bundle of 1–forms – e.g. by extension by zero on H(P), cf. below. The above definition of a connection may be re–stated as follows: there is a natural short exact bundle sequence

$$0 \longrightarrow V(P) \xrightarrow{i} T(P) \xrightarrow{\pi_*} \pi^* T(M) \longrightarrow 0 \qquad (4.4.2)$$

over P where π_* is induced by $T\pi : T(P) \longrightarrow T(M)$. The choice of an invariant summand H(P) of V(P) now amounts to the choice of an equivariant bundle map $\lambda: \pi^* T(M) \longrightarrow T(P)$ (whose image then is H(P)!), i.e. to an <u>equivariant splitting</u> of the sequence (4.4.2).

There is another short exact bundle sequence, this time over M, whose splittings correspond to the connections on P; this sequence was first used by Atiyah in the holomorphic case (cf. [A]), but is easily adapted to the current situation:

We begin by letting E_x, $x \in M$, be the vector space of <u>invariant vector fields</u> along the fibre $P_x \subset P$,i.e. $E_x = \Gamma(P_x, T(P))^G$. Since an invariant vector field is determined by its value at any fixed point of P_x, the vector space E_x is finite–dimensional ($\cong T_p(P_x)$ for some fixed $p \in P_x$). Set $\mathbb{E} = \cup E_x$ with the obvious projection $\rho: \mathbb{E} \longrightarrow M$. Then $\mathbb{E}$ is a vector bundle over M: local trivializations of $\mathbb{E}$ may be obtained, e.g., from local sections s of P by means of the isomorphisms $E_x \cong T_{s(x)}(P_x)$, etc. Again there is a natural bundle map $\pi_*: \mathbb{E} \longrightarrow T(M)$ induced by $T\pi$, well–defined since $T\pi \cdot R_g = T\pi$. One then verifies that $\ker(\pi_*) \subset \mathbb{E}$ is naturally isomorphic to the adjoint bundle $P(\mathfrak{g})$ of P, whence the short exact sequence

$$0 \longrightarrow P(\mathfrak{g}) \longrightarrow \mathbb{E} \xrightarrow{\pi_*} T(M) \longrightarrow 0 \qquad (4.4.3)$$

and a connection on P now corresponds to a splitting of this sequence – and conversely: the equivariance requirement in (4.4.2) is replaced by the very construction of (4.4.3) and thus "becomes automatic". These observations also yield a simple <u>existence proof</u> for connections on P: the sequence (4.4.3) certainly splits (smoothly)!

<u>Remark</u>: It is not hard to see that (4.4.2) is obtained from (4.4.3) by pull–back ;under π: $\pi^* P(\mathfrak{g}) \cong P \times \mathfrak{g} \cong V(P)$ land $\pi^* \mathbb{E} \cong T(P)$ under the map which assigns to $(p,X) \in (\pi^* \mathbb{E})_p = (p) \times E_x$ the value $X(p) \in T_p(P)$ of the invariant field X.

It is clear from definition 4.4.1 that a vector field X on P has a unique decomposition

$$X = X^v + X^h \tag{4.4.4}$$

with $X^v \in \Gamma(V(P)) := \mathfrak{X}^v(P)$ its vertical part, $X^h \in \Gamma(H(P)) := \mathfrak{X}^h(P)$ its horizontal part. This decomposition is induced by a pair of bundle maps $T(P) \longrightarrow T(P)$, say ρ^v and ρ^h, with the following properties:

ρ^v and ρ^h are equivariant idempotents, $\rho^v + \rho^h = 1$ and $\mathrm{im}(\rho^v) = V(P)$.

Conversely, such a pair allows the definition of a bundle H(P) either as $\ker(\rho^v)$ or as $\mathrm{im}(\rho^h)$, etc. all this is standard linear algebra.

(4.4.4) can be used to obtain another description of connections on P which is extremely useful in practice; we arrive at it in the following manner:

Given $X \in \mathfrak{X}(P)$, X^v is vertical and hence of the form Z_f for a unique $f \in C^\infty(P,\mathfrak{g})$. We define a $\mathfrak{g}$–valued 1–form ω on P by setting

$$\omega(X) = f; \tag{4.4.5}$$

in particular, $\omega(X^h) = 0$ and, in fact, $\ker(\omega) = H(P)$. On the other hand, if $X = Z_h$ is a fundamental vector field (given by $h \in \mathfrak{g}$, then $\omega(X) = h$. Moreover, it is immediate from (4.4.5) that $r_g^*\omega = \mathrm{Ad}(g^{-1})\omega$. With this, we give the following definition:

<u>Definition 4.4.6</u> A <u>connection form</u> on P is a $\mathfrak{g}$–valued 1–form ω which satisfies the following conditions:

(C1) $\omega(Z_h) = h$ for $h \in \mathfrak{g}$;

(C2) $r_g^*\omega = \mathrm{Ad}(g^{-1})\omega$ for $g \in G$.

In this case, one obtains a horizontal bundle by setting

$$H(P) = \ker(\omega) \subset T(P); \tag{4.4.7}$$

due to (C2) this is an invariant subbundle and (C1) can be used to show that $T(P) = V(P) \oplus H(P)$. In particular, formula (4.4.5) yields the following description of the vertical component X^v of an arbitrary vector field $X \in \mathfrak{X}(P)$:

$$X^v = Z_{\omega(X)}, \tag{4.4.8}$$

often very useful. Accordingly, X^h is given by $X^h = X - Z_{\omega(X)}$.

The constructions indicated in (4.4.5) and (4.4.7) establish a bijection between invariant "horizontal" bundles H(P) and connection forms on P and we commonly identify connections with their connection forms.

The set $\mathcal{C} = \mathcal{C}_P \subset A^1_G(P,\mathfrak{g})$ of connection forms is not a subspace since (C1) is not a homogeneous linear condition; it is, however, an affine space: if $\omega_i \in \mathcal{C}$ and if $\lambda_i \in \mathbb{R}$ are such that $\Sigma\lambda_i = 1$, then $\Sigma\lambda_i\, \omega_i \in \mathcal{C}$. We obtain a more explicit description of $\mathcal{C}$ by noting that (C1) implies that if ω', $\omega \in \mathcal{C}$, then $\omega' - \omega$ is horizontal. By (C2), it also is equivariant. Thus, given $\omega \in \mathcal{C}$, every connection form can be written in the form $\omega+\mu$, μ an equivariant horizontal form:

$$\mathcal{C} = \omega + A^1_{h,G}(P,\mathfrak{g}). \tag{4.4.9}$$

As was pointed out earlier, the vector space $A^1_{h,G}(P,\mathfrak{g})$ may be replaced by $A^1(M,P(\mathfrak{g}))$ since π^* yields an isomorphism of the latter onto the former. In this sense,

$$\mathcal{C} \cong \omega + A^1(M,P(\mathfrak{g})). \tag{4.4.10}$$

This also follows from (4.4.3) quite directly: any two right inverses of π^* differ by a bundle map $T(M) \longrightarrow \ker(\pi_*) = P(\mathfrak{g})$, i.e. by a $P(\mathfrak{g})$–valued 1–form. Note here that the descriptions of $\mathcal{C}$ we just obtained are meaningful since we know from earlier that $\mathcal{C} \neq \phi$!

Remarks:

(i) Connection forms play the role of "vector potentials" (or "gauge potentials") in (non–abelian) gauge theories and thus have acquired great importance in mathematical Physics; we shall return to this aspect of the theory of connections in later chapters and here add only the following remarks:

Suppose that (U_α) is an open cover of M such that $P|U_\alpha$ is trivial for each α; we may assume that the trivialization is given by a section s_α of P over U_α; let also $(g_{\alpha\beta})$ be the 1–cocycle representing P with respect to (U_α), so that $g_{\alpha\beta} \in C^\infty(U_\alpha \cap U_\beta, G)$ with the usual properties. If now ω is a connection form on P, set $\omega_\alpha = s_\alpha^* \omega$, so that $\omega_\alpha \in A^1(U,\mathfrak{g})$. The "transformation law" for the forms ω_α then is this: over $U_\alpha \cap U_\beta$,

$$\omega_\beta = \mathrm{Ad}(g_{\alpha\beta}^{-1})\omega_\alpha + g_{\alpha\beta}^{-1}\, dg_{\alpha\beta},$$

as a direct calculation shows. In this expression, $g^{-1}dg$ is a standard abuse of notations for the left logarithmic differential, i.e. the principal part of $L_{g^{-1}}Tg$ or, which is the same, for $g^*\theta$, θ the canonical left invariant 1–form on G. If $\Omega_\alpha = s_\alpha^*\Omega$, Ω the curvature form of ω, cf. below, then $\Omega_\beta = \mathrm{Ad}(g_{\alpha\beta}^{-1})\Omega_\alpha$ by a similar calculation using the Maurer–Cartan theorem (cf. below). Conversely, any family (ω_α) of 1–forms $\omega_\alpha \in A^1(U,\mathfrak{g})$ satisfying the given "transformation law" yields a connection form ω on P by "patching" and it is in this form that connections initially arose as vector potentials in the literature; for an early account, cf. e.g. [W,Y].

(ii) Suppose that $P = M \times G$ is the product bundle. The principal right action of G then is given by $(x,g)\cdot h = (x,gh)$. Since a connection ω is equivariant, it is uniquely determined by the values $\omega(x,e)$, for $\omega(x,g) = r_g^*\omega(x,e) = \mathrm{Ad}(g^{-1})\omega(x,e)$. By (C1), moreover, $\omega(x,e)(0,h) = h$ for $h \in \mathfrak{g}$. On the other hand, if $\epsilon\colon M \longrightarrow M \times G$ is the identity section $\epsilon(x) = (x,e)$, then $(\epsilon^*\omega)(x)u = \omega(x,e)(u,0)$ for $u \in T_x(M)$. Here $\epsilon^*\omega$ lies in

$A^1(M,\mathfrak{g})$ and we see that

$$\omega(x,e)(u,h) = (\epsilon^* \omega)(x)u + h.$$

Conversely, given $\beta \in A^1(M,\mathfrak{g})$, we define

$$\omega(x,e)(u,h) = \beta(x)u + h$$

and obtain a connection on M×G. With this, one possesses a complete description of the connections on a product bundle M×G.

(iii) The "vertical part" $\omega|V(P)$ of a connection ω is independent of the choice of ω by (C1): it is the "vertical form" α_P determined, as an element of $\Gamma(V^*(P))$, by (C1). α_P is not a 1–form on P since it is defined only on V(P) and not on all of T(P). In fact, a connection form ω is precisely an <u>equivariant extension</u> of α_P((C2)!) to a 1–form on P.

Let ω be a connection and H(P) its horizontal bundle on P. Then one can assign to a (vector–valued) differential form α, say of degree p, a "horizontal part" α^h by

$$\alpha^h(X_1,...,X_p) := \alpha(X_1^h,...,X_p^h) \tag{4.4.11}$$

where $X_i \in \mathfrak{X}(P)$. α^h then is a horizontal form in the standard sense. Also, if α already is horizontal, then $\alpha = \alpha^h$ as one immediately verifies. Furthermore, if X is a horizontal vector field, then so is $R_g X$ since H(P) is invariant and one concludes that if α is an equivariant E–valued p–form (with respect to some representation of G in E), then so is α^h. Accordingly, $\alpha \longrightarrow \alpha^h$ is a projection of $A^p(P,E)$ onto $A^p_h(P,E)$ which restricts to a projection of $A^p_G(P,E)$ onto $A^p_{h,G}(P,E)$ under the appropriate assumptions.

Observe that, in general, $d\alpha^h$ is not horizontal; on the other hand, the explicit definition of d shows that $(d\alpha^h)^h = (d\alpha)^h$ and we now define the covariant exterior derivative of a form α by

$$D_\omega \alpha := (d\alpha)^h . \tag{4.4.12}$$

Combining this with the remarks just made shows, in particular, that

$$D_\omega: A^p_{h,G}(P,E) \to A^{p+1}_{h,G}(P,E) \tag{4.4.13}$$

for $p \geq 0$. However, in general $D^2_\omega \neq 0$ as we shall see shortly. Before investigating this claim, let us add a useful observation:

The connection ω is given by an equivariant splitting $\lambda = \lambda_\omega$ of the short exact sequence (4.4.2), λ_ω a bundle map $\pi^* T(M) \to T(P)$. If now X is a vector field on M, then $\pi^* X$ is a G–invariant section of $\pi^* T(M)$ which therefore maps to the invariant section $\lambda_\omega \pi^* X$ of $H(P) \subset T(P)$; we denote this section by $\kappa_\omega X$ is the unique invariant horizontal vector field on P which is π–related to X. κ may also be obtained in a different manner, without direct appeal to (4.4.2): ω induces a splitting $\mathfrak{X}(P) = \mathfrak{X}^v(P) \oplus \mathfrak{X}^h(P)$ and the same splitting restricts to the modules $\mathfrak{X}_I(P)$, $\mathfrak{X}^v_I(P)$ and $\mathfrak{X}^h_I(P)$ of invariant vector fields. Then one shows directly that $T\pi$ induces a surjective module homomorphism (over $C^\infty(M)$)

$$\pi_*: \mathfrak{X}_I(P) \to \mathfrak{X}(M) \tag{4.4.14}$$

whose kernel evidently is $\mathfrak{X}^v_I(P)$. Therefore, π_* induces an isomorphism $\mathfrak{X}^h_I(P) \cong \mathfrak{X}(M)$ whose inverse is the horizontal lift κ.

<u>Remark</u>: Let $\omega+\mu$ be any other connection on P. Then formula (4.4.17) immediately implies that $D_{\omega+\mu}\beta = D_\omega \beta + \lambda^\cdot(\mu) \wedge \beta$; in particular, if $s \in C^\infty_C(P,E)$, then $D_{\omega+\mu}s = D_\omega s + \lambda^\cdot(\mu)s$. This is the reason why one often abbreviates the result in the form $D_{\omega+\mu} = D_\omega + \lambda^\cdot(\mu)$.

Let now α be an E–valued p–form on M, E a fixed vector space. Then $\pi^* \alpha \in A^p_h(P,E)$ and one easily verifies that $\alpha(X_1,\ldots,X_p) = (\pi^* \alpha)(\kappa X_1,\ldots,\kappa X_p)$, $X_i \in \mathfrak{X}(M)$. More generally, let $\mathbb{E}$ be a vector bundle over M with fibre E which is associated with P under a representation $\lambda: G \to GL(E)$ and denote by $\rho_{\mathbb{E}_p}$ the map $C^\infty_G(P,E) \to \Gamma(\mathbb{E})$, inverse of π^*. If $\alpha \in A^p(M,\mathbb{E})$ and therefore $\pi^* \alpha \in A^p_{h,G}(P,E)$, one again verifies that

$$\alpha(X_1,\ldots,X_p) = \rho_E(\pi^* \alpha(\kappa X_1,\ldots,\kappa X_p)). \tag{4.4.15}$$

This formula is more explicit than the earlier isomorphism, but requires the choice of a connection on P.

We next return to the covariant exterior derivative D_ω induced by the connection ω and wish to describe the maps in (4.4.13) more explicitly; we assume that G operates on E by means of the representation λ and denote by $\lambda^{\cdot}$ the induced representation of $\mathfrak{g}$ which is a Lie algebra homomorphism $\mathfrak{g} \longrightarrow gl(E)$ ($= T_e\lambda$ when $\mathfrak{g}$ is identified with $T_e(G)$, etc.). As a consequence, we now have the bilinear map $(h,y) \longrightarrow \lambda^{\cdot}(h)y$ of $\mathfrak{g}\times E$ into E which – by general principles – induces a pairing $A^{\cdot}(P,\mathfrak{g}) \times A^{\cdot}(P,E) \longrightarrow A^{\cdot}(P,E)$, mapping (α,β) to an E–valued form we denote by $\lambda^{\cdot}(\alpha) \wedge \beta$. If $h \in \mathfrak{g}$ and if Z_h is the corresponding fundamental vector field, then a simple calculation shows that the Lie derivative $\mathcal{L}(Z_h)\beta$ of the <u>equivariant</u> form β coincides with $-\lambda^{\cdot}(h) \wedge \beta$. As a consequence, we obtain the following result:

<u>Proposition 4.4.16</u> For any $\beta \in A^{\cdot}_{h,G}(P,E)$ and $\omega \in \mathcal{C}$,

$$D_\omega\beta = d\beta + \lambda^{\cdot}(\omega) \wedge \beta. \tag{4.4.17}$$

For the proof, we show firstly that the right–hand side is horizontal: Since β is horizontal, $\mathcal{L}(Z_h)\beta = i(Z_h)d\beta$ and $i(Z_h)\lambda^{\cdot}(\omega)\wedge\beta)$ becomes $(i(Z_h)\lambda^{\cdot}(\omega)\wedge\beta = \lambda^{\cdot}(iZ_h)\lambda^{\cdot}(\omega))\wedge\beta = \lambda^{\cdot}(i(Z_h)\omega)\wedge\beta = \lambda^{\cdot}(h)\wedge\beta$. Thus, by the remark preceding the proposition, $i(Z_h)(d\beta + \lambda^{\cdot}(\omega)\wedge\beta) = \mathcal{L}(Z_h)\beta + \lambda^{\cdot}(h)\wedge\beta = 0$, showing the claim. Therefore $d\beta + \lambda^{\cdot}(\omega)\wedge\beta = (d\beta + \lambda^{\cdot}(\omega)\wedge\beta)^h = D_\omega\beta + \lambda^{\cdot}(\omega^h)\wedge\beta^h = D_\omega\beta$ because of $\omega^h = 0$. This completes the proof.

We shall need the following special case: $E = \mathfrak{g}$ and $\lambda = \mathrm{Ad}$, so that $\lambda^{\cdot} = \mathrm{ad}$: $\mathrm{ad}(h)k = [h,k]$. In this case, observe that $\mathrm{ad}(\alpha)\beta = \frac{1}{2}[\alpha,\beta]$, a form which often also is denoted by $\alpha\wedge\beta$. For later use, we state the special case explicitly:

Corollary: If $\beta \in A^{\cdot}_{h,G}(P,\mathfrak{g})$, then

$$D_\omega\beta + d\beta + \tfrac{1}{2}[\omega,\beta]. \tag{4.4.18}$$

After these generalities, we now introduce the notion of the curvature of a connection:

Definition 4.4.19 The curvature Ω of the connection ω is the 2–form

$$\Omega = D_\omega\omega. \tag{4.4.20}$$

The following basic results then are quite immediate:

Proposition 4.4.21 For any connection ω on P and its curvature Ω, the following identities hold:

(i) $\Omega = d\omega + \frac{1}{2}[\omega,\omega]$ (Maurer–Cartan formula)

(ii) $D_\omega\Omega = 0$ (Bianchi identity).

To verify (i), observe that $\mathcal{L}(Z_h)\omega = i(Z_h)d\omega$ equals $-\mathrm{ad}(h)\omega$ since ω is equivariant. Moreover,

$$i(Z_h)[\omega,\omega] = 2[i(Z_h)\omega,\omega] = 2[h,\omega] = 2\mathrm{ad}(h)\omega,$$

whence $i(Z_h)(d\omega + \frac{1}{2}[\omega,\omega]) = 0$: the right–hand side is horizontal. Moreover, $[\omega,\omega]^h = [\omega^h,\omega^h] = 0$ and therefore

$$d\omega + \tfrac{1}{2}[\omega,\omega] = (d\omega + \tfrac{1}{2}[\omega,\omega])^h = [\Omega,\omega^h] = 0.$$

Remark: Equation (ii) sometimes appears as part of the "Yang–Mills equations" in the Physics literature. It is important to bear in mind that it is satisfied for any connection (vector potential) and hence does not contribute anything to the solution of the Yang–Mills problem!

The Maurer–Cartan formula now also provides more information on the sequence of operators D_ω introduced in (4.4.13):

Proposition 4.4.22 Let G act on E by means of $\lambda:G \longrightarrow GL(E)$. Then, for any form $\alpha \in A^{\cdot}_{h,G}(P,E)$,

$$D^2_\omega \alpha = \lambda^{\cdot}(\Omega)\wedge\alpha. \tag{4.4.23}$$

The proof consists of an explicit calculation: $D^2_\omega \alpha = D_\omega(d\alpha + \lambda^{\cdot}(\omega)\wedge\alpha)$ by (4.4.17); this equals

$$\begin{aligned} &d^2\alpha + d(\lambda^{\cdot}(\omega)\wedge\lambda) + \lambda^{\cdot}(\omega)\wedge d\alpha + \lambda^{\cdot}(\omega)\wedge\lambda^{\cdot}(\omega)\wedge\alpha \\ &= \lambda^{\cdot}(d\omega)\wedge\alpha - \lambda^{\cdot}(\omega)\wedge d\alpha + \lambda^{\cdot}(\omega)\wedge\alpha + \tfrac{1}{2}\lambda^{\cdot}[\omega,\omega]\wedge\alpha = \lambda^{\cdot}(\Omega)\wedge\alpha. \end{aligned}$$

In this sense, the curvature Ω "measures" the extent to which the sequence $(A^p_{h,G}(P,E),D_\omega)$ fails to be a complex: it is one iff. $\lambda^{\cdot}(\Omega) = 0$ which for a locally faithful representation λ amounts to $\Omega = 0$; in such a case, the connection ω is said to be flat. One should not here that, in general, P will not admit any flat connections.

It is immediate from definition 4.4.19 and the properties of ω that the curvature Ω is an equivariant horizontal 2–form, i.e. lies in $A^2_{h,G}(P,\mathfrak{g})$. Therefore, there exists a uniquely determined 2–form

$$F_\omega \in A^2(M,P(\mathfrak{g}))$$

such that

$$\pi^* F_\omega = \Omega. \tag{4.4.24}$$

This form still is referred to as the curvature of ω.

Remarks:

(i) In (free) Yang–Mills theories, F_ω is the field induced by the vector potential ω, cf. below. In the classical case $G = U(1)$, etc., F_ω is the Maxwell 2–form.

(ii) The Atiyah sequence (4.4.3) leads to the corresponding short exact sequence of section modules

$$0 \longrightarrow \Gamma(P(\mathfrak{g})) \longrightarrow \Gamma(\mathbb{E}) \xrightarrow{\pi_*} \mathfrak{X}(M) \longrightarrow 0$$

in which the middle term is naturally isomorphic to $\mathfrak{X}_I(P)$; in particular, one again finds the surjective Lie algebra homomorphism π_* of (3.4.14) whose kernel $\Gamma(P(\mathfrak{g}))$ is an ideal isomorphic to $\mathfrak{X}_I^v(P)$ under $\Gamma(\mathbb{E}) \cong \mathfrak{X}_I(P)$. Let now μ be a splitting of the Atiyah sequence: μ is a module homomorphism $\mathfrak{X}(M) \longrightarrow \Gamma(\mathbb{E})$ with $\pi_*\mu = 1$ and corresponds to a connection form on P. It makes sense to form $\mu[X,Y] - [\mu X,\mu Y]$ and it is clear that this expression lies in the ideal $\Gamma(P(\mathfrak{g}))$; the construction shows that it is a skew–symmetric $C^\infty(M)$–bilinear map $\mathfrak{X}(M)\times\mathfrak{X}(M) \longrightarrow \Gamma(P(\mathfrak{g}))$, i.e. a $P(\mathfrak{g})$–valued 2–form on M. This form is precisely F_ω and in this view, the curvature of ω "measures" the extent to which μ fails to be a Lie algebra homomorphism.

(iii) There is yet another interpretation of the curvature of ω: Recall that $d\omega(X,Y) = X\omega(Y) - Y\omega(X) - \omega([X,Y])$. Since $\Omega = D_\omega \omega$ is the horizontal part of this and ω vanishes on horizontal vector fields, one sees that $\Omega(X,Y) = \Omega(X^h,Y^h) = -\omega([X^h,Y^h])$. This vanishes if and only if $[X^h,Y^h]$ is horizontal. One concludes that $\Omega = 0$ is equivalent with the integrability of the horizontal bundle $H(P) \subset T(P)$. Thus, if ω is flat, then $H(P)$ defines a foliation of P which is transverse to the fibration $\pi:P \longrightarrow M$ and has leaf dimension equal to dim(M).

At this point, we deal briefly with the special case of a principal G–bundle $\pi:P \longrightarrow M$ with abelian structure group G. Due to the fact that now the inner automorphisms of G and Ad are trivial (and ad = 0), the preceding considerations simplify considerably:

The natural bundles $\mathcal{G}$ and $P(\mathfrak{g})$ are trivial: $\mathcal{G} \cong M\times G$ and $P(\mathfrak{g}) \cong M\times\mathfrak{g}$. Accordingly, $\mathcal{G}_P \cong C^\infty(M,G)$ and $\Gamma(P(\mathfrak{g})) \cong C^\infty(M,\mathfrak{g})$. Moreover, equivariant $\mathfrak{g}$–valued forms on P now are invariant: $r_g^*\alpha = \alpha$. In particular, a connection form ω is an invariant $\mathfrak{g}$–valued 1–form such that $i(Z_h)\omega = h$ for $h \in \mathfrak{g}$ and its curvature form Ω is an invariant horizontal 1–form with values in $\mathfrak{g}$. Accordingly, Ω is of the form $\Omega = \pi^*F_\omega$ for some $\mathfrak{g}$–valued 2–form F_ω on M. The earlier formulas for D_ω simplify and, in particular, the Maurer–Cartan formula reduces to $\Omega = d\omega$. $\pi^*F_\omega = \Omega$ then implies that $dF_\omega = 0$ (since π^* commutes with d).

Next we add some remarks on mappings of connections and end this section with a brief outline of the construction of products of connections.

Let $\pi':Q \longrightarrow M$ be a principal H–bundle and suppose that $\varphi:Q \longrightarrow P$ is a strong bundle map with respect to the group homomorphism $\sigma:H \longrightarrow G$. Since $\varphi\cdot r_h = r_{\sigma(h)}\cdot\varphi$ for $h \in H$ and $Ad\cdot\sigma = \dot{\sigma}\circ Ad$, the pull–back $\varphi^*\alpha$ of an equivariant $\mathfrak{g}$–valued form α is still equivariant (with values in $\mathfrak{g}$!). In general, however, if ω is a connection on P, $\varphi^*\omega$ will not yield a connection on Q. Nevertheless, there are the following results:

(A) Suppose that σ is locally faithful, i.e. that $\dot{\sigma}$ is injective $\mathfrak{h} \longrightarrow \mathfrak{g}$. Given $\omega \in \mathcal{C}_P$, there exists a unique $\omega' \in \mathcal{C}_Q$ such that

$$\varphi^* \omega = \dot{\sigma}\omega'. \tag{4.4.25}$$

Moreover, $\varphi^* \Omega = \dot{\sigma}\Omega'$.

Since $\dot{\sigma}$ is injective, the uniqueness is clear, as is the H–equivariance because of $r_h^* \dot{\sigma}\omega' = \dot{\sigma}\, r_h^* \omega'$. Next, for any $h \in \mathfrak{g}$, $T\varphi \cdot Z_h = Z_{\dot{\sigma}h}$ as follows immediately from the definitions. Therefore, $i(Z_h)\varphi^* \omega = i(Z_{\dot{\sigma}h})\omega = \dot{\sigma}h$. Letting for the moment ρ be the inverse of $\dot{\sigma}$ on $\mathrm{im}(\dot{\sigma}) \subset \mathfrak{g}$, one concludes that $\rho\varphi^* \omega$ is a connection ω' on Q which satisfies (4.4.25) by construction.

With only minor changes, the result still holds if φ is a bundle map over some diffeomorphism f of M, cf. [K,N], ch. 2, sec. 6.

(B) The result holds, in particular, when $\dot{\sigma}$ is an isomorphism, i.e. if H is locally isomorphic to G. An important special case is that of a <u>covering group</u> H of G ($\dot{\sigma}$ is an isomorphism and σ is surjective); we shall return to such a case later.

(C) $\pi' = \pi \cdot \varphi$ shows that $T\pi' = T\pi \cdot T\varphi$, i.e. that

$$V(Q) = \ker(T\pi') = (T\varphi)^{-1}(V(P));$$

in particular, $\ker(T\varphi) \subset V(Q)$. Suppose now that ω' is a connection on Q and that H(Q) is its horizontal bundle, $\mathrm{rank}(H(Q)) = \dim(M)$. We conclude that $T\varphi | H(Q)$ is injective on each fibre. Since $T\varphi$ also map V(Q) onto $V(P)/\mathrm{im}(\varphi)$, $T\varphi(H(Q))$ is a summand of $V(P)|\mathrm{im}(\varphi)$ in $T(P)|\mathrm{im}(\varphi)$. For $h \in H$, $T\varphi \cdot R_h = R_{\sigma(h)} \cdot T\varphi$, i.e. $T\varphi(H(Q))$ is invariant under H operating on $\mathrm{im}(\varphi)$ via σ.

If $p \in P$ is arbitrary, there exist $q \in Q$ and $g \in G$ such that $p = \varphi(q)g$ and we define $H_p(P)$ to be $R_g T_g \varphi(H_g(Q))$; one easily shows that this is independent of the choice of q and g.

Thus, $H(P) = \bigcup_p H_p(P)$ is an invariant subbundle of T(P) such that $T(P) = V(P) \oplus H(P)$ and therefore now defines a connection ω on P, "image under φ of ω'", $= \varphi_* \omega'$, and it is immediate that $\overset{*}{\varphi}\,\omega = \dot{\sigma}\omega'$, etc.

The considerations of (C) again apply with obvious minor changes to the case where φ is a bundle map over some diffeomorphism f of M.

(D) In the situation of (C), suppose that $H \subset G$ is an immersed group, that σ is the inclusion and that therefore $\varphi: Q \longrightarrow P$ is a reduction of P to the principal H–bundle Q. Then a connection ω on P is said to be <u>reducible</u> to a connection ω' on Q if it is obtained from ω' by the process of (C), i.e. if $\overset{*}{\varphi}\,\omega = \omega'$ (omitting $\dot{\sigma}$, the inclusion $\mathfrak{h} \subset \mathfrak{g}$). ω is irreducible if it cannot be obtained in this manner from any reduction of P to a principal H–bundle, H a proper closed subgroup of G.

In the current situation, φ is an injective immersion into P and $\overset{*}{\varphi}\,\omega$ thus is "the restriction of ω to Q". $T_q\varphi$ maps $H_q(Q)$ isomorphically onto $H_q(P)$ and in this sense this latter space is "tangent to Q" for each $q \in Q$. This necessary condition for the reducibility of ω also is sufficient:

If for every $q \in Q$ the horizontal space $H_q(P)$ of ω is tangent to Q, the ω is reducible to a connection ω' on the immersed bundle Q. It is obvious how to define ω': its horizontal space at q simply is $H_q(P)$.

(E) Combining the last result with proposition 4.4.16 leads to the following considerations:

Let $H \subset G$ be a closed subgroup, $P/H \to M$ the associated bundle with fibre G/H, μ: $P \to P/H$ the natural map. Choose a connection ω on P and let H(P) be its horizontal bundle. Since μ is a submersion, $T\mu$ maps T(P) onto T(P/H) and, as is easily seen, V(P) onto V(P/H) since the fibre at x of P/H simply is P_x/H. Moreover, $T\mu(H(P))$ is a direct summand of V(P/H) which we denote by H(P/H):

$$T(P/H) = V(P/H) \oplus H(P/H), \tag{4.4.26}$$

cf. below for more details. In this manner, ω induces a "horizontal bundle" om P/H.

If now s is a (local) section of P/H, we call it parallel or horizontal with respect to ω if for every x in the domain of s,

$$T_x s(T_x(M)) \subset H_{s(x)}(P/H);$$

cf. below again for more detail. With this, we can formulate the next result:

With the current notations, suppose that s is a global section of P/H and let Q be the principal H–bundle, reduction of P, defined by s. A connection ω on P is reducible to a connection ω' on Q if and only if s is ω–horizontal.

Note that here Q is a submanifold of P since H is closed. Suppose now that ω is reducible to a connection ω' on Q. By (D), $H_q(Q) = H_q(P)$ for $q \in Q$ and s(M) is a (closed) submanifold of P/H, diffeomorphic to M under s, so that $T_{s(x)}(s(M)) = T_x s(T_x(M))$. Choose $q \in Q$ with $\mu(q) = s(x)$. Then $T_q\mu(H_q(P))$ is contained in $T_{s(x)}(s(M))$ since μ maps Q onto s(M); for dimension reasons, the two now are equal and we conclude $H_{s(x)}(P/H) = T_{s(x)}(T_x(M))$: s is ω–horizontal.

Conversely, suppose that s is ω–parallel. In this case, $T_x s(T_x(M)) = H_{s(x)}(P/H)$ again for dimension reasons. On the other hand, $\mu : P \longrightarrow P/H$ is a submersion, hence certainly transversal over the submanifold s(M). Since $Q = \mu^{-1}(s(M))$, $T_q(Q) = (T_q\mu)^{-1}(T_{\mu(q)}(s(M)))$, cf. ch.I. But this is $(T_q\mu)^{-1}(H_{\mu(q)}(P/H)$ and thus contains $H_q(P)$; therefore $H_q(P) \subset T_q(Q)$ and the claim follows from (D): there is a connection ω' on Q such that $\omega | Q = \omega'$.

Observe now that $P/H \cong P\times_G(G/H)$, a bundle associated with P. This suggests that at least some parts of the above considerations might carry over to arbitrary associated bundles of P. This, indeed, is the case:

(F) Suppose that G operates from the left on the manifold F by means of $(g,y) \longrightarrow gy$. Form the associated bundle $\mathbb{F} = P\times_G F = (P\times F)/G$ and let ρ: $P \times F \longrightarrow \mathbb{F}$ be the natural map used earlier (Sec. 4.3). Identify $T(P\times F)$ with $T(P)\times T(F)$ in the usual way (cf. ch.I and the literature); then $T\rho$ is a surjection $T(P)\times T(F) \longrightarrow T(\mathbb{F})$ (since ρ naturally is a submersion) and is, in fact, a bundle map over ρ. The claim is the following:

<u>Proposition 4.4.27</u> Choose a connection ω on P and let H(P) be its horizontal bundle. If V(P), V($\mathbb{F}$) denote the naturally defined vertical bundles, then

$$T\rho(V(P) \times T(F)) = V(\mathbb{F}),$$

$$H(\mathbb{F}) := T_\rho(H(P) \times o) \text{ is a subbundle of } T(\mathbb{F}) \text{ and}$$

$$T(\mathbb{F}) = V(\mathbb{F}) \oplus H(\mathbb{F}). \tag{4.4.28}$$

In this manner, ω induces "horizontal bundles" on all associated bundles of P and it therefore is meaningful to speak of horizontal sections of such bundles.

For the proof, let $pr_1:P\times F \longrightarrow P$ be the obvious map and $\mu:\mathbb{F} \longrightarrow M$ the bundle projection. Then $\pi\cdot pr_1 = \mu\cdot\rho$ by construction and thus $T\pi\cdot Tpr_1 = T\mu\cdot T\rho$. In addition, $Tpr_1:T(P)\times T(F) \longrightarrow T(P)$ is the obvious first projection again whose kernel is $o\times T(F)$. Since $\ker(T\pi) = V(P)$, we conclude that $\ker(T\mu\cdot T\rho) = V(P)\times T(F)$, whence $T_\rho(V(P)\times T(F)) \subset V(\mathbb{F}) = \ker(T_\mu)$ as well as $\ker(T_\rho) \subset V(P) \times T(F)$. Since $V(\mathbb{F})_y = T_y(F_{\mu(y)})$, $\mathrm{rank}(V(\mathbb{F}))$ is $2\dim(F)$. Moreover, since $T(P)\times T(F) = (V(P)\times T(F))\oplus(H(P)\times 0)$ and $\ker(T\rho) \subset V(P) \times T(F)$, one sees that $T(\mathbb{F}) = T\rho(V(P)\times T(F))\oplus H(\mathbb{F})$. It is easy to show that $H(\mathbb{F})$ has constant rank, hence now also the first summand. In fact, the rank of $H(\mathbb{F})$ is $\dim(M)$ and so the first part has rank $2\dim(F)$. Since it is contained in $V(\mathbb{F})$, it now must coincide with this bundle and we are done.

Remarks:

(i) Induced horizontal bundles in the sense of the proposition just established will be used later again especially in the case of vector bundles associated with P and will lead to an explicit description of the covariant derivative (linear connection) induced in such a bundle by the choice of $\omega \in \mathcal{C}_P$, cf. sec. 4.1.

(ii) In some applications, it is useful to possess an explicit description of $\ker(T\rho)$ which may be obtained as follows:

The group G acts on the left on F and hence, there is an induced map $\mathfrak{g} \longrightarrow \mathfrak{X}(F)$; we denote the image of $h \in \mathfrak{g}$ under this map by Y_h. Explicitly: $Y_h(y) = \frac{d}{dt}(g_t y)|_{t=0}$ where $g_t = \exp(th)$.

The same 1–parameter group g_t also operates on P and, for a fixed p in P, defines a curve $t \longrightarrow pg_t$ whose tangent vector at $t = 0$ is $Z_h(p)$, by definition. There now also are the two curves $(pg_t y)$ and $(p, g_t y)$ in $P \times F$ for any choice of $p \in P$, $y \in F$, with the obvious tangent vectors at $t = 0$. Since $\rho(pg_t, y) = \rho(p, g_t y)$ for all t, we conclude that $T\rho(Z_h(p), o_y) = T\rho(o_p, Y_h(y))$. In other words:

For $h \in \mathfrak{g}$, $(p,y) \in P \times F$, $(Z_h(p), -Y_h(y)) \in \ker(T\rho)$.

Since $Z_h(p) = 0$ implies $h = 0$, the map $h \longrightarrow (Z_h(p), -Y_h(y))$ is a linear injection whose image therefore has dimension $= \dim(\mathfrak{g}) = \dim(G)$ and this is, in particular, constant on $P \times F$: these images form a sub<u>bundle</u> of rank $= \dim(G)$ of $T(P \times F)$. On the other hand, $(P \times F, \rho)$ is a principal G–bundle whose vertical bundle $\ker(T\rho)$ has rank $\dim(G)$. With this, we are done:

$$\ker(T\rho) = \{(Z_h(p), -Y_h(y)) \mid h \in \mathfrak{g},\ (p,y) \in P \times F\}.$$

The natural trivialization of this vertical bundle is given by the map $h \longrightarrow (Z_h, -Y_h)$ and we have obtained an explicit description of $\ker(T\rho)$, as desired.

We close this section with a few remarks on "products of connections" in the following context:

Let $\pi : P \longrightarrow M$ and $\rho : Q \longrightarrow N$ be principal bundles with group G,H, resp. Then $P \times Q$ is a principal $G \times H$–bundle over $M \times N$ with bundle projection $\pi \times \rho$, as is readily verified. The tangent bundle of $P \times Q$ is identified, as usual, with $T(P) \times T(Q)$. If $(a,b) \in \mathfrak{g} \times G$ and if Z_a, Z_b are the fundamental vector fields on P and Q induced by a and b, then $Z_a \times Z_b = Z_{(a,b)}$ is the fundamental vector field through (a,b).

Given the connections ω on P and ψ on Q, it is immediate that the $\mathfrak{g}\times\mathcal{G}$–valued 1–form $\omega\times\psi$ is a connection of P×Q, the <u>product</u> of ω and ψ. Let T(P) = V(P) ⊕ H(P), T(Q) = V(Q) ⊕ H(Q) be the splittings defined by ω and Ψ, respectively. Then it is clear that T(P×Q) = T(P)×T(Q) = (V(P)×V(Q) ⊕ (H(P)×H(Q)) and also that

$$V(P\times Q) = V(P)\times V(Q). \tag{4.4.29}$$

Consequently,

$$H(P\times Q) = H(P) \times H(Q) \tag{4.4.30}$$

is a horizontal bundle since the invariance is clear. This is, in fact, the horizontal bundle of $\omega\times\Psi$ since $(\omega\times\Psi)(u,v) = o$ iff. $\omega u = o$ and $\Psi v = o$ ($u \in T(P)$, $v\in T(Q)$).

Next, recall that if α and β are differential forms on P and Q of the same degree, then $d(\alpha\times\beta) = d\alpha\times d\beta$. Accordingly, (3.4.30) implies that

$$D_{\omega\times\Psi}(\alpha\times\beta) = D_{\omega}\alpha\times D_{\Psi}\beta. \tag{4.4.31}$$

In particular, the <u>curvature</u> of $\omega\times\psi$ is $\Omega\times\psi$. Observe also that if we write $\mathfrak{g}\oplus\mathfrak{h}$, instead of $\mathfrak{g}\times\mathfrak{h}$, then $\alpha\times\beta$ becomes the form $\overset{*}{pr_1}\alpha + \overset{*}{pr_2}\beta$ where pr_i are the projections $P\times Q \longrightarrow P,Q$. This is the notation e.g. of [K,N], ch.2, sec. 6; our curvature formula then is in agreement with the one given there.

Suppose now that N = M and let $\Delta : M \longrightarrow M\times M$ be the diagonal embedding. The pull–back bundle $\overset{*}{\Delta}(P\times Q)$ is denoted by $P\times_M Q$ and is the (fibred) product of P and Q over M. If the pull–back of $\omega\times\psi$ is denoted by $\omega+\psi$, then $\omega+\psi$ is a connection on $P\times_M Q$ and its curvature is $\Omega+\Psi$. We shall return to these matters later.

4.5 The special case $G \longrightarrow G/H$

As a preliminary consideration, recall from 4.1 the group Aut(P) of equivariant diffeomorphisms of P. It was shown there that V(P) is invariant under the $\alpha \in \text{Aut}(P)$. If now $\omega \in \mathcal{C}_P$ and $\alpha \in \text{Aut}(P)$, then $\alpha^* \omega$ again is equivariant because of $\alpha \cdot r_g = r_g \cdot \alpha$ and the invariance of V(P) implies that $(\alpha^* \omega)(Z_h) = \omega(Z_h) = h$: $\alpha^* \omega$ is a connection again. In other words:

There is a natural right action $\mathcal{C}_P \times \text{Aut}(P) \longrightarrow \mathcal{C}_P$, given by the pullback map $(\omega, \alpha) \longrightarrow \alpha^* \omega$. This action, restricted to the gauge group $\mathcal{G}_P$ or certain extensions of this "minimal" gauge group, is of great importance in Yang–Mills theories, cf. the appropriate sections of later chapters.

For current purposes, we merely need the following: The connection ω is said to be invariant under $\alpha \in \text{Aut}(P)$ if $\alpha^* \omega = \omega$ and one then also says that α "preserves the connection ω". Evidently, the set of such automorphisms α is a subgroup of Aut(P), the isotropy of ω in the natural action of Aut(P) on $\mathcal{C}_P$.

For the remainder of this section, we assume that the principal bundle under consideration is of the form

$$\pi : G \longrightarrow G/H \tag{4.5.1}$$

where H, the structure group of the bundle, is a closed subgroup of G. The Lie algebras of G and H are denoted by $\mathfrak{g}$ and $\mathfrak{h}$ and we recall that we often tacitly identify $\mathfrak{g}$ (= vector space of left–invariant vector fields with $T_e(G)$ under <u>left</u> translations. Since the left translations by elements of G commute with right translations by the elements of H, G is naturally realized as a group of bundle automorphisms of $G \longrightarrow G/H$ and it is the aim of this section to describe in a very explicit form the connections on the bundle

$G \longrightarrow G/H$, especially in the case where G admits an <u>invariant connection</u>, i.e. one which is invariant under left translations by G. The following simple observation will be fundamental for the sequel: A connection ω on $G \longrightarrow G/H$ is invariant iff. ω is a $\mathfrak{g}$–valued <u>left–invariant</u> 1–form satisfying (C1) and (C2) of definition 4.4.6. Such a form ω then is, in particular, an element of $L(\mathfrak{g},\mathfrak{g}) \cong \mathfrak{g}^* \otimes \mathfrak{g}$ and is uniquely determined by its values on the left–invariant vector fields.

The first result is the following:

<u>Proposition 4.5.2</u> There exists an invariant connection ω on $G \longrightarrow G/H$ if and only if $\mathfrak{h} \subset \mathfrak{g}$ admits a (vector space) summand $\mathfrak{m}$ such that $Ad(H)(\mathfrak{m}) = \mathfrak{m}$, i.e. iff. G/H is reductive.

First of all, suppose that ω is an invariant connection and interpret as a linear map $\mathfrak{g} \longrightarrow \mathfrak{g}$. Set $\mathfrak{m} = \ker(\omega)$. Then $\mathfrak{g} = \mathfrak{h} \oplus \mathfrak{m}$: Indeed, it is easy to see that the fundamental vector fields Z_h, $h \in \mathfrak{h}$, coincide with the left–invariant vector fields given by $\mathfrak{h}$; the vertical tangent space $V_e(G)$ is thus identified with $\mathfrak{h}$, whereas $H_e(G)$ is $\ker(\omega(e))$, identified with $\ker(\omega) = \mathfrak{m}$. Next, if X is a horizontal field, so is $R_h X$ for any $h \in H$ and now, if X is the left–invariant field through $k \in \mathfrak{m}$, then $R_h X$ is defined by $Ad(h^{-1})k$; since this must again be horizontal, we conclude that $Ad(H)(\mathfrak{m}) \subset \mathfrak{m}$.

Conversely, suppose that G/H is reductive and let $\mathfrak{m}$ be an invariant summand of $\mathfrak{h}$: $\mathfrak{g} = \mathfrak{h} \oplus \mathfrak{m}$, $Ad(H)(\mathfrak{m}) \subset \mathfrak{m}$. Let $p: \mathfrak{g} \longrightarrow \mathfrak{h}$ be the projection onto $\mathfrak{h}$ with kernel $\mathfrak{m}$. This defines a left–invariant $\mathfrak{g}$–valued 1–form ω which is a connection since $p|\mathfrak{h} = 1$ (whence (C2)) and p commutes with Ad(H) by the choice of the splitting of $\mathfrak{g}$ (whence (C1)). This proves the proposition.

For the rest of this section, we assume that G/H is reductive and that $\mathfrak{g} = \mathfrak{h} \oplus \mathfrak{m}$ is a fixed Ad(H)–invariant splitting; we denote by ω_0 the invariant connection defined by the projection $p: \mathfrak{g} \longrightarrow \mathfrak{h}$, $\ker(p) = \mathfrak{m}$, and refer to it as the "canonical connection" determined by the given splitting. Any connection ω on $G \longrightarrow G/H$ now is of the form $\omega_0 + \beta$ with $\beta \in A^1_{h,H}(G,\mathfrak{h}) \cong A^1(G/H,P(\mathfrak{h}))$. This latter is the section module of the bundle $T^*(G/H) \otimes P(\mathfrak{h})$ and both $T^*(G/H)$ and $P(\mathfrak{h})$ are homogeneous, so that the remarks made at the end of section 4.3 imply that

$$A^1_{h,H}(G,\mathfrak{h}) \cong C^\infty_H(G,\mathfrak{m}^* \otimes \mathfrak{h}) \tag{4.5.3}$$

where H acts on $\mathfrak{m}^* \otimes \mathfrak{h}$ by means of $\mathrm{Ad}^* \otimes \mathrm{Ad}$ or, re–writing $\mathfrak{m}^* \otimes \mathfrak{h}$ as $L(\mathfrak{m},\mathfrak{h})$, by $\varphi \longrightarrow \mathrm{Ad}(h)\cdot\varphi\cdot\mathrm{Ad}(h^{-1})$; thus, we may also write

$$A^1_{h,H}(G,\mathfrak{h}) \cong C^\infty_H(G,L(\mathfrak{m},\mathfrak{h})). \tag{4.5.4}$$

A brief computation using the Maurer–Cartan formula shows that if $h,k \subset \mathfrak{h}$, the curvature form Ω_0 is given by

$$\Omega_0(h,k) = [ph,pk] - p[h,k] \tag{4.4.5}$$

at least at e. Since Ω_0 is a left–invariant form, it is completely determined by this identity and if both h and k lie in $\mathfrak{m}$, then the expression reduces to

$$\Omega_0(h,k) = -p[h,k] \tag{4.5.6}$$

and it is instructive to compare this last formula with one derived in the remarks following proposition 4.4.22.

At this point, let D_0 be the covariant exterior derivative of ω_0 on a function space $C^\infty(G,E)$ or $C^\infty_H(G,E)$ (if a representation $\lambda: H \longrightarrow GL(E)$ is chosen). A brief computation shows that for a left invariant vector field X with $X_e = h \in \mathfrak{m}$,

$$D_0 f\cdot X = Xf. \tag{4.5.7}$$

An arbitrary connection ω on $G \longrightarrow G/H$ is of the form $\omega = \omega_0 + \alpha$ with $\alpha \in C^{\infty}_{H}(G;,L(\mathfrak{m},\mathfrak{h}))$, cf. (4.5.4). Let $E = \mathfrak{g}$, $\mathfrak{h}$ or $\mathfrak{m}$ – on which H operates by means of Ad. Formula (4.4.18) then implies that on $C^{\infty}_{h,H}(G,\mathfrak{h})$,

$$D_{\omega} = D_0 + \mathrm{ad}(\alpha). \tag{4.5.8}$$

Next, the connection ω is invariant if and only if α is left–invariant, i.e. now iff. α is constant: $\alpha \in L(\mathfrak{m},\mathfrak{h})$ with $\mathrm{Ad}(h)\alpha = \alpha\,\mathrm{Ad}(h)$, $h \in H$ because of the equivariance of α. Thus, α has to lie in the subspace $L(\mathfrak{m},\mathfrak{h})^H = (\mathfrak{m}^* \otimes \mathfrak{h})^H$ of invariants under H and we conclude that the affine space of invariant connections is of the form

$$\omega_0 + (\mathfrak{m}^* \otimes \mathfrak{h})^H \tag{4.5.9}$$

and is, in particular, finite–dimensional.

Let $q = 1 - p$ be the projection $\mathfrak{g} \longrightarrow \mathfrak{m}$ with kernel $\mathfrak{h}$ and let $\omega = \omega_0 + \alpha$ be an invariant connection. α is constant and we use the Maurer–Cartan formula together with the bilinearity of [,] and the fact that for 1–forms α and β, $[\alpha,\beta] = [\beta,\alpha]$ to compute the curvature of ω (at e which is sufficient since o is left–invariant); the result is

$$\Omega(h,k) = -p[h,k] + [\alpha h,\alpha k] - \alpha q[h,k] \tag{4.5.10}$$

for $h,k \in \mathfrak{m}$.

A slight generalization of all this may be useful: Suppose that λ is a representation of H in E (a homomorphism $H \longrightarrow GL(E)$) and let $\lambda^{\cdot}\mathfrak{h} \longrightarrow gl(E)$ be the induced representation of the Lie algebra $\mathfrak{h}$. Given $\alpha \in L(\mathfrak{m},\mathfrak{h})^H$, set $\beta = \lambda^{\cdot}\alpha$, so that $\beta \in L(\mathfrak{m},gl(E))^H$. Then we find for $\omega = \omega_0 + \alpha$ and any $f \in C^{\infty}{}_{h,H}(G,E)$,

$$D_{\omega}f = (D_0 + \beta)f. \tag{4.5.11}$$

Moreover, the "curvature" $R_{\omega} = D^2_{\omega}$ satisfies:

$$R_{\omega}(h,k) = -\lambda p[h,k] + [\beta h,\beta k] - \beta q[h,k] \tag{4.5.12}$$

for h,k $\in \mathfrak{m}$ because of (4.4.22) and (4.5.11).

Similar formulas will hold for E–valued (equivariant horizontal) differential forms of degree > 0, but will not be given here. We shall return to these matters below in sec. 4.7.

We close the section with a few remarks on an especially important special case of reductive homogeneous spaces, the so–called symmetric spaces:

Assume that the Lie group G possesses an involutive automorphism: there exists $\sigma \in \mathrm{Aut}(G)$ with $\sigma^2 = 1$. Let G_σ be the –evidently closed – subgroup of fixed points of σ:

$$g \in G_\sigma \text{ iff. } \sigma(g) = g. \tag{4.5.13}$$

The derivative $\dot{\sigma} \in \mathrm{Aut}(\mathfrak{g})$ also is involutive: $\dot{\sigma}^2 = 1$, and hence $\mathfrak{g}$ splits into the direct sum of the (± 1)–eigenspaces of $\dot{\sigma}$,

$$\mathfrak{h}_\sigma = \{h \in \mathfrak{g} \mid \dot{\sigma}h = h\},$$
$$\mathfrak{m} = \{h \in \mathfrak{g} \mid \dot{\sigma}h = -h\}.$$

Then $\mathfrak{g} = \mathfrak{h}_\sigma \oplus \mathfrak{m}$, $\mathfrak{h}_\sigma$ is a subalgebra and $[\mathfrak{h}_\sigma,\mathfrak{m}] \subset \mathfrak{m}$, $[\mathfrak{m},\mathfrak{m}] \subset \mathfrak{h}_\sigma$, as one easily verifies.

By general principles, G_σ is a Lie subgroup whose Lie algebra is $\mathfrak{h}_\sigma$. Let K denote the identity component of G_σ. Since K is connected, $[\mathfrak{h}_\sigma,\mathfrak{m}] \subset \mathfrak{m}$ is equivalent with $\mathrm{Ad}(K)(\mathfrak{m}) \subset \mathfrak{m}$; in other words: G/K is reductive. Such a space G/K is called a symmetric space (with connected fibre).

In this case, the connection ω_0 is uniquely determined by σ and is the "symmetric space connection". The projection p: $\mathfrak{g} \longrightarrow \mathfrak{h}_\sigma$ with kernel $\mathfrak{m}$ is given by $p = \frac{1}{2}(1 + \dot{\sigma})$ and so q is $\frac{1}{2}(1 - \dot{\sigma})$. The earlier curvature formulas now can be simplified somewhat: If h,k $\in \mathfrak{m}$, then $[h,k] \in \mathfrak{h}_\sigma$ and hence $p[h,k] = [h,k]$. Accordingly,

$$\Omega_0(h,k) = -[h,k] \tag{4.5.14}$$

for h,k $\in \mathfrak{m}$. Moreover, $[\mathfrak{m},\mathfrak{m}] \subset \mathfrak{h}_\sigma$ implies $q[ph,k] = 0$ for h,k $\in \mathfrak{m}$. Therefore, the last terms of the right–hand sides of (3.5.11) and (3.5.13) vanish and we are left with the following formula:

$$R_\omega(h,k) = -\lambda[h,k] + [\beta h,\beta k] \tag{4.5.15}$$

for $\omega = \omega_0 + \alpha$ invariant, $\beta = \lambda\alpha$.

For examples, we refer to [K,N] and [G,H,V]; for information on the differential geometry of symmetric spaces to the standard reference [He].

4.6 Invariant connections on principal bundles

In this section, we outline a generalization of the considerations of 4.5: this time, the principal bundle $P \to M$ under consideration need not be of the form $G \to G/K$. The basic assumptions made here are the following:

π: $P \to M$ is a principal G–bundle; H is a Lie group acting on the left on M; the action $H \times M \to M$ is written simply as $(h,x) \to hx$. More restrictive conditions will be imposed as needed, but to begin with we merely assume the following:

(i) The left H–action on M lifts to an action $H \times P \to P$ by bundle automorphisms. More explicitly, this means that H acts form the left on P in such a way that $h(pg) = (hp)g$ for $h \in H$, $p \in P$ and $g \in G$, and that $\pi(hp) = h\pi(p)$.

Following standard procedures, we next define a left action of H in certain function spaces on P: Let E be (finite–dimensional) manifold and denote by $C^\infty(P,E)$ the space of smooth maps $P \longrightarrow E$. For $f \in C^\infty(P,E)$ and $h \in H$, define $hf \in C^\infty(P,E)$ by $(hf)(p) = f(h^{-1}p)$, $p \in P$. This construction generalizes the "left regular representation" of a group in the space of smooth functions on the group. Suppose, moreover, that G acts on the left on E by "automorphisms of E", $(g,e) \longrightarrow \lambda(g)e$; let $C^\infty_G(P,E)$ denote the space of equivariant smooth maps $P \longrightarrow E$. Then $C^\infty_G(P,E)$ is H–invariant as a brief verification shows. In fact, one has the following result:

<u>Proposition 4.6.1</u> Suppose that G acts on the left on E by automorphisms of E. Then the left H–action on P lifts to an action by bundle automorphism of the associated bundle $\mathbb{E} = P \times_G E$.

Indeed, writing again [p,e] for the element of $\mathbb{E}$ represented by (p,e), set $h[p,e] = [hp,e]$. Any other representative of [p,e] is of the form $(pg,\lambda(g)^{-1}e)$ and so $(h(pg),\lambda(g)^{-1}e) = ((hp)g,\lambda(g)^{-1}e) \sim (hp,e)$; hence the action of H is well–defined and it is obvious that it covers the action of H on M.

In other words: Bundles associated with P are "H–bundles" over M. Since $\Gamma(\mathbb{E}) \cong C^\infty_G(P,E)$, the above action of H on $C^\infty_G(P,E)$ induces one on $\Gamma(\mathbb{E})$ and it is easy to see that this is the expected action: H acts on the sections of $\mathbb{E}$ by $(hs)(x) = h(s(h^{-1}x))$.

Under the action of H on M, M splits into the disjoint union of the orbits of its points. Let $x_0 \in M$ and denote its orbit by $\mathcal{O} := Hx_0$. Then $\mathcal{O}$ becomes a smooth manifold when identified with H/K, K the isotropy group of x_0; $\mathcal{O}$ need not be a submanifold of M, but in any case the natural map $H/K \longrightarrow M$ is an injective immersion with image $\mathcal{O}$. By pullback, P induces a principal G–bundle P_K on H/K on which H again operates (from the left) by bundle automorphisms covering the left translations of H/K. In this case, H operates transitively on the set of fibres of P_K.

In a vague sense, now, P "is composed" of such bundles P_K, pairwise disjoint. In fact, if the orbits of H are submanifolds (eg. when H is compact), then this is literally true. Note also that H acts transitively on the set of fibres of P if and only if it is transitive on the base M of P. At any rate, to a large extent, it suffices to consider the case $M = H/K$ and we therefore impose the following condition on the action of H:

(ii) H acts transitively on M, i.e. $M = H/K$ for some closed subgroup K of H. Again, let $0 \in H/K$ be the neutral coset and choose a base point $p_0 \in P_0$. Since K leaves 0 fixed, it maps P_0 into itself. Accordingly, for each $k \in K$, there is unique $\lambda(k) \in G$ such that

$$kp_0 = p_0 \, \lambda(k) \tag{4.6.2}$$

and it is immediate that the map λ: $K \longrightarrow G$ obtained in this manner is a (smooth) homomorphism, often called the isotropy representation of K in G. Note that the isotropy representation of K is unique up to conjugation: if $q \in P_0$ is any other point in the "neutral fibre", then $q = p_0 g$ for a unique $g \in G$. Thus, if $\lambda'(k)g$ is given by $hq = q\lambda'(k)$, then $hq = hp_0 g = p_0\lambda(k)g = qg^{-1}\lambda(k)g$ and hence $\lambda'(k) = g^{-1}\lambda(k)g$.

The isotropy representation also defines the bundle $H{\times_K}G$ over H/K where K acts on G by left translations via λ. One then easily obtains the following result:

Proposition 4.6.3 The choice of a base point $p_0 \in P_0$ induces a (strong) bundle isomorphism

$$P = H{\times_K}G \tag{4.6.4}$$

by the map $hp_0 g \longrightarrow [h,g]$. This is an H–equivariant isomorphism of principal

G–bundles, showing now that P is associated with the principal K–bundle $H \longrightarrow H/K$ under the isotropy representation of K in G.

Under this isomorphism, the left H–action on P corresponds to the obvious such action on $H \times_K G$. In particular, one sees that any bundle associated with P also is associated with $H \longrightarrow H/K$, i.e. is a homogeneous bundle over H/K; in fact, the homogeneous bundles obtained in this way are the ones which "admit a G–structure" as defined by P.

Next we introduce a special class of <u>connections</u> on P.

<u>Definition 4.6.5</u> A connection ω is said to be <u>invariant</u> if

$$h^* \omega = \omega \tag{4.6.6}$$

for all $h \in H$.

Now let L_h be the derivative of $p \longrightarrow hp$ (or $x \longrightarrow hx$ on H/K) and suppose that ω *is* an invariant connection on P. Clearly, then, if u is a horizontal vector at p, $L_h u$ is horizontal at hp. Next, assume that the horizontal bundle H(P) of the connection ω is H–invariant. We claim that ω then is invariant: Observe, first of all, that the equivariance of the projection ρ: $P \longrightarrow H/K$ implies that the vertical bundle V(P) always is H–invariant. Let $p \in P$, $u \in T_P(P)$ and write $u = u_1 + u_2$ where u_1 is vertical, u_2 horizontal. Then for $h \in H$, $(h^*\omega)(p)u = \omega(hp)(L_h u) = \omega(hp)(L_h u_1 + L_h u_2)$. Since u_2 is horizontal, so is $L_h u_2$ by assumption and hence the last identity reduces to $\omega(hp)L_h u_1 = (h^*\omega)(p)u_1$. Since h is a bundle automorphism, $h^*\omega$ is a connection and so agrees with the "canonical vertical from" of P on V(P) (cf. also below); in particular, $h^*\omega = \omega$ on V(P). We are done: if H(P) is H–invariant, so is ω. Summarizing, we obtain:

Proposition 4.6.7 A connection P is H–invariant if and only if its horizontal bundle is H–invariant.

4.7 Linear connections in vector bundles

Let E be a (finite–dimensional) vector space and consider the product bundle $M \times E \longrightarrow M$ with fibre E. Its sections are naturally identified with the elements of $C^\infty(M,E)$ and it is clear that any such map s has a derivative ds which is an element of $A^1(M,E) = A^1(M) \otimes E$. This operation of differentiation generalizes somewhat while still retaining two crucial properties: The bundle $L(M \times E)$ is identified with $M \times L(E)$ and its sections now simply are smooth maps $M \longrightarrow L(E)$. Given such a map γ, define the operator $d+\gamma$ in the obvious manner:

$$((d+\gamma)s)(x) = ds(x) + \gamma(x)s(x). \tag{4.7.1}$$

Clearly, $(d+\gamma)s \in A^1(M,E)$ and the operator has the following two properties: it is additive in s and, if $f \in C^\infty(M)$, then $(d+\gamma)(fs) = df \cdot s + f(d+\gamma)s$. Conversely, any map $D: C^\infty(M,E) \longrightarrow A^1(M,E)$ with these last two properties is easily seen to be of the form $d + \gamma$ with γ in $C^\infty(M,L(E))$.

If $\mathbb{E} \longrightarrow M$ is a non–trivial vector bundle, then (exterior) differentiation of its sections is, in general, meaningless; however, the last two properties of the "differentiation operators" just mentioned still retain meaning and single out an important class of operators:

Definition 4.7.2 A linear connection in the vector bundle $\mathbb{E} \longrightarrow M$ is an additive map $\nabla : \Gamma(\mathbb{E}) \longrightarrow A^1(M,\mathbb{E})$ which satisfies

$$\nabla(fs) = df \cdot s + f\nabla s \tag{4.7.3}$$

for $s \in \Gamma(\mathbb{E})$ and $f \in C^\infty(M)$.

According to this definition, linear connections are a special class of first order differential operators $\mathbb{E} \longrightarrow T^*(M) \otimes \mathbb{E}$. An equivalent description is given as follows: ∇ is a bilinear map $\mathcal{X}(M) \times \Gamma(\mathbb{E}) \longrightarrow \Gamma(\mathbb{E})$ such that

(i) $\nabla_X s$ is $C^\infty(M)$–linear in $X \in \mathcal{X}(M)$;

(ii) $\nabla_X(fs) = (Xf)s + f\nabla_X s$ for $s \in \Gamma(\mathbb{E})$ and $f \in C^\infty(M)$.

We shall also write ∇_X for the (R–linear) map $s \longrightarrow \nabla_X s$ of $\Gamma(\mathbb{E})$ into itself. Note here that since $X \longrightarrow \nabla_X s$ is a differential form, the value $(\nabla_X s)(x)$ depends only on $X(x)$, not on the vector field X; in other words: if $u \in T_x(M)$, then there is a well–defined $\nabla_u s(x) \in E_x$, independent of any extension of u to a vector field on M.

Given two linear connections ∇_1,∇_2 in $\mathbb{E}$, it is immediate from (4.7.3) that $(\nabla_1-\nabla_2)s$ is $C^\infty(M)$–linear also in s and hence determines an $L(\mathbb{E})$–valued 1–form: $\nabla_1-\nabla_2 \in A^1(M,L(\mathbb{E}))$. Conversely, if ∇ is a linear connection in E and β is an $L(\mathbb{E})$–valued 1–form, then $\nabla + \beta$ is readily seen to be a linear connection. This analogue of (4.7.1) establishes the following claim:

Proposition 4.7.4 Let ∇ be a linear connection in $\mathbb{E}$. Then the set $\mathcal{C}_{\mathbb{E}}$ of all such connections is an affine space with tangent vector space $A^1(M,L(\mathbb{E}))$:

$$\mathcal{C}_{\mathbb{E}} = \nabla + A^1(M,L(\mathbb{E})). \tag{4.7.5}$$

As we shall see shortly, the analogy with (4.4.9) is no accident. First of all, however, we extend the notion of a covariant derivative to all of $A^{\cdot}(M,\mathbb{E}) = \bigoplus_{p\geq 0} A^p(M,\mathbb{E})$ as well as to certain other bundles associated with $\mathbb{E}$:

For any $p > 0$, the module $A^p(M,\mathbb{E})$ is naturally isomorphic to the tensor product $A^p(M) \otimes_{C^\infty} \Gamma(\mathbb{E})$ and thus is generated as a $C^\infty(M)$–module by the decomposable elements $\alpha \otimes s$, $\alpha \in A^p(M)$ and $s \in \Gamma(\mathbb{E})$. For such form, we define ∇ by

$$\nabla(\alpha \otimes s) = d\alpha\cdot s + (-1)^p \alpha = \nabla_s. \tag{4.7.6}$$

It is quite immediate that this map extends to an operator

$$A^p(M,\mathbb{E}) \longrightarrow A^{p+1}(M,\mathbb{E}) \tag{4.7.7}$$

which satisfies $\nabla(f\beta) = df\cdot\beta + (-1)^p f\,\nabla\beta$ ($f \in C^\infty(M)$). In this way, one finally obtains a map $\nabla : A^{\cdot}(M,\mathbb{E}) \longrightarrow A^{\cdot}(M,\mathbb{E})$ of degree 1 satisfying the "Leibniz rule" just repeated. This "extension" of ∇ is called the covariant exterior derivative induced by the given linear connection in E. Note, however, that $(A^{\cdot}(M,\mathbb{E}),\nabla)$ will not be a complex, in general. The obstruction to $\nabla^2 = 0$ is measured by what is called the curvature of ∇:

Definition 4.7.8 If ∇ is a linear connection in E, its curvature R_∇ is defined by

$$R_\nabla(X,Y)s = [\nabla_X,\nabla_Y]s - \nabla_{[X,Y]}s \tag{4.7.9}$$

and is a differential form on M:

$$R_\nabla \in A^2(M,L(\mathbb{E})). \tag{4.7.10}$$

Moreover, for any $\beta \in A^{\cdot}(M,\mathbb{E})$,

$$\nabla^2\beta = R_\nabla = \beta. \tag{4.7.11}$$

Finally, the connection ∇ is said to be flat if $R_\nabla = 0$. Formula (4.7.10) makes sense since it is immediate that $R_\nabla(X,Y)s$ is C^∞–linear also in s; (4.7.11) is easy to verify if $\beta = \alpha \otimes s$ and then clearly extends to all of $A^{\cdot}(M,\mathbb{E})$.

One also concludes that $(A^{\bullet}(M,\mathbb{E}),\nabla)$ is a complex if and only if ∇ is flat. An example of a flat connection is $\nabla = d$ in case $\mathbb{E} = M \times E$. On the other hand, if $\mathbb{E}$ is non–trivial, it will not admit flat connections, as a rule; in fact, the cohomology of M will have certain elements, "characteristic classes" of $\mathbb{E}$, which are naturally associated with $\mathbb{E}$ and will have to vanish if $\mathbb{E}$ posseses a flat connection.

Remarks: Since $\mathbb{E}$ is locally trivial, any connection in $\mathbb{E}$ will locally look like $d + \gamma$. In this sense, γ plays the role of the "Christoffel symbols" of classical Riemannian geometry. In fact, the theory of linear connections is a (far–reaching!) generalization of some fundamental work, at first still in the "metric" context, of Hermann Weyl [cf. his Raum, Zeit und Materie", 4th ed., 1920] and of Elie Cartan [cf. e.g. C.R. Acad. Sci. Paris, t.174, 1922] and it is in these notes as well as in Cartan's memoir "Sur les variétés à connexion affine..." that one already encounters the term "connection", not yet used by Weyl. All these considerations were limited to $\mathbb{E} = T(M)$, depending on a more geometric notion of "parallel transport" of tangent vectors on a (pseudo–) Riemannian manifold. We shall add some details on this fundamentally important case later and, as usual, refer to the literature for more information. The definition of a connection adopted here seems to be due essentially to J.L. Koszul and has become fairly standard. Since ∇ then is a differential operator, it "localizes" properly and could also be defined using sheaf–theoretic methods, an approach occasionally taken in the literature.

There is another way of defining the covariant exterior derivative of an $\mathbb{E}$–valued form, perhaps more explicit than (4.7.6), which once more points out the analogy with "ordinary" exterior differentiation:

Let $\beta \in A^p(M,\mathbb{E})$ and $X_i \in \mathfrak{X}(M)$, $0 \le i \le p$. Then

$$(\nabla\beta)(X_0,\dots,X_p) = \Sigma(-1)^i\nabla_{X_i}(\beta(X_0,\dots,\hat{X}_i,\dots,X_p))$$

$$+ \sum_{i<j}(-1)^{i+j}\beta([X_i,X_j],\dots,\hat{X}_i,\dots,\hat{X}_j,\dots,X_p). \tag{4.7.12}$$

Once again, this formula permits an easy verification e.g. of (4.7.11).

Next we briefly indicate the "extension" of ∇ to some associated bundles, etc. This is routine linear algebra and the very simple proofs are omitted here.

Let ∇ be a linear connection in $\mathbb{E}$. Then there is a unique connection ∇^* in the dual bundle $\mathbb{E}^*$ such that, if $<\ ,\ >$ is the natural pairing of $\mathbb{E}^*$ and $\mathbb{E}$ and if $u \in \Gamma(\mathbb{E}^*)$, $s \in \Gamma(\mathbb{E})$, then $d<u,s> = <\nabla^* u,s> + <u,\nabla_s>$; ∇^* is the "dual connection" of ∇ and will be understood in $\mathbb{E}^*$ unless otherwise mentioned. Similarly, for any $p > 1$, there is a unique connection ∇ in $\otimes^p\mathbb{E}$ such that for $s_i \in \Gamma(\mathbb{E})$, $1 \leq i \leq p$,

$$\nabla(s_1 \otimes \cdots \otimes s_p) = \Sigma\, s_1 \otimes \cdots \otimes \nabla s_i \otimes \cdots \otimes s_p.$$

The same construction, using ∇^* in lieu of ∇, applies of course to $\otimes^p\mathbb{E}^*$ and the two connections so obtained are duals of each other: if $\Phi \in \Gamma(\otimes^p E^*)$ and the s_i are in $\Gamma(\mathbb{E})$, then

$$d(\Phi(s_1,\dots,s_p)) = (\nabla\Phi)(s_1,\dots,s_p) + \Sigma\Phi(s_1,\dots,\nabla s_i,\dots s_p).$$

Somewhat more generally, if ∇_1, ∇_2 are linear connections in the vector bundles $\mathbb{E}$ and $\mathbb{F}$, resp., there is a natural induced connection in $\mathbb{E} \otimes \mathbb{F}$ (the "tensor product connection") defined by $\nabla_1 \otimes 1 + 1 \otimes \nabla_2$. In particular, recalling that $L(\mathbb{E},\mathbb{F}) \cong \mathbb{E}^* \otimes \mathbb{F}$, there is an induced connection in $L(\mathbb{E},\mathbb{F})$ which is characterized by the property that $\nabla_2(\varphi s) = (\nabla\varphi)s + \varphi(\nabla_1 s)$; for more detail. Note also that there will be an induced connection in $\mathbb{E} \oplus \mathbb{F}$, given by $\nabla_1 + \nabla_2$, etc. The curvatures of these connections all are quite easy to compute and we merely indicate the following special cases: The curvature of $\nabla_1 \otimes 1 + \otimes \nabla_2$ is $R_1 \otimes 1 + 1 \otimes R_2$, the one of ∇^* is ${}^tR$ (t = transpose of a linear map, here interpreted as a map $L(\mathbb{E}) \longrightarrow L(\mathbb{E}^*)$). In particular, the curvature of the connection induced in $L(\mathbb{E},\mathbb{F})$ satisfies $\hat{R}\varphi = R_2 \circ \varphi + \varphi \circ R_1$.

Given $\nabla \in \mathcal{C}_{\mathbb{E}}$, let ∇ also denote the induced connection in $L(\mathbb{E}) = \mathbb{E}^* \otimes \mathbb{E}$ as well as its covariant exterior derivative in $A^{\cdot}(M,L(\mathbb{E}))$. Let also $s \in \Gamma(\mathbb{E})$. Then

$$\nabla^3 s = \nabla(\nabla^2 s) = \nabla(R_\nabla \cdot s) = (\nabla R_\nabla)\cdot s + R_\nabla = \nabla s;$$

on the other hand,

$$\nabla^3 s = \nabla^2(\cdot s) = R_\nabla \,\hat{\mathcal{C}}\, \nabla s.$$

This holds for all s and so comparison of the two terms yields the Bianchi identity in the form

$$\nabla R_\nabla = 0 \tag{4.7.13}$$

for any linear connection in $\mathbb{E}$.

Remark: The existence of linear connections in $\mathbb{E}$ can be obtained either by an argument using local triviality and a partition of unity on M—or by later considerations which amount to a reduction of the problem to the corresponding question concerning connections on principal bundles, answered earlier. Accordingly, we postpone the argument, but point out there that in any case

$$\mathcal{C}_{\mathbb{E}} \neq \phi. \tag{4.7.14}$$

To end this section, we add some remarks on the "local" description of linear connections:

Suppose that Ψ_U: $\mathbb{E}|U \cong U \times E$ is a trivialization of $\mathbb{E}$ over the open subset $U \subset M$. Then $\Gamma(U,\mathbb{E}) = \Gamma(\mathbb{E}|U) \cong \mathcal{C}^\infty(U,E)$ and thus, exterior differentiation d: $C^\infty(U,E) \longrightarrow A^1(U,E)$ induces a linear connection in $\mathbb{E}|U$. By earlier arguments, any other connection "is" of the form $d + \gamma_U$ for some $\gamma_U \in A^1(U,L(\mathbb{E}))$. Somewhat more explicitly, suppose that $(s_i)_{1\leq i\leq r}$ is a local frame for $\mathbb{E}$ of the form $s_i = \Psi_U^{-1} e_i$ where (e_i) is a basis of the vector space E. The sections of $\mathbb{E}|U$ then are the linear combinations $s = \Sigma f^i s_i$, $f^i \in C^\infty(U)$, and the isomorphism $\Gamma(\mathbb{E}|U) \cong C^\infty(U,E)$ maps $\Sigma f^i s_i$ to $\Sigma f^i e_i$; here e_i is identified with the constant map $U \longrightarrow e_i \in E$. If ∇ is linear connection in E (or in $E|U$), then

$$\nabla s_i = \Sigma\, \gamma_i^j \cdot s_j \tag{4.7.15}$$

with $\gamma_i^j \in A^1(U)$. Thus, for $s = \Sigma f^i s_i$, $\nabla s = \sum_i df^i \cdot s_i + \sum_{i,j} f^i \gamma_i^j \cdot s_j$. This shows that (γ_i^j) is the matrix representation of γ_U with respect to the given basis (e_i) of E.

We omit here the derivation of the "transformation law" for the local connection forms γ_U since this will be done implicitly in later considerations.

Suppose furthermore that U also carries a coordinate chart and let (x^α), $1 \leq \alpha \leq m = \dim(M)$, be local coordinates. Then the partial derivatives $\partial_\alpha = \partial/\partial x^\alpha$ form a frame for T(M) over U whose dual frame is (dx^α). The 1–forms γ_i^j then yield scalar functions $\gamma_{i\alpha}^j$ by means of $\gamma_i^j = \sum_\kappa \gamma_{i\alpha}^j\, dx^\alpha$, i.e. by setting $\gamma_{i\alpha}^j = \gamma_i^j(\partial_\alpha)$. Also, ∇s "is" an E–valued 1–form and we thus obtain the scalars

$$(\nabla s)_\alpha^i = \partial_\alpha f^i + \sum_j \gamma_{j\alpha}^i f^j$$

for $s = \Sigma f^i s_i$. The curvature R of ∇ has components $R_{j\alpha\beta}^i$ with respect to the frame (s_i) and the local coordinates (x^α); these are obtained in the usual manner by means of

$$R(\partial_\alpha, \partial_\beta) s_j = \sum_i R_{j\alpha\beta}^i\, s_i\,.$$

Note that $R_{j\alpha\beta}^i$ is skew–symmetric in α,β since R is a 2–form. A straight–forward computation then shows that

$$R_{j\alpha\beta}^i = \partial_\alpha \gamma_{j\beta}^i - \partial_\beta \gamma_{j\alpha}^i + \sum_k (\gamma_{k\alpha}^i \gamma_{j\beta}^k - \gamma_{k\beta}^i \gamma_{j\alpha}^k),$$

a classical formula known from Riemannian geometry (where the $\gamma_{j\alpha}^i$ are the Christoffel symbols which represent the Levi–Cività connection of the given Riemannian structure).

4.8 Connection forms and linear connections

We now establish the relations between linear connections in vector bundles and connections on principal bundles as described in the preceding sections. The following first step is quite immediate on the grounds of what was done earlier:

Suppose that $\pi: P \longrightarrow M$ is a principal G–bundle, $\lambda: G \longrightarrow GL(E)$ a representation of G in the (real or complex) vector space E and that $\mathbb{E} = P\times_G E$ is the associated vector bundle. As explained earlier, in this case $\Gamma(\mathbb{E}) \cong C^\infty_G(P,E)$, $A^1(M,\mathbb{E}) \cong A^1_{h,G}(P,E)$, $A^1(M,L(\mathbb{E})) \cong A^1_{h,G}(P,L(E))$, etc.

Let ω be a connection on P. If $s \in C^\infty_G(P,E)$, then $ds \in A^1_G(P,E)$ and therefore

$$D_\omega s := (ds)^h \in A^1_{h,G}(P,E) \tag{4.8.1}$$

is well–defined, $(ds)^h$ again denoting the ω–horizontal part of ds. Evidently, then D_ω yields a linear map

$$D_\omega : \Gamma(\mathbb{E}) \longrightarrow A^1(M,\mathbb{E}) \tag{4.8.2}$$

and it is easily verified that D_ω is a <u>linear connection</u> in $\mathbb{E}$, said to be induced by ω. One thus obtains a natural map $\mathcal{C}_P \longrightarrow \mathcal{C}_{\mathbb{E}}$ which, in general, is neither injective nor surjective. Note that we continue to use the symbol D_ω instead of the more cumbersome ∇_ω (which would be more consistent with the preceding section!).

Next, fix $\omega_0 \in \mathcal{C}_P$ and denote the induced linear connection in $\mathbb{E}$ by D_0. Any other connection will be of the form $D_0 + \alpha$ for some form α in $A^1(M,L(\mathbb{E}))$; since this space is isomorphic to $A^1_{h,G}(P,L(E))$, one readily concludes

Proposition 4.8.3 The choice of ω_0 establishes an affine isomorphism

$$D_0 + A^1_{h,G}(P,L(E)) \cong \mathcal{C}_{\mathbb{E}} . \tag{4.8.4}$$

Under this isomorphism, the linear connections induced by the members of $\mathcal{C}_P$ correspond to the affine subspace $D_0 + \lambda^\cdot(A_{h,G}(P,\mathfrak{g}))$.

In the second claim, $\lambda^\cdot$ again denotes the derivative of λ, i.e. the induced algebra homomorphism $\mathfrak{g} \longrightarrow gl(E)$ and $gl(E)$ is the Lie algebra $(L(E), [\ ,\ ])$. The result is, of course, to be expected, but is often useful in applications where a more explicit description of the linear connections in $\mathbb{E}$ is required (e.g. in homogeneous vector bundles).

Remark: If P is interpreted as a "G–structure in $\mathbb{E}$", then the connections induced by the ones of P are the "admissible connections" for this structure. This is the case e.g. for reductions P of the frame bundle to 0(E)–resp. U(E)–principal bundles ($\mathbb{E}$ then is a Riemannian resp. hermitian vector bundle). In this case, the admissible connections are precisely the "metric connections".

Corollary 4.8.5 The construction $\omega \longrightarrow D_\omega$ yields a natural affine bijection between the connections on the linear frame bundle $\mathcal{F}(\mathbb{E})$ and $\mathcal{C}_{\mathbb{E}}$.

Indeed, in this case, $\lambda = \text{id}: GL(E) \longrightarrow GL(E)$ and the claim is immediate.

In the general case, it follows easily from earlier considerations that the curvature tensor R_ω of a linear connection of the form D_ω is computed in terms of $F_\omega \in A^2(M,P(\mathfrak{g}))$ by

$$R_\omega = \lambda^\cdot(F_\omega) \tag{4.8.6}$$

where $\lambda^\cdot$ denotes the bundle map $P(\mathfrak{g}) \longrightarrow L(\mathbb{E})$ defined by $\lambda^\cdot: \mathfrak{g} \longrightarrow gl(E)$.

Let now $q: P \times E \longrightarrow P \times_G E = \mathbb{E}$ be the projection (which is a principal G–bundle) and let $\epsilon: \mathbb{E} \longrightarrow M$ be the bundle projection. Choose a connection ω on P and let H(P) be its horizontal bundle. Since $T(P \times E) = T(P) \times T(E)$ in a natural way, $H(P) \times (0)$ is a subbundle of $T(P \times E)$ and it follows easily from section 4.4 that Tq is fibrewise injective on this subbundle, the image $H(\mathbb{E})$ is a subbundle of $T(\mathbb{E})$ and that $T(\mathbb{E}) = V(\mathbb{E}) \oplus H(\mathbb{E})$; $H(\mathbb{E})$ is the <u>horizontal bundle</u> of the connection D_ω.

On the other hand, ϵ induces a fibrewise surjective bundle map $T\epsilon: T(\mathbb{E}) \longrightarrow T(M)$ over ϵ, hence a bundle surjection $\epsilon_*: T(\mathbb{E}) \longrightarrow \epsilon^*(M)$ whose kernel obviously is $V(\mathbb{E})$; thus, there is the following short exact sequence <u>over</u> $\mathbb{E}$:

$$0 \longrightarrow V(\mathbb{E}) \longrightarrow T(\mathbb{E}) \xrightarrow{\epsilon_*} \epsilon^* T(M) \longrightarrow 0. \tag{4.8.7}$$

From $T(\mathbb{E}) = V(\mathbb{E}) \oplus H(\mathbb{E})$, we next conclude:

<u>Proposition 4.8.8</u> The connection $\omega \in \mathcal{C}_P$ induces a splitting $\mu = \mu_\omega$ of the sequence (4.8.7), $\mu: \epsilon^* T(M) \longrightarrow T(\mathbb{E})$.

Indeed, $\epsilon_* | H(\mathbb{E})$ is an isomorphism onto $\epsilon^* T(M)$ and so, $\mu = (\epsilon_* | H(\mathbb{E}))^{-1}$ will serve the purpose.

The converse of proposition 4.8.8 is false: A splitting of (4.8.7) need not arise from a linear connection even when $P = \mathcal{F}(\mathbb{E})$ and it is somewhat involved to characterize the splittings, i.e. those horizontal bundles which arise from linear connections.

The following analogue of the horizontal lift map for principal bundle is almost immediate:

Corollary 4.8.9 The connection $\omega \in \mathcal{C}_P$ defines a $C^\infty(M)$–linear injection $\hat{\mu}: \mathfrak{X}(M) \longrightarrow \Gamma(H(\mathbb{E}))$ which to each vector field $X \in \mathfrak{X}(M)$ assigns the unique horizontal vector field $\hat{\mu}X$ on $\mathbb{E}$ such that $\hat{\mu}X \underset{\epsilon}{\sim} X$. $\hat{\mu}X$ is the $(D_\omega-)$ horizontal lift of X.

It is to be noted here that the assertions of proposition 4.8.8 and corollary 4.8.9 do not really require $H(\mathbb{E})$ to be the horizontal bundle of a linear connection in $\mathbb{E}$: it is sufficient to choose any supplement $H(\mathbb{E})$ of $V(\mathbb{E})$ – for instance the orthogonal $V(\mathbb{E})^\perp$ for any Riemannian structure on the manifold $\mathbb{E}$.

The remainder of this section is devoted to a brief sketch of the "classical" special case of connections on M, by which we mean linear connections in T(M); we also refer back to the appropriate sections of ch. 2. The conventions now are these: M is a manifold modelled on E, so that E is the fibre of T(M); we write P for the linear frame bundle $\mathcal{F}(T(M))$ of M and G for GL(E), etc., in order to simplify the notations somewhat. Recall that now $L(T(<M)) \cong T^*(M) \otimes T(M)$ and therefore,

$$\mathrm{End}(T(M)) = \Gamma(L(T(M)) \cong A^1(M,T(M)). \tag{4.8.10}$$

In particular, the identity section of L(T(M)) yields a T(M)–valued 1–form ι.

Definition 4.8.11 Let D be a connection on M. The torsion $T = T_D$ of D is defined to be

$$T = D\iota, \tag{4.8.12}$$

the covariant exterior derivative of ι.

Clearly, $T \in A^2(M,T(M))$ and it is easy to derive the earlier formula for T, cf. ch. 2: $D\iota(X,Y) = D_X\iota(Y) - D_{\mathcal{G}}(X) - \iota([X,Y])$, i.e.:

$$T(X,Y) = D_XY - D_YX - X,Y \tag{4.8.13}$$

for $X,Y \in \mathfrak{X}(M)$.

On the other hand, the 1–form ι lifts to an E–valued 1–form $\theta \in A^1_{h,G}(P,E)$. If ω is the connection form of D (cf. corollary 4.8.5), then:

Proposition 4.8.14 The 2–form T is realized on P by the form

$$\Theta_\omega = D_\omega \theta \in A^2_{h,G}(P,E).$$

The existence of the form T (resp. Θ) is very special to the tangent bundle T(M) (resp. P and its reductions): There is no analogue for arbitrary vector bundles since $T^*(M) \otimes \mathbb{E}$ will not have a "privileged" section. The 1–form θ has a more explicit description if one interprets $P = \mathcal{F}_M$ as the bundle of isomorphisms $E \longrightarrow T_x(M)$, i.e. as the bundle Is(M×E,T(M)): Let $u \in P$, $V \in T_u(P)$. Then

$$\theta(u)v = u^{-1}(T_\mu \pi \cdot v), \tag{4.8.15}$$

the form introduced earlier, cf. (4.2.1). The verification can easily be done locally or, if desired, globally by using e.g. the horizontal lift of any connection on M, and it is left as an exercise.

A connection $D = D_\omega$ on M i called symmetric (or "torsionfree") if $T_D = 0$. In the general case, it is easily seen that $\hat{D}_X Y = D_X Y - \frac{1}{2}T(X,Y)$ defines a new connection whose torsion vanishes, the so–called "symmetric part" of D. Since it is obvious that $\hat{D}_X X = D_X X$, the two connections have the same geodesics, by the way.

For the following basic result, we also refer back to ch. 2:

Theorem 4.8.16 (Ricci) Let g be a pseudo–Riemannian structure on M. Then there exists a unique connection D on M satisfying $Dg = 0$ (compatibility with g) and $T_D = 0$ (symmetric), called the Levi–Cività connection of (M,g).

Proof: The two conditions imposed on D amount to the following:

(i) $X(g(Y,Z)) = g(D_X Y,Z) + g(Y,D_X Z)$

(ii) $D_X Y - D_Y X = [X,Y]$

for any three vector fields X,Y and Z on M. If there is such a connection, a brief calculation shows that it has to satisfy the following identity:

$$g(D_X Y,Z) = \tfrac{1}{2}\{Xg(Y,Z) + Yg(X,Z) - Zg(X,Y) + g([X,Y],Z) + g([Z,X],Y) - g([Y,Z],X\} \tag{4.8.17}$$

In particular, D is uniquely determined since g is non–degenerate.

Conversely, this identity defines $D_X Y$ for $X,Y \in \mathfrak{X}(M)$ and it is again the result of easy calculation that D then is, indeed a connection on M. This establishes the theorem.

Definition 4.8.18 The curvature of the pseudo–Riemannian manifold (M,g) is the curvature of its Levi–Cività connection:

$$R(X,Y)Z = [D_X,D_Y] - D_{[X,Y]} \tag{4.8.19}$$

and so is an L(T(M))–valued 2–form on M whose values are skew–symmetric with respect to g.

Formula (4.7.16) gives the expression of R in local coordinates where now the components of the local connection form are the Christoffel symbols Γ^i_{jk} of the metric (g_{ij}). Explicitly, they are given by

$$\Gamma^i_{jk} = \frac{1}{2}\sum_{\ell} g^{il}(\partial g_{jl}/\partial k_k + \partial g_{kl}/\partial x_j - \partial g_{jk}/\partial x_l) \tag{4.8.20}$$

as follows directly from (4.8.17).

Since R takes values in the skew–symmetric endomorphisms of T(M), the expression

$$R^{\cdot}(U,V,X,Y) = g(R(U,V)X,Y) \tag{4.8.21}$$

is skew–symmetric not only in U,V, but also in X,Y and hence may be interpreted as a bilinear form on $\Lambda^2 T(M)$. It is called the <u>Riemann tensor</u> of (M,g); a number of identities involving $R^{\cdot}$ may be found in ch. 2 as well as in the literature, to which we also refer for such notions as the Ricci tensor and the scalar curvature of (M,g).

<u>Remark</u>: In local coordinates, $R^{\cdot}$ is obtained from R "by lowering an index". This process, well known in the classical tensor formalism, and its inverse ("raising indices") have a global meaning which we illustrate in the following simplest cases:

(a) If $X \in \mathcal{X}(M)$, a tensor of type (1,0), then i(X)g is a 1–form, i.e. a tensor of type (0,1), and this correspondence is a $C^\infty(M)$–isomorphism of $\mathfrak{X}(M)$ onto A^1 as a consequence of the Riesz–Fischer type isomorphism $T(M) \cong T^*(M)$ induced by g.

(b) If $A \in End(T(M))$ is a tensor of type (1,1), one can associate with it a tensor of type (0,2), i.e. a bilinear form on T(M), by means of $\breve{A}(X,Y) = g(AX,Y)$. This yields a $C^\infty(M)$–module isomorphism of $End(T(M)) = A^1(M,T(M))$ onto $\Gamma(\otimes^2 T^*(M))$, the "module of bilinear forms on M".

It is clear from these examples how one proceeds in general.

References, Chapter 4

[A]	Atiyah, M. F.	Complex Analytic Connections in Fibre Bundles, Trans. Amer. Math. Soc. 85, (1957), pp. 181–207.
[B]	Bourbaki, N.	Élements de Mathematique, Variétés différentielles et analytiques, Fasciule XXXIII, Paragraph 1 à 7, (1967), Fasciule XXXVI, Paragraph 8 à 15, (1971), Hermann, Paris.
[Bo]	Bott, R.	Homogeneous Vector Bundles, Ann. of Math. (2), 66, (1957), pp. 203–248.
[D,I]	Dieudonné, J.	Foundations of Modern Analysis (vol I), Academic Press, New York and London, (1969).
[G,H,V]	Greub, W. Halperin,S. Vanstone, R.	Connections, Curvature and Cohomology, vol. II, Academic Press, New York, (1973).
[He]	Helgason, S.	Differential Geometry, Lie Groups, and Symmetric Spaces, Pure and Applied Math. Academic Press, 80, New York, (1978).
[H]	Hirzebruch, F.	Topological Methods in Algebraic Geometry, Springer Verlag, Berlin, (1966).
[K,N]	Kobayashi, S. Nomizu, K.	Foundations of Differential Geometry, Interscience Publishers: New York, (1968).
[S]	Steenrod, N.E.	The Topology of Fibre Bundles, Princeton Mathematical Series 14, Princeton University Press, Princeton, (1951).
[St]	Sternberg, S.	Lectures on Differential Geometry, Prentice Hall, Englewood Cliffs, N.J., (1964).
[W,Y]	Wu, T.T Yang, C. N.	Concept of Nonintegrable Phase Factors and Global Formulation of Gauge Fields, Phys. Rev. D (3) 12, (1975), no 2, pp. 107–112.

Chapter 5

FUNCTION SPACE

In this chapter we present some elementary material on globally defined spaces of functions. These spaces play an essential role in later chapters, especially in chapter nine. Naturally we need the basic concepts of general topology and functional analysis. We refer to [Ke] and [Heu] respectively as introductory texts. Throughout the chapter the reader is left with numerous routine calculations. But first we review some basic examples of function spaces used in functional analysis. The domain of the functions will not be a compact manifold but rather open subspaces of Euclidean spaces. These spaces play a fundamental role in the field of partial differential equations defined locally.

5.1 Space of functions and distributions

a) C(M,E)

We start with the space of all continuous functions defined on a compact set K or more general on some compact topological space M. The values of these functions are assumed to be in a normed vector space E which is given over the reals (or the complex numbers).

Let $\| \cdot \|$ be a norm on E. Denote the collection of all continuous E–valued functions on M by C(M,E) and observe that it is a vector space under the pointwise defined operations. Given any $f \in C(M,E)$ then

$$\| \cdot \| \circ f$$

is a real valued continuous function.

The norm $\| \cdot \|$ on E yields a norm on C(M,E), again denoted by $\| \cdot \|$, which is given by

$$\|f\| := \sup_{p\in M}\|f(p)\| .$$

If E is complete (in this case E is called a Banach space) then C(M,E) is a Banach space too. Completeness is defined via Cauchy sequences in the usual way. In the sequel E is supposed to be a Banach space.

b) $C^k(0,E)$, $k = 0,1,2,...$

Let 0 be an open subset of $\mathbb{R}^n$. In the following we will endow $C^k(0,E)$, the vector space of all k–times continuously differentiable functions from 0 into E with a locally convex topology. k = 0 means continuous. Such topology is given by seminorms. To define these seminorms consider a non negative integer $\ell \leq k$ and a compact subset $K \subset 0$. For each $f \in C^k(0,E)$ the lth derivative $D^\ell f$ of f maps 0 continuously into the Banach space $L^\ell(\mathbb{R}^n,E)$ (consisting of all ℓ–linear maps from the ℓ–fold product of $\mathbb{R}^n$ into E and being equipped with the usual norm, [cf.D,I]). The number

$$\tau_{K,k}(f) := \sup_{p\in K}\{\|f(p)\|,\|Df(p)\|,...,\|D^kf(p)\|\}$$

exists. Assigning to each f the real number $\tau_{K,k}(f)$ yields the seminorm $\tau_{K,k}: C^k(0,E) \longrightarrow \mathbb{R}$. The collection

$$\{\tau_{K,k} \mid K \subset 0 \text{ compact}\}$$

defines a locally convex topology on $C^k(0,E)$ which in addition is complete. A neighbourhood of zero is given by finite intersections of sets of the form

$$U_{\epsilon,\tau_{K,k}}(0) := \{f \in C^k(0,E) \mid \tau_{K,k}(f) < \epsilon\}$$

where ϵ is a positive real number. However countably many of these seminorms are enough to define the locally convex topology just described, simply because 0 can be exhausted by countably many compact subsets. Thus $C^k(0,E)$ is a complete metrizable locally convex topological vector space. Such a space is called a Fréchet space (cf. [Ja]).

c) $C^\infty(0,E)$

$C^\infty(0,E)$, the collection of all smooth functions from 0 to E, is a linear subspace of $C^k(0,E)$ for all non negative integers k. We equip $C^\infty(0,E)$ thus with the coarsest among all topologies for which the inclusion of $C^\infty(0,E)$ into $C^k(0,E)$ is continuous. Again $C^\infty(0,E)$ is a Fréchet space, namely the inverse limit of all $C^k(0,E)$. A major observation is that $C^\infty(0,E)$ is a dense subset of $C^k(0,E)$ for all $k = 0,1,2,...$

d) $C^k(K,E)$, $k = 0,1,...,\infty$

Let $f \in C^k(0,E)$. The <u>support</u> supp(f) of f is the closure of the set

$$\{p \in 0 \mid f(p) \neq 0\}.$$

By $C^k(K,E)$ we mean the vector space of all k–times continuously differentiable functions from 0 to E which have their support in K. Here $K \subset 0$ is a compact subset. Again we define the seminorm $\tau_{K,k}$ which in this case is a norm. Thus $C^k(K,E)$ is a Banach space iff k is finite. If $k = \infty$, then we equip $C^\infty(K,E)$ with the topology generated by $\tau_{K,0}, \tau_{K,1}, ...$ and obtain a Fréchet space.

e) $\mathcal{D}^k(0,E)$, $k = 0,1,...,\infty$

The collection

$$\mathcal{D}^k(0,E) := \bigcup_{K \subset 0} C^k(K,E) \subset C^k(0,E)$$

with K compact, is certainly a linear space. Each member of this union is a Fréchet space (whether k is finite or not). We thus can endow $\mathcal{D}^k(0,E)$ with the direct limit topology (cf. [Ja],[Tr]). The resulting space is no longer a Fréchet space (cf. [Ja]). It is a so called <u>LF–space</u>. Also in this case $\mathcal{D}^\infty(0,E) \subset \mathcal{D}^k(0,E)$ is a dense subset for all non negative integers k. $\mathcal{D}^\infty(0,E)$ is called the space of test functions on 0 with values in E.

Let us point out, that any differential operator of order k

$$D: \mathcal{D}^k(0,\mathbb{R}) \longrightarrow \mathcal{D}^0(0,\mathbb{R})$$

is continuous. In fact $\mathcal{D}^k(0,\mathbb{R})$ carries the coarsest among all topologies for which all differential operators of order k are continuous. In the sequel $\mathcal{D}^0(0,\mathbb{R})$ will be simply denoted by $\mathcal{D}(0,\mathbb{R})$.

f) $L^p(0)$, $p = 1,2,...$

Of another nature than the function spaces considered up to now are the L^p–spaces (cf. [Ja]). These sort of spaces will help us, loosely speaking, to enrich the structure of the function spaces introduced above. For simplicity choose E to be $\mathbb{R}$.

Again let $0 \subset \mathbb{R}^n$ be an open set equipped with the Lebesgue measure. By $\mathcal{L}^1(0)$ we denote the vector space of all real valued functions f for which $|f|$ is integrable. Observe that

$$f \longmapsto \int |f| d\mu$$

is a seminorm on $\mathcal{L}^1(0)$. Let $L^1(0)$ be the factor space

$$\mathcal{L}^1(0) / \{f \in \mathcal{L}^1(0) \; \int |f| d\mu = 0\}$$

Assigning to each equivalence class $\tilde{f} \in L^1(0)$ of $f \in \mathcal{L}^1(0)$ the non negative number $\int |f| d\mu$ we obtain a norm on $L^1(0)$ called $\|f\|_1$. The relation of $L^1(0)$ with $D(0,\mathbb{R})$ is somewhat described by the following (cf. [Tr]):

Theorem 5.1.1 $L^1(0)$ equipped with $\|\cdot\|_1$ is a Banach space containing the vector space of all continuous functions with compact support as a dense subset.

By $L^p(0)$ for $p \geq 1$ we mean the vector space of equivalence classes with respect to the above defined equivalence relation of all those measurable functions $f : 0 \longrightarrow \mathbb{R}$ for which

$$\left[\int |f|^p \, d\mu \right]^{1/p}$$

exists. The map

$$\tilde{f} \longmapsto \left[\int |f|^p d\mu \right]^{1/p}$$

with $f \in \tilde{f}$ is a norm under which $L^p(0)$ is a Banach space again.

Next we turn our attention to distributions. We do not intend to give an introduction to the theory of distributions at all. A vast literature on this subject will provide the reader with a variety of excellent presentations. As references we state [Tr] and [C,W,D].

Here we rather review the most basic notions only to define the Sobolev spaces. These type of spaces also will appear as parameter spaces of certain kinds of manifolds of maps.

By a distribution we mean a continuous scalar valued linear map on $\mathcal{D}^\infty(0,\mathbb{R})$, the space of all $\mathbb{R}$–valued smooth functions with compact support.

To prepare the introduction of an important type of distributions consider with $f,g \in \mathcal{D}^\infty(0,\mathbb{R})$ the pairing

$$\langle f,g\rangle := \int f \cdot g \, d\mu$$

where μ is the Lebesgue measure. If f is fixed, then $g \longmapsto \langle f,g\rangle$ is a continuous linear $\mathbb{R}$–valued functional, called T_f, that is a distribution.

Observe that $<f,g>$ also makes sense if f is continuous and $g \in \mathcal{D}^\infty(O,\mathbb{R})$, thus every continuous function can be viewed as a distribution. More general this is true for $f \in \mathcal{L}^p(O)$. Not all distributions are of this form however. A counter example is the Dirac measure:

Let $O = \mathbb{R}$ and $a \in \mathbb{R}$. Define

$$\delta_a\colon \mathcal{D}^\infty(\mathbb{R},\mathbb{R}) \longrightarrow \mathbb{R}$$

by $\delta_a(f) = f(a)$. Clearly δ_a is continuous. To show that it is not generated by a function f, consider

$$g_\epsilon(x) = \begin{cases} \exp(-\epsilon^2/(\epsilon^2 - x^2)^2) & \text{for all } x \in B_\epsilon \\ 0 & \text{for all } x \notin B_\epsilon \end{cases}$$

a member of $\mathcal{D}^\infty(O,\mathbb{R})$, where $B_\epsilon \subset O$ is the open ball with center zero and of radius ϵ. Clearly if $a = 0$

$$\delta_0(g_\epsilon) = e^{-1}.$$

On the other hand if f is locally integrable we let

$$<f,g_\epsilon>: = \int f\cdot g_\epsilon d\mu.$$

Since

$$|<f,g_\epsilon>| \le \int |f|\ |g_\epsilon| d\mu$$

the number $<f,g_\epsilon>$ has to tend to zero for ϵ approaching zero whereas it tends to e^{-1} in case δ_0 were represented by f.

A first classification of distributions is the one given by the order. To define this notion consider the continuous inclusion

$$\mathcal{D}^\infty(O,\mathbb{R}) \subset \mathcal{D}^k(O,\mathbb{R})$$

which maps onto a dense subspace for all non negative integers k. Then the dual spaces (the vector spaces of all continuous $\mathbb{R}$–valued functionals) are hence related by

$$\mathcal{D}^\infty(O,\mathbb{R})' \supset \mathcal{D}^k(O,\mathbb{R})'.$$

The distributions of order k are those in $\mathcal{D}^k(0,\mathbb{R})'$. Here k can be any non negative integer.

Next we introduce the derivative of a distribution based on the following observation. For any two functions $f,g \in \mathcal{D}^\infty(0,\mathbb{R})$ and a coordinate function x^r on $\mathbb{R}^n$ we have (cf. [Tr])

$$<\frac{\partial}{\partial x^r} f,g> = \int \frac{\partial}{\partial x^r} f\cdot g \, d\mu$$

$$= \int \frac{\partial(f\cdot g)}{\partial x^r} d\mu - \int f\cdot\frac{\partial}{\partial x^r} g \, d\mu$$

$$= -\int f\cdot\frac{\partial}{\partial x^r} g \, d\mu$$

$$= -<f,\frac{\partial}{\partial x^r} g>.$$

This motivates the following definition: Let $T \in \mathcal{D}^\infty(0,\mathbb{R})'$. We define

$$\left[\frac{\partial}{\partial x^r} T\right](g) := -T\left[\frac{\partial}{\partial x^r} g\right]$$

for all $g \in \mathcal{D}^\infty(0,\mathbb{R})$. This notion generalizes in an obvious way to any multiindex $\alpha = (\alpha_1,\dots,\alpha_n)$: Let

$$\left[\frac{\partial}{\partial x}\right]^\alpha = \left[\frac{\partial}{\partial x^1}\right]^{\alpha_1} \cdots \left[\frac{\partial}{\partial x^n}\right]^{\alpha_n}.$$

Then

$$\left[\left[\frac{\partial}{\partial x}\right]^\alpha T\right](g) := (-1)^{|\alpha|} T\left[\left[\frac{\partial}{\partial x}\right]^\alpha g\right].$$

With respect to the above classification by the order we have

Theorem 5.1.2 Every distribution of finite order is a finite sum of derivatives of continuous functions.

For the proof see [Tr] e.g.

The Sobolev spaces are now defined as follows:

$H^{p,s}(0,\mathbb{R})$ is the vector space of all those elements in $L^p(0)$ whose partial derivatives up to order s are in $L^p(0)$ where $p > 1$ and $s \geq 1$ are integers.

$H^{p,s}(0,\mathbb{R})$ carries a natural norm given by

$$\|f\| := \left[\sum_{|\alpha| \leq s} \int \left| \left[\frac{\partial}{\partial x}\right]^{\alpha} f \right|^p d\mu \right]^{\frac{1}{p}}$$

and we have (cf. [Tr])

Theorem 5.1.3 $H^{p,s}(0,\mathbb{R})$ is a Banach space whose dual is $H^{p,s}(0,\mathbb{R})$, where p and q are related by $\frac{1}{p} + \frac{1}{q} = 1$. Thus $H^{2,s}(0,\mathbb{R})$ is a Hilbert space.

Instead of $H^{2,s}(0,\mathbb{R})$ we just write $H^s(0,\mathbb{R})$. What sorts of elements $H^{p,s}(0,\mathbb{R})$ consists of is said in the following famous theorem proved in [Tr]:

Theorem 5.1.4 $H^{p,s}(0,\mathbb{R}) \subset C^k(0,\mathbb{R})$ for $k \leq s - \frac{1}{p}$.

Observe that $C^\infty(0,\mathbb{R}) \subset H^{p,s}(0,\mathbb{R})$ is not dense if $0 \neq \mathbb{R}^n$ however $\mathcal{D}^\infty(\mathbb{R}^n,\mathbb{R}) \subset H^{p,s}(\mathbb{R}^n,\mathbb{R})$ is dense (cf. [C,W,D]).

These notions, introduced up to now, make perfect sense if $\mathbb{R}$ is replaced by $\mathbb{C}$. The corresponding theorems then still hold.

If E is finite dimensional vector space we denote by $H^{p,s}(0,E)$ the direct sum

$$\bigoplus_{i=1}^{\ell} H^{p,s}(0,\mathbb{R})$$

where $e_1,\ldots,e_\ell$ is a basis of E and $\mathbb{R}\cdot e_i$ is identified with $\mathbb{R}$ for all $i = 1,\ldots,\ell$. Observe however, that $H^{p,s}(0,E)$ can be introduced by using an invariant measure on E and proceed as in the case of $H^{p,s}(0,\mathbb{R})$. The concept of a distribution will be generalized later to what is called a current.

5.2 Globally defined function spaces such as $C_\varphi(M,\mathbb{E})$, $C^k(M,N)$, $C^\infty(M,N)$, $\Gamma^0(\varphi^*\mathbb{E})$, $\Gamma^k\mathbb{E}$, $\Gamma\mathbb{E}$, $H^k\mathbb{E}$

In this section M denotes a compact (smooth) manifold and $\mathbb{E}$ a finite dimensional vector bundle with typical fibre E over a manifold N.
Assume that $\mathbb{E}$ carries a bundle norm

$$\| \ \| : \mathbb{E} \longrightarrow \mathbb{R}$$

This means that $\| \ \|$ is continuous and when restricted to a fibre $\mathbb{E}_p$ of $\mathbb{E}$ yields a norm of $\mathbb{E}_p$. A bundle norm is obtained e.g. by taking a Riemannian structure $<,>$ on $\mathbb{E}$ and setting

$$\|x_p\| = \sqrt{<x_p,x_p>}$$

for all $x_p \in \mathbb{E}$.

a) $C_\varphi(M,\mathbb{E})$

The collection $C(M,\mathbb{E})$ is certainly not a vector space. However, if we fix a continuous map $\varphi : M \longrightarrow N$ and consider $C_\varphi(M,\mathbb{E})$, the collection of all $f \in C(M,\mathbb{E})$ for which $\pi \circ f = \varphi$, where $\pi : \mathbb{E} \longrightarrow N$ is the projection, then $C_\varphi(M,\mathbb{E})$ is a vector space, in fact it can be naturally identified with a space of sections: Denoting $\Gamma^\circ(\mathbb{F})$ the space of continuous sections of a vector bundle $\mathbb{F}$ over M we easily establish a natural isomorphism

$$\Gamma^\circ(\varphi^*\mathbb{E}) \longrightarrow C_\varphi(M,\mathbb{E})$$

sending any $s \in \Gamma^o(\varphi^* \mathbb{E})$ into $f : M \longrightarrow \mathbb{E}$ given by $f(p) = s(p) \in \mathbb{E}_{\varphi(p)}$ for all $p \in M$. Again $C_\varphi(M,\mathbb{E})$ carries a natural norm, the sup–norm, assigning to each $f \in C_\varphi(M,\mathbb{E})$ the real number

$$\|f\| = \sup_{p \in M} \|f(p)\|.$$

The following is easily verified:

Proposition 5.2.1 Given a continuous map φ from a compact manifold M into a manifold N and $\mathbb{E}$ a vector bundle over N. The sup–norm turns $C_\varphi(M,\mathbb{E})$, which is naturally isomorphic to $\Gamma^o(\varphi^* \mathbb{E})$, into a Banach space.

b) $C^k(M,N)$

In this subsection we will endow $C^k(M,N)$ with Whitney's C^k–topology. To this end we form the system of sets

$$\{\{f \in C^k(M,N) \mid j^\ell(f)(M) \subset 0\} \mid 0 \subset N \text{ open}, 0 \le \ell \le k\}.$$

This system is a subbasis of open sets on $C^k(M,N)$. The topology defined by the above subbasis is called Whitney's C^k–topology. Since N is second countable the C^k–topology on $C^k(M,N)$ admits a countable basis. Moreover it is Hausdorff. Hence by a simple version of a metrization theorem based on the uniformity associated with the C^k–topology $C^k(M,N)$ is a metrizable space (cf. [Go,Gui]).

Next let $N = \mathbb{E}$ where $\mathbb{E}$ is a smooth vector bundle over M with projection π. The space $\Gamma^k \mathbb{E}$ of all k–times differentiable sections of $\mathbb{E}$ is a closed vector space. In fact it is closed in $C^k(M,\mathbb{E})$.

We will show next that $\Gamma^k\mathbb{E}$ is a Banach space. We start with a general remark:

Assume that $\mathbb{F}_1$ and $\mathbb{F}_2$, two topological vector bundles over M, are equipped with bundle norms $\| \|_1$ and $\| \|_2$ respectively. We denote the vector bundle of all strong continuous bundle maps from $\mathbb{F}_1$ to $\mathbb{F}_2$ by $L^o(\mathbb{F}_1,\mathbb{F}_2)$. Let $K = \{v \in \mathbb{F}_1 \mid \|v\|_1 \leq 1\}$. It is not hard to see that

$$\| \|_L : L^o(\mathbb{F}_1,\mathbb{F}_2) \longrightarrow \mathbb{R}$$

determined by $\|\varphi\|_L = \sup_{v \in K} \|\varphi(v)\|_2$ is a bundle norm again. We will write $\| \|$ instead of $\| \|_L$ if no confusion arises.

Next let ∇ and $\tilde{\nabla}$ be connections on M and on $\mathbb{E}$, that is in TM and in $T\mathbb{E}$, respectively. Assume $k \geq 1$ and consider any bundle map

$$(\alpha,f)\colon TM \longrightarrow T\mathbb{E}$$

of class C^k, with $f \in C^k(M,\mathbb{E})$ defined by $f(p) = \pi_{\mathbb{E}}\, \alpha(v_p)$ for all $p \in M$ and all $v_p \in T_pM$; $\pi_{\mathbb{E}}$ the projection $T\mathbb{E} \longrightarrow \mathbb{E}$. Regarding α as a strong bundle map

$$\alpha\colon TM \longrightarrow f^*T\mathbb{E}$$

we define in accordance with section 2.10

$$\nabla_X(\alpha)Y = \tilde{\nabla}_X(\alpha \circ Y) - \alpha(\nabla_X Y)$$

for all $X,Y \in \Gamma TM$. Here $\tilde{\nabla}$ denotes also the connection in $f^*T\mathbb{E}$ induced by $\tilde{\nabla}$ in $T\mathbb{E}$. Now observe that for fixed $X \in \Gamma TM$ this yields a strong bundle map

$$\nabla_X(\alpha)\colon TM \longrightarrow f^*T\mathbb{E}$$

of class C^{k-1}. Hence

$$\nabla(\alpha)\colon TM \longrightarrow L^{k-1}(TM,f^*T\mathbb{E})$$

is a strong C^{k-1}–bundle map, where $L^{k-1}(TM,f^*T\mathbb{E})$ denotes the vector bundle of all strong bundle maps $TM \longrightarrow f^*T\mathbb{E}$ of class C^{k-1}. Moreover, if $k \geq 2$, then $L^{k-1}(TM,f^*T\mathbb{E})$ carries a natural connection ∇^L determined by ∇ and $\tilde{\nabla}$ (cf. section 2.10). Thus we may define the strong C^{k-2}–bundle map

$$\nabla(\nabla(\alpha)): TM \longrightarrow L^{k-2}(TM,L^{k-1}(TM,f^*(T\mathbb{E})),$$

called $\nabla^2(\alpha)$. Proceeding by induction we define $\nabla^\ell(\alpha)$ for each non-negative integer $\ell \leq k$. (Set $\nabla^o(\alpha) = \alpha$).

Now we redefine Whitney's C^k-topology on $\Gamma^k\mathbb{E}$ by means of a norm $\| \|^k$ as follows: For k = 0 we choose a bundle norm on $\mathbb{E}$ and set

$$\|s\|^o = \sup_{p \in M} \|s(p)\| \qquad \forall s \in \Gamma^o\mathbb{E} .$$

In case $k \geq 1$ any section $s \in \Gamma^k\mathbb{E}$ yields the bundle map

$$(Ts,s): TM \longrightarrow T\mathbb{E}$$

which is of class C^{k-1}. Thus

$$Ts: TM \longrightarrow s^*T\mathbb{E}$$

is a strong C^{k-1} bundle map. Specifying bundle norms on TM and on $T\mathbb{E}$ (hence on $s^*T\mathbb{E}$) and connections ∇ and $\tilde{\nabla}$ as above we define $\| \|^k$ by

$$\|s\|^k = \sup\{\|s\|^o, \|Ts\|_L, \|\nabla(Ts)\|_L, \ldots, \|\nabla^{k-1}(Ts)\|_L\}$$

for all $s \in \Gamma^k\mathbb{E}$. This is a norm on $\Gamma^k\mathbb{E}$, called the <u>C^k-norm</u>, which turns $\Gamma^k\mathbb{E}$ into a Banach space, a routine exercise which is left to the reader.

We could have proceeded alternatively as follows: First we extend the notion of an ℓ-jet of a section in $\Gamma\mathbb{E}$ in the obvious manner to sections in $\Gamma^k\mathbb{E}$ for $k \geq \ell$. The resulting jet bundle is isomorphic to $J^\ell\mathbb{E}$. Its typical fibre is again

$$F: = E \oplus \bigoplus_{s=1}^{\ell} S^s(\mathbb{R}^m,E)$$

as in proposition 2.8.2, with E the typical fibre of $\mathbb{E}$ and $m = \dim M$. We fix norms on E and on $\mathbb{R}^m$, take the sup-norms on $S^s(\mathbb{R}^m,E)$ over the unit ball in the s-fold cartesian product of $\mathbb{R}^m$, and consider the product norm $\| \|$ on all of F. Given a local trivialization

$$\psi: \pi^{-1}(U_\alpha) \longrightarrow U_\alpha \times F$$

of $J^\ell\mathbb{E}$ over $U_\alpha \subset M$ the map $\| \| \circ pr_F \circ \psi$ is a bundle norm on $\pi^{-1}(U_\alpha)$, where π is the projection $J^\ell\mathbb{E} \longrightarrow M$ and pr_F the projection of $U_\alpha \times F$ onto the second factor. Now we take a partition of unity subordinated to a cover of M over which $J^\ell\mathbb{E}$ trivializes to define a bundle norm on $J^\ell\mathbb{E}$ in the obvious manner. Hence $\Gamma^o J^\ell\mathbb{E}$ is a Banach space under the sup–norm as seen earlier. Call this norm ${}^\ell\| \,\|$. Given $s \in \Gamma^k\mathbb{E}$ we define

$$\|s\|_k = \sup_{0\leq \ell\leq k} {}^\ell\|j^\ell(s)\|.$$

Under this norm $\Gamma^k\mathbb{E}$ is a Banach space again and due to the closed graph theorem it is isomorphic to $(\Gamma^k\mathbb{E}, \| \,\|^k)$. Therefore we prove easily the following:

<u>Proposition 5.2.2</u> $\Gamma^k\mathbb{E}$ is a Banach space of which the topology is Whitney's C^k–topology inherited from $C^k(M,\mathbb{E})$. Moreover it is a modul over $C^k(M,\mathbb{R})$ via the pointwise defined multiplication which is continuous.

c) $H^k\mathbb{E}$

First we will define the notion of a square integrable section s with values in a vector bundle $\mathbb{E}$ given over a compact orientable manifold M of dimension m. On M we fix a volume form μ, i.e. a nowhere vanishing section into $\Lambda^m T^* M$. On $\mathbb{E}$ we choose a positive definite quadratic structure $<\,,\,>$. Any two sections $s_1,s_2 \in \Gamma^o\mathbb{E}$ yield a continuous function

$$<s_1,s_2> : M \longrightarrow \mathbb{R}$$

assigning $<s_1(p),s_2(p)>$ to any $p \in M$. Thus

$$<<s_1,s_2>>: = \int_M <s_1,s_2>\mu$$

is a positive definite quadratic form on $\Gamma^o\mathbb{E}$. Let $H^o(\mathbb{E})$ denote the completion of $\Gamma^o\mathbb{E}$ under $<<,>>$. Obviously $H^o(\mathbb{E})$ is a Hilbert space. Let next $k \geq 1$. Observe that

$$j^k : \Gamma^k\mathbb{E} \longrightarrow \Gamma^o J^k\mathbb{E} \subset H^o J^k\mathbb{E}$$

is an injection. Hence $<<,>>$ defined on $H^o J^k\mathbb{E}$ pulls back to $\Gamma^k\mathbb{E}$. Denote this scalar product again by $<<,>>$. The <u>Sobolev space</u> $H^{2,k}(\mathbb{E})$ is defined to be the completion of

$\Gamma^k\mathbb{E}$ with respect to <<,>>. $H^{2,k}(\mathbb{E})$ is a Hilbert space usually denoted just by $H^k(\mathbb{E})$ if no confusion arises.

d) $C^\infty(M,N)$, $\Gamma\mathbb{E}$

Next we treat Whitney's C^∞–topology on $C^\infty(M,N)$, the space of all smooth maps from M to N, as well as on $\Gamma\mathbb{E}$, the space of smooth sections of M into a vector bundle $\mathbb{E}$ over M. Both topological spaces will be obtained by means of some of the spaces introduced above.

As usual for a Fréchet setting we require M to be compact. The sets $C^\infty(M,N)$ and $\Gamma\mathbb{E}$ are naturally represented as intersections, namely

$$C^\infty(M,N) = \bigcap_{k=0}^{\infty} C^k(M,N) \text{ and } \Gamma\mathbb{E} = \bigcap_{k=0}^{\infty} \Gamma^k\mathbb{E}\,.$$

The C^∞–topologies on $C^\infty(M,N)$ and on $\Gamma\mathbb{E}$ respectively are defined via the canonical injections

$$i^k : C^\infty(M,N) \longrightarrow C^k(M,N) \text{ and } i^k\!: \Gamma\mathbb{E} \longrightarrow \Gamma^k\mathbb{E},$$

namely as the coarsest among all topologies (cf. [Ke]) on $C^\infty(M,N)$ and $\Gamma\mathbb{E}$ respectively for which i^k with $k = 0,1...$ are continuous. Thus both $C^\infty(M,N)$ and ΓE are hence the projective limits over all $C^k(M,N)$ and $\Gamma^k\mathbb{E}$ respectively. Evidently $\Gamma\mathbb{E}$ is a subspace of $C^\infty(M,\mathbb{E})$. Since $C^k(M,\mathbb{E})$ is metrizable for each k both $C^\infty(M,N)$ and $\Gamma\mathbb{E}$ are metric spaces. Evidently $\| \;\|_k$ and $\| \;\|^k$ are continuous maps on $\Gamma\mathbb{E}$ for each $k = 0,...$ It is a matter of routine to show the following

Proposition 5.2.3 Let M be compact, N a manifold and $\mathbb{E}$ a vector bundle over M. The set $C^\infty(M,N)$ with Whitney's C^∞–topology is a metrizable space. $\Gamma\mathbb{E}$ is a complete metrizable locally convex topological vector space, i.e. a Fréchet space. The pointwise defined multiplication

$$C^\infty(M,\mathbb{R}) \times \Gamma\mathbb{E} \longrightarrow \Gamma\mathbb{E}$$

$$(f,s) \longmapsto f\cdot s$$

is continuous. Moreover $C^\infty(M,N) \subset C^k(M,N)$ and $\Gamma\mathbb{E} \subset \Gamma^k E$ are dense subspaces for all $k = 0,1,\ldots$

Let next $\mathbb{F}$ be another vector bundle over M and

$$\varphi : \mathbb{E} \longrightarrow \mathbb{F}$$

a smooth map which preserves the fibres i.e. which maps each fibre of $\mathbb{E}$ into a fibre of $\mathbb{F}$ (it need not to be linear). This map defines thus a smooth map $f : M \longrightarrow M$ by sending $p \in M$ into $\pi' \mid \varphi(\mathbb{E}_p)$ where π' is the projection of $\mathbb{F}$ to M. Hence we have a uniquely determined smooth strong map

$$\tilde{\varphi} : \mathbb{E} \longrightarrow f^*\mathbb{F} .$$

"Strong" again means, that $\tilde{\varphi}$ induces the identity on M. This map yields for each $k = 0,\ldots$ the modul homomorphism

$$\tilde{\varphi}_* : \Gamma^k\mathbb{E} \longrightarrow \Gamma^k f^*\mathbb{F}$$

$$s \longmapsto \tilde{\varphi} \circ s$$

which clearly is continuous as the reader verifies via a routine check. Hence the restriction

$$\tilde{\varphi}_* : \Gamma\mathbb{E} \longrightarrow \Gamma f^*\mathbb{F}$$

is continuous if both domain and range carry the C^∞–topology.

The map φ allows a partial derivative along the fibres of $\mathbb{E}$ called the <u>fibre derivative</u> of φ given as follows: Let $\sigma: \mathbb{R} \longrightarrow \mathbb{E}_p$ be a smooth curve with values in the fibre $\mathbb{E}_p$ of $\mathbb{E}$ over $p \in M$ for which we assume furthermore that $\sigma(0) = v_p$ and $\dot{\sigma}(0) = w_p$. Set

$$\mathbb{F}\varphi(v_p)(w_p) = \frac{d}{dt}\, \varphi \circ \sigma(t)|_0 .$$

Obviously $\mathbb{F}\varphi$ when expressed in local coordinates coincides with the partial derivative along the fibre variables. The map

$$\mathbb{F}\varphi(v_p): \mathbb{E}_p \longrightarrow \mathbb{F}_{f(p)}$$

is clearly linear. Thus if $s \in \Gamma^k\mathbb{E}$ is fixed, the map $\mathbb{F}\varphi(s(p))$ is linear for each $p \in M$ and each $k \geq 0$. Hence if we set $\mathbb{F}\varphi(s)(w_p) := \mathbb{F}\varphi(s(p))(w_p)$ for all $p \in M$ and all $w_p \in \mathbb{E}_p$ both

$$\mathbb{F}\varphi(s) : \mathbb{E} \longrightarrow \mathbb{F} \text{ and } \mathbb{F}\tilde{\varphi}(s): \mathbb{E} \longrightarrow f^*\mathbb{F}$$

are bundle maps, saying in particular that

$$\mathbb{F}\tilde{\varphi}(s) \in L^k(\mathbb{E}, f^*\mathbb{F})$$

for each $s \in \Gamma^k\mathbb{E}$, where $L^k(\mathbb{E}, f^*\mathbb{F})$ denotes the vector bundle of all strong C^k–bundle maps of $\mathbb{E}$ into $f^*\mathbb{F}$. Clearly the map

$$\mathbb{F}\tilde{\varphi}(s)_* : \Gamma^k\mathbb{E} \longrightarrow \Gamma^k f^*\mathbb{F}$$

$$s' \longmapsto \mathbb{F}\tilde{\varphi}(s)(s')$$

is continuous with respect to the C^k–topology for all k. The fibre derivative of φ and $\tilde{\varphi}$ respectively are of special relevance in the analysis on manifolds of maps as we will observe in the next few sections.

Appendix: <u>Currents</u>

Distributions, introduced in sec.5.1 as the elements of the topological dual $\mathcal{D}(U,\mathbb{R})'$ for open domains U in Euclidean spaces, generalize to manifolds in a rather obvious manner. However, it is useful to go even further and introduce at the same time – at essentially no extra expense! – the analogous notions using differential forms of all orders rather than just the "test functions". This leads to the notion of a <u>current</u> on a manifold M, due to de Rham, cf. [dR], which we now outline very briefly:

Just as in the case of functions, the <u>support</u> of a differential form ω is the closure of the set $\{x \in M \mid \omega(x) \neq 0\}$. We denote by $A^p_c(M)$ the module of smooth p–forms with compact support; for $p = 0$, this is the space of "test functions" on M. For each compact set $K \subset M$, $A^p_K(M)$ is the space of those smooth p–forms whose support is contained in K and it is clear that

$$A^p_c(M) = \bigcup_K A^p_K(M).$$

Adapting the methods outlined earlier, one constructs a Fréchet topology in each $A^p_K(M)$ and equips the union $A^p_c(M)$ of these spaces with the inductive limit topology (which makes it an LF–space).

<u>Definition 5.2.4</u> The space of <u>p–currents</u> on an n–dimensional smooth manifold M is defined to be

$$K^p(M) = A^{n-p}_c(M)',$$

the space of continuous linear forms on $A^{n-p}_c(M)$.

One motivation for the choice of this grading of currents is the following: Suppose that M is oriented. Let α be any smooth p–form. Clearly, then, for any $\beta \in A^{n-p}_c(M)$, we have $\alpha \wedge \beta \in A^n_c(M)$ and hence, the integral $T_\alpha(\beta) = \int \alpha \wedge \beta$ is well–defined. It also is obvious that this is a linear and continuous function of β, i.e. a p–current: $T_\alpha \in K^p(M)$. This construction yields a linear injection $A^p(M) \longrightarrow K^p(M)$: every p–form "is" a p–current! This holds, by the way, for p–forms which are not necessarily smooth; continuity, e.g., certainly is sufficient for the convergence of the integral we just used.

Moreover, one now can extend the differential d: $A^p(M) \longrightarrow A^{p+1}(M)$, i.e. exterior differentiation, to all of $K^p(M)$. This is done in the following way: First of all, exterior differentiation d: $A^{n-p-1}_c(M) \longrightarrow A^{n-p}_c(M)$ certainly is continuous and linear, hence: its transpose ${}^t d$ maps dual into dual, i.e. maps $K^p(M)$ into $K^{p+1}(M)$. However, since we wish the embedding $A^p(M) \longrightarrow K^p(M)$ to preserve exterior differentiation, ${}^t d$ is not quite the "right choice". Instead, one has to replace it by $d = (-1)^{p+1} \cdot {}^t d$. The reason is that: For $\alpha \in A^p(M)$ and $\beta \in A_c^{n-p+1}(M)$, $d(\alpha\wedge\beta) = d\alpha\wedge\beta + (-1)^p \alpha\wedge d\beta$ and, by Stokes theorem, $\int d(\alpha\wedge\beta) = 0$. Therefore,

$$T_{d\alpha}(\beta) = \int d\alpha\wedge\beta = (-1)^{p+1}\int \alpha\wedge d\beta = (-1)^{p+1} T_\alpha(d\beta).$$

Remark: Even though the spaces $K^p(M)$ are, in some sense, "much larger than $A^p(M)$", one can establish the very important result that the cohomology of the complex $(K^p(M),d)_{p\geq 0}$ is naturally isomorphic to the de Rham cohomology of M. We omit all the details and refer to the literature.

5.3 Remarks on Calculus

The manifolds of maps we shall have to deal with will, in general, be modelled on non–normable Fréchet spaces and hence, a brief introduction to Calculus in such spaces seems required. Unfortunately, while there are two rather classical notions of differentiability for maps between Banach spaces, namely Gâteaux– (= directional) and Fréchet – differentiability whose relations mirror those of the finite–dimensional case, there is no longer any such "simple" situation once the spaces in question are not normable any more: There are numerous forms of differential calculus in general locally convex spaces, adapted to specific types of problems or to the authors' tastes, and their interaction is not all that easy to formulate, cf. e.g. [Kel]. One way out of the problem is to concentrate on a "simplest possible" concept of a C^1–map instead of the now somewhat awkward question of differentiability at a point. There is no real loss of generality in such an approach since in practice, one usually will be interested in smooth (C^∞) maps, at least for elementary purposes. The easiest approach would seem to be to generalize the notion of a Gâteaux–C^1 map to maps defined in open subsets of a locally convex space E, with values in another such space F. In practice, it often will be possible to obtain "stronger" differentiability after the fact: one verifies that the directional derivative obtained actually satisfies considerably stronger continuity conditions. In fact, it is important to bear in mind that the problem of "finding derivatives" often can be reduced to finite–dimensional considerations which will at least provide a possible candidate for a derivative; one then shows, e.g., that the expression so obtained satisfies the conditions stated below – and with this, differentiability is established.

The definition to choose appears to be the following one which is discussed briefly in [Ha], somewhat more extensively in [Mi]:

<u>Definition 5.3.1</u> Let E,F be locally convex T_2 spaces, $U \subset E$ open. The map $f: U \longrightarrow F$ is said to be C^1 in U if the following conditions are satisfied:

(i) For $x \in U$ and $h \in E$, $\lim_{t\to 0} \frac{1}{t}(f(x+th) - f(x)) := Df(x)h$ exists.

(ii) The map $(x,h) \longmapsto Df(x)h$ is continuous from $U \times E$ to F.

One now shows that $h \longmapsto Df(x)h$ is <u>linear</u> (and by (ii) continuous) in h, so that the <u>derivative</u> $Df(x)$ lies in $L(E,F)$, the space of continuous linear maps $E \longrightarrow F$. Observe here that if E is infinite–dimensional, then there is no locally convex topology in $L(E,F)$ such that (ii) is equivalent with the continuity of the derived map $Df: U \longrightarrow L(E,F)$.

Let us now assume that F is complete. In this case, one obtains the following form of the fundamental theorem of Calculus:

If $f: U \longrightarrow F$ is C^1, then for $x, x+h \in U$,

$$(*) \qquad f(x+h) - f(x) = \int_0^1 Df(x+th)h \, dt.$$

A word of explanation might be in order: The integral on the right is the ordinary "Riemann integral" obtained as the limit of the filter base of all "Riemann sums" for the obviously continuous map $t \longrightarrow Df(x+th)h$ from $[0,1]$ to F. Using a defining family of seminorms on F, it is easy to show that one is dealing with a Cauchy filterbase which therefore converges in F. The elementary integral thus obtained has all the usual properties (and is a very special case of the so–called Pettis–integral). For the proof of (*), one considers the real–valued function $g(t) = \langle u, f(x+th)\rangle$, u a continuous linear form on F, and applies the elementary theorem to these functions, etc.

As a consequence of (*), one obtains the following weak form of a "théorème des accroissements finis":

For any continuous seminorm p on F,

$$p(f(x+h) - f(x)) \leq \sup_{0 \leq t \leq 1} p(Df(x+th)h).$$

Moreover, (ii) implies that given p and $\epsilon \geq 0$, there is a (convex balanced) neighbourhood V of $0 \in E$ such that $p(Df(x+th)h) < \epsilon$ for $h \in V$ and $0 \leq t \leq 1$. One now concludes from the above inequality that a C^1–map $f: U \longrightarrow F$ is continuous in U.

As always, higher derivatives are defined by iteration: Assume that $f: U \longrightarrow F$ is C^1. We define

$$D^2f(x)(h,k) = D(Df(x)k)h = \lim_{t \to 0} \tfrac{1}{t}(Df(x+th)k - Df(x)k),$$

if this limit exists, and f will be C^2 in U if $D^2f(x)$ exists for every $x \in U$ and is such that $(x,h,k) \longmapsto D^2f(x)(h,k)$ is a continuous map $U \times E \times E \longrightarrow F$. One continues in this manner:

$$D^{k+1}f(x)(h,h_1,\ldots,h_k) = D(D^kf(x)(h_1,\ldots,h_k))h$$

with this by now obvious continuity requirement. The map f is C^∞(smooth) if it is C^k for all k. Examples of smooth functions are e.g., the continuous multilinear maps from E to F, hence also the polynomial maps.

Observe that if f is e.g. C^2, then $D^2f(x)$ is a <u>symmetric</u> bilinear map $E \times E \longrightarrow F$; similarly for higher derivatives. We refer for all detail to [Ha], I.3 and to [Mi].

To return to the "practical" question of establishing differentiability of a given map f: One often can proceed, as mentioned earlier, by a kind of "reduction to the finite–dimensional case". In the current context, this may be achieved as follows:

Given f: $U \longrightarrow F$, let $u \in F'$ and now consider $f_{u,h}$: $t \longrightarrow \langle u, f(x+th)\rangle$. This is defined for small $|t|$ and one (formally) differentiates this function at $t = 0$. If now it is possible to find $\varphi(x,h) \in F$, <u>independent of</u> u, such that $\frac{d}{dt} f_{u,h}(t)|_{t=0} = \langle u, \varphi(x,h)\rangle$, this for all u in some total subset of F', then the map $h \longrightarrow \varphi(x,h)$ is the only possible candidate for Df(x)h; one then has to verify (i) and (ii) in order to prove the differentiability of f – by construction. As remarked earlier this approach implicitly uses a special case of the chain rule: It so happens that the form of calculus sketched here does indeed admit a chain rule in its usual form (including for C^k–maps, any k, cf. [Ha], loc. cit.).

While the most elementary part of Calculus thus is seen to proceed routinely, this is definitely not the case for the first major theorem required in many applications, namely a form of an <u>inverse function theorem</u>. In the general case, the difficulties one encounters appear to be insurmountable, so that a suitable class of spaces and maps has to be found for which some such theorem can be established which will be a reasonable analogue of the standard theorem in Banach spaces (cf. e.g. [La], [D,I]). It is precisely the goal of Hamilton's article [Ha] to do this and it is clear that space limitations do not even permit us to sketch the solution. The only thing we can do, therefore, is to strongly recommend the study of this article at least to the extent needed to comprehend the statement and use of the Nash–Moser theorem. Some applications in our context will be mentioned later, cf. sec. 5.5.

We apply next this calculus to

$$\varphi_* : \Gamma^k\mathbb{E} \longrightarrow \Gamma^k\mathbb{F}$$

given by $\varphi_*(s) = \varphi \circ s$ for all $s \in \Gamma^k(\mathbb{E})$, where $\varphi: \mathbb{E} \longrightarrow \mathbb{F}$ is a smooth fibre preserving map of two vector bundles $\mathbb{E}$ and $\mathbb{F}$ both given over a compact manifold M. In addition we assume that φ is strong, i.e. it induces the identity on M. From the previous section we know that φ_* is continuous. In fact φ_* is differentiable, which is expressed by the so called <u>Ω–lemma</u> (cf. [Pa]):

<u>Lemma 5.3.2</u> Let $\mathbb{E}$ and $\mathbb{F}$ be any two vector bundles. For any smooth map $\varphi: \mathbb{E} \longrightarrow \mathbb{F}$ preserving fibres and inducing the identity on M

$$\varphi_* : \Gamma^k\mathbb{E} \longrightarrow \Gamma^k\mathbb{F}$$

is C^1 for any $k = 0,1,\ldots,\infty$. Indeed if $s,s' \in \Gamma^k\mathbb{E}$

$$D\varphi_*(s)(s') = \mathbb{F}\varphi(s)(s').$$

Outline of the proof: Let $\| \|_{\mathbb{E}}$ and $\| \|_{\mathbb{F}}$ be bundle norms on $\mathbb{E}$ and $\mathbb{F}$ respectively. First we treat the case $k = 0$. We have to verify conditions (i) and (ii) of definition 5.3.1. If $s,s' \in \Gamma^o\mathbb{E}$ and $t \in \mathbb{R}$, we have for all $p \in M$

$$\tfrac{1}{t}(\varphi_*(s+ts') - \varphi_*(s))(p) = \tfrac{1}{t}(\varphi(s(p) + ts'(p)) - \varphi(s(p))).$$

Therefore

$$\|\tfrac{1}{t}(\varphi_*(s+ts') - \varphi_*(s)) - \mathbb{F}\varphi(s)(s')\|^\circ$$

$$= \sup_{p\in M} \|\tfrac{1}{t}(\varphi(s(p)+ts'(p)) - \varphi(s(p))) - \mathbb{F}\varphi(s(p))(s'(p))\|_{\mathbb{F}} \cdot$$

Passing to the limit as t tends to zero yields

$$\lim_{t\to 0} \tfrac{1}{t}(\varphi_*(s+ts') - \varphi_*(s)) = \mathbb{F}\varphi(s)(s').$$

To verify condition (ii) one shows

$$\|d\varphi_*(s)(s')\|^\circ = \|\mathbb{F}\varphi(s)(s')\|^\circ \leq \|\mathbb{F}\varphi(s)\|_L \cdot \|s'\|^\circ .$$

Now $\|\mathbb{F}\varphi(s)\|_L$ is bounded in suitable small neighbourhoods which follows from the smoothness of φ.

Next we consider the case $k = \ell$. To this end we choose bundle norms on TM, $T\mathbb{E}$ and on $T\mathbb{F}$ and show that for all $s,s' \in \Gamma^1\mathbb{E}$

$$\sup_{v\in K} \|T(\tfrac{1}{t}(\varphi_*(s+ts') - \varphi_*(s) - \mathbb{F}\varphi(s)(s'))(v)\|_{T\mathbb{F}}$$

tends to zero if $t \longrightarrow 0$, where $K = \{v \in TM \mid \|v\|_{TM} \leq 1\}$. Therefore we regard $T\mathbb{E}$ as a vector bundle over TM with projection $T\pi$, where $\pi: \mathbb{E} \longrightarrow M$ is the given projection. Let us denote scalar multiplication with respect to this bundle structure by $*$, i.e.

$$\lambda * a = T\mu_\lambda(a)$$

for all $\lambda \in \mathbb{R}$ and $a \in T\mathbb{E}$, where $\mu_\lambda: \mathbb{E} \longrightarrow \mathbb{E}$ is given by $\mu_\lambda(e) = \lambda e$ for all $e \in \mathbb{E}$. Analogously we consider $T\mathbb{F}$ as a vector bundle over TM. Then we have

$$T(\tfrac{1}{t}(\varphi_*(s+ts') - \varphi_*(s)) - \mathbb{F}\varphi(s)(s'))(v)$$

$$= \tfrac{1}{t}*(T\varphi(Ts(v) + t*Ts'(v)) - T\varphi(Ts(v))) - T(\mathbb{F}\varphi(s)(s'))(v)$$

$$= \tfrac{1}{t}*(T\varphi(Ts(v)+t*Ts'(v)) - T\varphi(Ts(v))) - \mathbb{F}T\varphi(Ts(v))(Ts'(v)).$$

Now we are in a position similar as above and proceed analogously by taking the supremum over all $v \in K$. Then the smoothness of φ implies that (ii) holds for the C^1–topology. The next step involves $\nabla(T\varphi)$. This we define by

$$\nabla^{\mathbb{F}}_X(T(\varphi\circ s))Y = \nabla_X(T\varphi)TsY + T\varphi(\nabla^{\mathbb{E}}_X(Ts)Y)$$

for all $X,Y \in \Gamma TM$, where $\nabla^{\mathbb{E}}$ and $\nabla^{\mathbb{F}}$ are the connections in $(\varphi\circ s)^* T\mathbb{F}$ respective in $s^* T\mathbb{E}$ given by fixed connections on $M,\mathbb{E}$ and $\mathbb{F}$. We leave the calculations to the reader and refer to [Mi] for a detailed proof of the Ω–lemma.

The above lemma will be applied naturally in the following situation which will reappear afterwards in connection with transition maps in the context of manifolds of maps. The application is based on the notion of a tubular neighbourhood:

Definition 5.3.3 Let N' be a submanifold of the smooth manifold N. A neighbourhood W of N' within N is called a tubular neighbourhood if there is a vector bundle $\mathbb{E}$ over N', an open neighbourhood U of $0_{\mathbb{E}}(N')$ where $0_{\mathbb{E}}$ is the zero section of $\mathbb{E}$ and a smooth diffeomorphism

$$\varphi : U \longrightarrow W$$

such that $\varphi \circ 0_{\mathbb{E}} = id_{N'}$. Moreover let $\Gamma^k U := \{s \in \Gamma^k \mathbb{E} \mid s(N') \subset U\}$

In this definition U could equivalently be replaced by the whole of $\mathbb{E}$. For the existence of tubular neighbourhoods we refer to [Hi,1].

Given a tubular neighbourhood W of a compact submanifold N' of a smooth manifold N, there is obviously for each $k = 0,\ldots,\infty$ a homeomorphism of $\Gamma^k U$ onto an open subset $C^k(N',W)$ namely

$$\varphi_* : \Gamma^k U \longrightarrow C^k(N',W)$$
$$s \longmapsto \varphi \circ s.$$

The $\Gamma^k U$ is equipped with the C^k–topology. Let $\mathbb{F}$ be another vector bundle over N' and V an open tubular neighbourhood of $0_{\mathbb{F}}(N') \subset \mathbb{F}$ together with a smooth diffeomorphism

$$\psi : V \longrightarrow W$$

satisfying $\psi \circ 0_{\mathbb{F}} = id_{N'}$. Assume that

$$\psi^{-1} \circ \varphi : U \longrightarrow V,$$

is fibrepreserving and thus induces the identity on N′. Hence lemma 5.3.2 applied to the iterations of derivatives yields immediately:

Lemma 5.3.4 If $\psi^{-1}\circ\varphi$ is fibrepreserving then it yields a C^1–map

$$(\psi^{-1}\circ\varphi)_*: \Gamma^k U \longrightarrow \Gamma^k V$$

for any $k = 0,\ldots,\infty$. In fact $(\psi^{-1}\circ\varphi)_*$ is smooth.

5.4 $C^k(M,N)$ as a manifold

In this section we will show that $C^k(M,N)$ carrying Whitney's C^k–topology is a manifold for all $k = 0,1,\ldots,\infty$. To construct a chart about $f \in C^k(M,N)$, we choose a smooth Riemannian metric g on N. This metric determines its spray (of section 3.4) and in turn its exponential map exp as introduced in section 3.3. The zero section $0 : N \longrightarrow TN$, the smooth vector field assuming the value zero everywhere, admits an open neighbourhood $0_z \subset TN$ of $0(N)$ on which for any $q \in M$

$$\exp_q: 0_z \cap T_qN \longrightarrow N$$

is a smooth diffeomorphism onto an open neighbourhood $V(q)$. This can easily be seen as follows. Let $0' \subset TN$ be an open neighbourhood of $0(N) \subset TN$ and let

$$\mathrm{Exp} : 0' \longrightarrow N \times N$$

$$v_q \longmapsto (q, \exp_q v_q).$$

This map is smooth. Clearly $\mathrm{Exp}\, 0' \supset \{(q,q) \in N \times N \mid q \in N\}$. Every value $(q,q) \in N \times N$ of Exp is regular. Thus Exp maps a neighbourhood $V_q \subset TN$ of $0(q)$ diffeomorphically onto a neighbourhood $W(q,q) \subset N \times N$. The set $0_z := \bigcup_{q\in N} V_q$ satisfies the above requirements for a tubular neighbourhood.

Pulling TN back to M by f yields f^*TN and the bundle map $\bar{f}: f^*TN \longrightarrow TN$ of class C^k as introduced in section 2.3. The set $\bar{f}^{-1}(0_z)$ is open in f^*TN. The map

$$\mathrm{Exp}_f: \bar{f}^{-1}(0_z) \longrightarrow M \times N$$
$$v_p \longmapsto (p, \exp_{f(p)} v_p)$$

is a C^k–diffeomorphism onto a neighbourhood of the graph of f in $M \times N$. The graph of f is given by

$$\text{graph } f: = \{(p,f(p)) \in M \times N \mid p \in M\}.$$

Now let us construct a chart about $f \in C^k(M,N)$. To this end let $0^*: M \longrightarrow f^*TN$ be the zero section and

$$U(0^*): = \{s \in \Gamma^k f^*TN \mid s(M) \subset \bar{f}^{-1}(0_z)\}.$$

This set is open in $\Gamma^k f^*TN$. On the other hand let

$$W(f) = \{h \in C^k(M,N) \mid h(p) \in \exp_{f(p)}(0_z \cap T_{f(p)}N)\}$$

a C^o–open neighbourhood of $f \in C^k(M,N)$. The map

$$C^k(M,N) \supset W(f) \xrightarrow{\phi} U(0^*) \subset \Gamma^k f^*TN$$
$$h \longmapsto \mathrm{Exp}_f^{-1} \circ (id_M \times h)$$

is certainly a bijective map which is also a homeomorphism with respect to the C^k–topology. Moreover $\phi(f) = 0^*$. Charts of the type $(W(f),\phi)$ are called <u>natural charts</u> of f. Clearly $\mathrm{Exp}_f(\bar{f}^{-1}(0_z))$ together with the smooth diffeomorphism

$$\mathrm{Exp}_f: \bar{f}^{-1}(0_z) \longrightarrow \mathrm{Exp}_f(\bar{f}^{-1}(0_z))$$

is a tubular neighbourhood of graph $f \subset M \times N$. Having two natural charts of f constructed with respect to different sprays say, yielding two different exponential maps inducing two different Exp_f and Exp'_f respectively, then by lemma 5.3.4 $(\mathrm{Exp}'_f \circ \mathrm{Exp}_f^{-1})_*$ (whenever defined) is smooth. The collection of all natural charts constructed about each map in $C^k(M,N)$ defines hence a smooth atlas on $C^k(M,N)$. As to whether the parameter spaces are finite or infinite dimensional by a smooth atlas we mean a covering collection of charts whose transition functions are smooth. The following now is evident:

Theorem 5.4.1 Given two (smooth) manifolds M and N of which M is compact. Then for each $k = 0,\ldots,\infty$

$$C^k(M,N)$$

is a smooth manifold modelled over Banach spaces in case k is finite or modelled over Fréchet spaces in case $k = \infty$. Accordingly $C^k(M,N)$ is either a Banach or a Fréchet manifold.

From now on we restrict us to $C^\infty(M,N)$. Most notions and constructions we will present for $C^\infty(M,N)$ follow for $C^k(M,N)$ accordingly. However one has to be aware of the derivative loss if passing to tangent manifolds and linearizations of C^k–maps.

The above construction of natural charts applies also in the case M is not compact (cf. [Mi]). However the compactness is essential for the following theorem (cf. [Gu] and [Mi]).

Theorem 5.4.2 Let M_1, M_2 and N be two smooth manifolds where M_1 and M_2 are compact. Then

$$a : C^\infty(M_1 \times M_2, N) \rightarrow C^\infty(M_1, C^\infty(M_2, N))$$
$$k \longmapsto (p \longmapsto k(p,.))$$

is a bijection.

Remark. $C^\infty(M_1, C^\infty(M_2, N))$ is a Fréchet manifold too since the above construction of charts is applicable also if N is a Fréchet manifold. The map a is smooth (cf. [Gu] and [Mi]).

Next we will treat the tangent manifold of $C^\infty(M,N)$. To do so we first define the notion of the tangent space for an open subset of a space of sections.

Let $\mathbb{E}$ be a vector bundle over a compact manifold M with projection π. Given any open set $A \subset \Gamma\mathbb{E}$ we define the tangent space T_sA at $s \in A$ by

$$T_sA := \Gamma\mathbb{E}.$$

Now assume $U \subset \mathbb{E}$ to be an open set with $\pi(U) = M$. Then

$$\Gamma U \subset \Gamma\mathbb{E}$$

is open in Whitney's C^∞–topology and therefore we set

$$T_s\Gamma U := \Gamma\mathbb{E}.$$

Given two smooth vector bundles $\mathbb{E}$ and $\mathbb{F}$ over the compact manifold M and M′. Let

$$\Psi : \mathbb{E} \longrightarrow \mathbb{F}$$

be a smooth strong fibre preserving map. Since $D\Psi(s) = \mathbb{F}\Psi \circ s$ for any $s \in \Gamma\mathbb{E}$ the map

$$(\Psi_*, \mathbb{F}\Psi) : \Gamma\mathbb{E} \times \Gamma\mathbb{E} \longrightarrow \Gamma\mathbb{F} \times \Gamma\mathbb{F}$$

is the tangent map of Ψ_*.

We define $T_fC^\infty(M,N)$ analogously as in the finite dimensional setting of chapter one: Any $v_f \in T_fC^\infty(M,N)$ can be represented by a natural chart W(f), its chart map ϕ and an appropriate section $s \in \Gamma f^*TN$, hence by a quadruple $(W(f),\phi,f,s)$. Thus $v_f = \overline{(W(f),\phi,f,s)}$, where the bar denotes the equivalence class defined as in the finite dimensional case.

The tangent manifold $TC^\infty(M,N)$ is again formed analogously as in the finite dimensional setting. It is a smooth manifold as the reader verifies routinely. However in this case we have much more at hand:

We can naturally represent $T_f C^\infty(M,N)$ as

$$\{Y \in C^\infty(M,TN) \mid \pi_N \circ Y = f\},$$

equipped with Whitney's C^∞–topology. This is a Fréchet space isomorphic to $\Gamma f^* TN$. To demonstrate this representation of $T_f C^\infty(M,N)$ more explicitely let

$$\sigma : I \longrightarrow W(f)$$

be a smooth curve defined on some open interval about zero such that

$$\sigma(0) = f.$$

Associated with this curve we define the smooth map

$$\tilde{\sigma} : I \times M \longrightarrow N$$

$$(t,p) \longmapsto \sigma(t)(p)$$

and let $\tilde{\sigma}_p : I \longrightarrow N$ be given by $\tilde{\sigma}_p(t) = \sigma(t)(p)$ for each fixed $p \in M$ and any $t \in I$. Clearly

$$D_1\tilde{\sigma}(0,p) = \dot{\tilde{\sigma}}_p(0) \in T_{f(p)}N \equiv T_{f(p)}N \times \{0\}.$$

Thus σ yields a smooth vector field along f namely

$$D_1\tilde{\sigma}(0,\ldots) : M \longrightarrow TN$$

$$p \longmapsto D_1\tilde{\sigma}(0,p)$$

which hence defines a smooth section $s := \bar{f}^{-1}\dot{\tilde{\sigma}} \in f^*TN$. On the other hand any smooth vector field $Y : M \longrightarrow TN$ along f is obtained via a curve such as σ. Indeed, let

$$\bar{Y} := \bar{f}^{-1} \circ Y$$

for any given Y along f. We set $\tau(t) := t \cdot \bar{Y}$ defined for all t near zero. The field $\bar{Y}$ is its derivative. The curve τ yields σ by letting $\sigma : \mathbb{R} \longrightarrow C^\infty(M,N)$ be given by $\sigma(t) := pr_N \circ Exp_f \circ \tau(t)$, where $pr_N : M \times N \longrightarrow N$ is the canonical projection onto N. Thus for all $p \in M$

$$D_1\tilde{\sigma}(0,p) = Y(p).$$

This procedure applied to any $f \in C^\infty(M,N)$ yields a natural bijection

$$T : TC^\infty(M,N) \longrightarrow C^\infty(M,TN).$$

As shown in [Mi] or [Schm] one has the following:

Theorem 5.4.3 For any pair of smooth manifolds M and N of which M is compact there is a smooth bijection

$$T : TC^{\infty}(M,N) \longrightarrow C^{\infty}(M,TN).$$

It assigns to any tangent vector $\xi_f \in TC^{\infty}(M,N)$ which is assumed to be determined by a smooth curve $\sigma : \mathbb{R} \longrightarrow C^{\infty}(M,N)$ satisfying $\sigma(0) = f$ and $\dot{\sigma}(0) = \xi$, the vector field

$$D_1\tilde{\sigma}(0,...) : M \longrightarrow TN$$

$$p \longmapsto D_1\tilde{\sigma}(0,p).$$

T does not depend on the particular curve chosen.

We conclude this section by computing two basic examples of tangent mappings. These linearizations are simple but useful. M,N,R denote smooth manifolds of which M is compact. Given $f \in C^{\infty}(N,R)$ we define

$$f_* : C^{\infty}(M,N) \longrightarrow C^{\infty}(M,R)$$

by $f_*(g) = f \circ g$ for all $g \in C^{\infty}(M,N)$. This map is smooth. To compute Tf_* at g applied to a tangent vector ξ we choose a smooth curve σ in $C^{\infty}(N,R)$ defined on an open neighbourhood of zero in $\mathbb{R}$ with $\sigma(0) = g$ and $\dot{\sigma}(0) = \xi$. Then

$$\begin{aligned}(Tf_*(g)(\xi))(p) &= \frac{d}{dt}(f\circ\sigma)(t)(p)\,|_{t=0} \\ &= Tf(\sigma(0)(p))(\dot{\sigma}(0)(p)) \\ &= Tf(g(p))(\xi(p))\end{aligned}$$

for any $p \in M$. Thus we have

Lemma 5.4.4 $Tf_*(g)(\xi)(p) = Tf(g(p)\xi(p)$ for all $\xi \in C^{\infty}(M,TN)$ and all $p \in M$.

Secondly we assume that M and N are compact and choose $f \in C^{\infty}(M,N)$. This time

$$f^* : C^{\infty}(N,R) \longrightarrow C^{\infty}(M,R)$$

maps any $g \in C^{\infty}(N,R)$ to $f^*(g) = g \circ f$. Again f^* is smooth and we verify

$$(Tf^*(g)(\xi))(p) = \frac{d}{dt}(\sigma(t) \circ f)(p)|_{t=0}$$
$$= \xi(f(p))$$

for all $p \in M$. Here σ denotes a curve in $C^{\infty}(N,R)$ with $\sigma(0) = g$ and $\dot{\sigma}(0) = \xi$. Thus we may state

Lemma 5.4.5 $Tf^*(g)(\xi) = \xi \circ f$ for all $\xi \in C^{\infty}(N,TR)$.

Of frequent use is Tev, where

$$ev : C^{\infty}(M,N) \times M \longrightarrow N$$

maps a pair (f,p) into $f(p)$. Again choose smooth curves (σ_1,σ_2) with $(\sigma_1(0),\sigma_2(0)) = (f,p)$ and

$$(\dot{\sigma}_1(0),\dot{\sigma}_2(0)) = (\xi,v) \in C^{\infty}(M,TN) \times TM.$$

Then

$$Tev(f,p)(\xi,v) = \frac{d}{dt}(\sigma_1(t),\sigma_2(t))|_{t=0}$$
$$= \dot{\sigma}_1(0)(\sigma_2(0)) + T\sigma_1(0)(\dot{\sigma}_2(0))$$
$$= \xi(p) + Tf(p)(v)$$

showing that

Lemma 5.4.6 $Tev(f,p)(\xi,v) = \xi(p) + Tf(p)(v)$ for $(f,p) \in C^{\infty}(M,N) \times M$, all $\xi \in C^{\infty}(M,TN)$, and all $v \in T_pM$ implying

$$T_1ev(f,p)(\xi,0) = \xi(p) \text{ and } T_2ev(f,p)(0,v) = Tf(p)(v).$$

Finally we state (cf. [Mi])

Theorem 5.4.7 Let M,N,R be smooth manifolds of which M and N are both compact. Then the composition mapping

$$C^\infty(M,N) \times C^\infty(R,N) \longrightarrow C^\infty(M,R)$$
$$(f,g) \longmapsto g\circ f$$

is smooth.

5.5 Examples of manifolds of maps and some tangent mappings

In this section we present some special examples which are of interest in general or else will be used later in this book. Associated with these examples we compute some linearizations of maps being closely related to the example itself. We already know that given a compact smooth manifold M, the collection $C^\infty(M,N)$ of all smooth maps from M into a smooth manifold N is a Fréchet manifold when endowed with Whitney's C^∞–topology. The first examples are submanifolds of $C^\infty(M,N)$.

a) I(M,N), E(M,N), Diff M, $\mathcal{M}$(m), derivatives of the metric and the Riemannian volume

By $I(M,N) \subset C^\infty(M,N)$ we denote the collection of all smooth immersions of M into N. A smooth immersion, just called an immersion, is an element $f \in C^\infty(M,N)$ such that

$$Tf(p) : T_pM \longrightarrow T_{f(p)}N$$

is injective for all $p \in M$. Clearly, if dim M > dim N then $I(M,N) = \phi$. The next theorem proved in [Hi,1] and [Hi,2] will establish I(M,N) as a Fréchet manifold.

Theorem 5.5.1 Let M and N be smooth connected manifolds, M compact. Then $I(M,N) \subset C^\infty(M,N)$ is open; it is connected and dense if $\dim N > 2\cdot\dim M$.

We close this subsection by referring to Nash's theorem and by computing the derivative of the Riemannian volume with respect to the metric as well as the derivative of the pull back of a Riemannian metric with respect to the embedding. First we observe that $\mathcal{M}(M)$, the collection of all smooth Riemannian metrics is an open set of $S^2(M)$, the Fréchet space of all smooth symmetric two tensors of M equipped with C^∞–topology. Hence $\mathcal{M}(M)$ is a Fréchet manifold.

We assume now that M is orientable. The Riemannian volume on an oriented Riemannian manifold M with nowhere vanishing volume form $\mu(g)$ is determined by the equation

$$\mu(g)(X_1,\ldots,X_m)\cdot\mu(g)(Y_1,\ldots,Y_m) = \det[g(X_i,Y_j)]$$

with $X_i,Y_j \in \Gamma TM$ and $[g(X_i,Y_j)]$ denotes the matrix with i as a column and j as a row index. Clearly

$$\mu : \mathcal{M}(M) \longrightarrow \Gamma\Lambda^m T^* M$$

is a smooth map. To compute the derivative $D(\mu)(g)$ of μ at g, choose $S \in S^2(M)$, $p \in M$ and a base $v_1,\ldots,v_m$ of T_pM. Differentiating the above equation (and multiplying by $\det[g(v_i,v_j)]^{-1/2}$) yields

$$D\mu(g)(S)(v_1,\ldots,v_m)\cdot\mu(g)(v_1,\ldots,v_m) =$$
$$\frac{1}{2}\,\mu(g)(p)(v_1,\ldots,v_m)\cdot \mathrm{tr}([S(p)(v_i,v_j)]\cdot[g(p)(v_i,v_j)]^{-1}).$$

Let $\mathrm{tr}_{g(p)}S(p)$ denote the trace of the symmetric bilinear map $S(p)\colon T_pM\times T_pM \longrightarrow \mathbb{R}$ with respect to $g(p)$, i.e.

$$\mathrm{tr}_{g(p)}S(p) = \mathrm{tr}\ A(p)$$

where $A(p)\colon T_pM \longrightarrow T_pM$ is the selfadjoint endomorphism defined by

$$S(p)(v,w) = g(p)(A(p)v,w)$$

for all $v,w \in T_pM$. By choosing an orthonormal basis $v_1,\ldots v_m$ of T_pM consisting of eigenvectors of $A(p)$ it is easily seen that the trace of the matrix

$$[S(p)(v_i,v_j)]\cdot[g(p)(v_i,v_j)]^{-1}$$

is equal to $tr_{g(p)}S(p)$. Thus $D\mu(g)$ is given by

Lemma 5.5.2 The derivative of the Riemannian volume μ at g in the direction of $S \in S^2(M)$ is

$$D\mu(g)(S) = \frac{1}{2}(tr_g S)\cdot\mu.$$

Now we suppose next that the metric on M is determined by an embedding. To this end we consider a smooth Riemannian manifold N with metric g. Any smooth embedding j of a smooth compact manifold M into N yields a Riemannian metric $j^*(g)$ given by

$$j^*g(X,Y) = g(TjX,TjY) \quad \forall\, X,Y \in \Gamma TM.$$

The collection of all smooth embeddings from M into N is denoted by E(M,N). We thus have a map

$$m : E(M,N) \longrightarrow \mathcal{M}(M),$$

which in general is not surjective, as seen below. However it is smooth as shown by [Pe].

We determine next the tangent mapping of m at $i \in E(M,N)$. With this end in mind we choose a smooth E(M,N)–valued curve σ defined on an open interval (–a,a) of reals. This curve is supposed to satisfy the following two conditions

i) $\sigma(0) = i$

ii) $\frac{d}{dt}\sigma(t)|_{t=0} = h$ for a given $h \in C^\infty(M,TN)$.

Clearly $h : M \longrightarrow TN$ is a smooth vector field along i, meaning that $\pi_N \circ h = i$. Then given any $p \in M$

$$\frac{d}{dt} m(\sigma(t))(X(p),Y(p))|_{t=0} = \frac{d}{dt} g(T\sigma(t)X(p),T\sigma(t)Y(p))|_{t=0}$$

$$= g(\nabla(g)_{\frac{d}{dt}} T\sigma(t)X(p),\, T\sigma(t)Y(p))|_{t=0} + g(T\sigma(t)X(p),\nabla(g)_{\frac{d}{dt}} T\sigma(t)Y(p))|_{t=0}$$

implying

$$\frac{d}{dt}\, m(\sigma(t))|_{t=0}(X,Y) = g(\nabla(g)_X h, TiY) + g(TiX, \nabla(g)_Y h).$$

Therefore we state:

<u>Lemma 5.5.3</u> Given any smooth Riemannian manifold N with metric g and any compact smooth manifold M. The tangent mapping Tm at $i \in E(M,N)$ is given by

$$Tm(i)(h)(X,Y) = g(\nabla(g)_X h, TiY) + g(TiX, \nabla(g)_Y h)$$

for each $h \in C^\infty(M,TN)$ with $\pi_N \circ h = i$ and any pair $X,Y \in \Gamma TM$. In case $N = \mathbb{R}^n$ and g coincide with a fixed scalar product $<,>$ then for any $h \in C^\infty(M,\mathbb{R}^n)$ the map $Th = (h,dh)$, where $dh : TM \longrightarrow \mathbb{R}^n$, is locally represented nothing else but the Fréchet derivative of h. Hence

$$Dm(j)(h) = \langle dj,dh \rangle + \langle dh,dj \rangle \, .$$

The map $m : E(M,\mathbb{R}^n) \longrightarrow \mathcal{M}(M)$ is not surjective in general. The following theorem of Nash gives a lower bound on n yielding the surjectivity of m.

<u>Theorem 5.5.4 (Nash)</u> The map $m : E(M,\mathbb{R}^n) \longrightarrow \mathcal{M}(M)$ is surjective if $n \geq \frac{1}{2} \dim M \cdot (3 \dim M + 11)$.

In fact Nash's theorem asserts much more (cf. [N]). The proof is based on an implicit function theorem. For such a deep theorem we refer at this point to [Sch] and especially to [Ha]. An interesting version is given in [Jac].

b) <u>Diff M for compact M; the Lie derivative, the derivative of the metric again, div Z</u>

In the case M is compact and connected $E(M,N) \subset C^\infty(M,N)$ is open. Let Diff M be the collection of all smooth diffeomorphisms of M. Clearly Diff M $\subset E(M,M)$. To show that equality holds observe that $Tj : TM \longrightarrow TM$ has maximal rank everywhere for any smooth embedding j. Thus $j(M) \subset M$ is open and hence $j(M) = M$. Thus Diff M = E(M,M) which shows furthermore that Diff M $\subset C^\infty(M,M)$ is an open subset.

Diff M is certainly a group with the composition comp as its multiplication. By theorem 5.4.7 this multiplication is smooth. Next consider the map

$$\text{inv: Diff } M \longrightarrow \text{Diff } M$$
$$g \longmapsto g^{-1}.$$

To show that this is also smooth it is enough to verify it at id. For given $f \in$ Diff M we have

g

$$g^{-1} = (f^{-1} \circ g)^{-1} \circ f^{-1}$$
$$= [(f^{-1})^* \circ \text{inv} \circ f_*^{-1}](g)$$

out of which we arrive via lemma 5.3.2 at the reduction mentioned.

The differentiability of inv at id can be verified by proceeding locally: Let $(\mathcal{W},\phi)$ be a chart at id associated with a given spray on M. The local representation of inv is given by $\iota : \phi(\mathcal{W}) \longrightarrow \phi(\mathcal{W})$

$$\iota := \phi \circ \text{inv} \circ \phi^{-1},$$

provided that inv $\phi(\mathcal{W}) \subset \phi(\mathcal{W})$. Secondly consider the smooth map

$$c := \phi \circ \text{comp} \circ (\phi^{-1} \times \phi^{-1}).$$

Since for $x \in \phi(\mathcal{W})$

$$c[x,\iota(x)] = \phi(\text{id}) = 0$$

we may proceed formally to obtain

$$D_1c[x,\iota(x)] + D_2c[x,\iota(x)] \cdot D\iota(x) = 0.$$

Thus

$$D\iota(x) = -D_2c[x,\iota(x)]^{-1} \circ D_1c[x,\iota(x)].$$

Certainly $D\iota(x)$: $\Gamma TM \longrightarrow \Gamma TM$ is a candidate for the derivative. A somewhat lengthly investigation establishes $D\iota$ as the derivative of ι as shown in [Mi] or [Gu].

The tangent space at id $\in$ Diff M is by theorem 5.4.3

$$T_{id}\mathrm{Diff}\ M = \Gamma TM.$$

The tangent space at $\varphi \in$ Diff M is constructed via the left translation L_φ determined by $\varphi \in$ Diff M:

$$L_\varphi : \mathrm{Diff}\ M \longrightarrow \mathrm{Diff}\ M$$

Let $\Psi(t)$ be a smooth parameterized family of diffeomorphisms such that $\Psi(0) = \mathrm{id}$ and $\Psi^{\cdot}(0) = X$. Then

$$\frac{d}{dt}(\varphi \circ \Psi(t))|_{t=0}\ TL_\varphi(\mathrm{id})\Psi^{\cdot}(0) = T\varphi_*(\mathrm{id})X = (T\varphi) \circ X.$$

Therefore

$$T_\varphi \mathrm{Diff}\ M = T_\varphi \Gamma TM = \{Y \in C^\infty(M,TM) \,|\, \pi_M \circ Y = \varphi\}.$$

The properties of the flow F_X of a $X \in \Gamma TM$ as mentioned in section 1.2 allow us to regard F_X as a one parameter group

$$F_X : \mathbb{R} \longrightarrow \mathrm{Diff}\ M.$$

The notion of a one parameter group in this infinite dimensional setting is defined in analogy to the one in case of Lie groups as presented in the first chapter. In fact F_X is smooth by theorem 5.4.2. Hence

$$\frac{d}{dt} F_X(t)|_{t=0} = X.$$

Next we study some natural operations of Diff M on spaces of smooth tensors. By a (smooth) tensor B of type r,s we mean a $C^\infty(M,\mathbb{R})$–multilinear map

$$B : (\Gamma TM)^r \longrightarrow (\Gamma TM)^s.$$

Such a map can be obviously identified with a unique map

$$\breve{B} : (\Gamma TM)^r \otimes (\Gamma T^*M)^s \longrightarrow \mathbb{R}.$$

The exponents r and s denote the r– and s–fold cartesian product of ΓTM and ΓT^*M respectively. By $(\Gamma TM)^0$ we mean $C^\infty(M,\mathbb{R})$. Given a tensor B of type r,s any map

$\varphi \in \text{Diff } M$ yields the pull back

$$\varphi^* B : (\Gamma TM)^r \to (\Gamma TM)^s$$

of B defined by

$$\varphi^* B(X_1,\ldots,X_r) = B(T\varphi X_1 \circ \varphi^{-1},\ldots,T\varphi X_r \circ \varphi^{-1}).$$

If φ is replaced by $F_X(t)$, the flow of $X \in \Gamma TM$ at $t \in \mathbb{R}$ then

$$F_X(t)^*(B)(X_1,\ldots,X_r) \in (\Gamma TM)^s$$

depends smoothly on t for any fixed choice of $(X_1,\ldots,X_r) \in (\Gamma TM)^r$. The Lie derivative $L_X(B)$ of B is then defined to be

$$L_X(B)(X_1,\ldots,X_r) := \frac{d}{dt}(F_X(t)^*(B)(X_1,\ldots X_r))|_{t=0}$$
$$- B(\frac{d}{dt} TF_X(t)X_1 \circ F_X(-t)|_{t=0},\ldots,\frac{d}{dt} TF_X(t)X_r \circ F_X(-t)|_{t=0}).$$

We leave it as an exercise to show the well known fact

$$\frac{d}{dt} TF_X(t)Z \circ F_X(-t)|_{t=0} = [X,Z]$$

for any $Z \in \Gamma TM$. Setting

$$L_X Z := \frac{d}{dt} TF_X(t)Z \circ F_X(-t) = [X,Z]$$

we therefore have

$$L_X(B)(X_1,\ldots,X_r) = \frac{d}{dt} F_X(t)^* B(X_1,\ldots,X_r)|_{t=0}$$
$$- B(L_X X_1,\ldots,L_X X_r) .$$

A particular consequence of this formula is the following one: Let α be a smooth r–form on M. For $X \in \Gamma TM$ we define the interior multiplication of X with α as the smooth $(r-1)$–form $i_X \alpha$ given by

$$(i_X \alpha)(X_1,\ldots,X_{r-1}) = \alpha(X,X_1,\ldots,X_{r-1})$$

for all $X_1,\ldots,X_{r-1} \in \Gamma TM$. Then

$$L_X(\alpha) = d i_X \alpha + i_X d\alpha.$$

Now we return to the derivative of the pull back of a fixed Riemannian metric g on N by an embedding $j \in E(M,N)$. First we assume $M = N$ and $h = X \in \Gamma TM$. In this case

$$Tm(id)(X)(X_1,X_2) = g(\nabla(g)_{X_1}X,X_2) + g(X_1,\nabla(g)_{X_2}X)$$

is the tangent map of

$$m : \text{Diff } M \longrightarrow \mathcal{M}(M)$$

at id applied to X. Since $\nabla(g)$ is torsion free the equation

$$[Z,Y] = \nabla(g)_Z Y - \nabla(g)_Y Z$$

holds, showing hence

$$Tm(id)(X)(X_1,X_2) = L_X(g)(X_1,X_2).$$

Now we turn back to the general case again. If $j \in E(M,N)$ and $h \in T_jE(M,N) = \{h \in C^\infty(M,TN) \mid \pi_N \circ h = j\}$ a pointwise orthogonal decomposition with respect to $TjTM \subset TN$ yields

$$h = TjX_h + h^\perp$$

with $X_h \in \Gamma TM$ and $h^\perp \in C^\infty(M,\nu(j))$ where $\nu(j)$ is the normal bundle of TjTM in TN. A simple calculation expresses Tm(j)(h) for any two $X_1,X_2 \in \Gamma TM$ by

$$Tm(j)(h)(X_1,X_2) = L_{X_h}(g)(X_1,X_2) + g(\nabla(g)_{X_1}h^\perp,TjX_2)$$
$$+ g(TjX_1,\nabla(g)_{X_2}h^\perp).$$

Since $g(\nabla(g)_{X_1}h^\perp,TjX_2) = -g(h^\perp,\nabla(g)_{X_1}TjX_2) = -g(h^\perp,Tj\,\nabla(m(j))_{X_1}X_2)$, by definition of $\nabla(m(j))$ as given in section 3.6, we immediately verify

$$g(\nabla(g)_{X_1}h^\perp,TjX_2) = g(\nabla(g)_{X_2}h^\perp,TjX_2).$$

Here $\nabla(m(j))$ is the Levi–Cività connection of m(j). In analogy to the second fundamental tensor introduced in section 3.6 we set

$$\mathscr{h}(h^\perp)(X_1,X_2) = g(\nabla(g)_{X_1}h^\perp,TjX_2).$$

Thus Tm(j)(h) is of the following form:

Proposition 5.5.5 The linearization of

$$m : E(M,N) \longrightarrow \mathcal{M}(M)$$

at any $j \in E(M,N)$ is given by

$$Tm(j)(h) = L_{X_h}(g) + 2\,\ell(h^\perp)$$

where h is pointwise orthogonally decomposed into

$$h = TjX_h + h^\perp$$

with $X_h \in \Gamma TM$ and $h^\perp \in C^\infty(M,\nu(j))$.

Let us take the trace on both sides of the equation for $Tm(j)(h)$. To do so we set

$$\operatorname{div} X := \operatorname{tr} \nabla(m(j))X \quad \forall\, X \in \Gamma TM$$

called the metric divergence of Z and

$$H(h^\perp) := \frac{1}{\dim M} \operatorname{tr}_{m(j)} \ell(h^\perp).$$

Then we have

Corollary 5.5.6 $\operatorname{tr}_{m(j)} Tm(j)(h) = 2(\operatorname{div} X_h + \dim M \cdot H(h^\perp)$.

Finally we remark that

$$\operatorname{tr} \nabla(g')Z \cdot \mu(g') = \operatorname{div} Z \cdot \mu(g') = L_Z \mu(g') \cdot \mu(g') \quad \forall\, Z \in \Gamma TM$$

for all Riemannian metrics g' on M, a calculation we leave as a simple exercise.

c) U(M,N), the principal Diff M–bundle E(M,N)

Let M and N be smooth (finite dimensional) connected manifolds of which M is compact. Following [Bi,Fi] we will show here that E(M,N) is a principal Diff M–bundle over a natural Fréchet manifold U(M,N). The latter one consists of all smooth submanifolds of N of diffeomorphism type M.

We begin by defining the action β of Diff M on E(M,N). Let

$$\beta\colon E(M,N) \times \text{Diff } M \longrightarrow E(M,N)$$

$$(i,g) \longmapsto i \circ g.$$

β is smooth by theorem 5.4.7. Again it is left to the reader to show that

$$U(M,N) := E(M,N)/\text{Diff } M$$

i.e. the quotient of the action β is a Hausdorff topological space. It easily can be verified by using the slice construction in the sequel. Denote the associated projection by π_U. The charts in U(M,N) will be constructed via the tubular neighborhoods (cf. [Hi,1]). To do so let $i \in E(M,N)$ be fixed and choose any Riemannian metric G on N. Denote by ν_i the normal bundle of i(M) and let O_i^ϵ the open neighbourhood of the zero section in ν_i consisting of all vectors $X \in \Gamma\nu_i$ with $\sqrt{G(X(p),X(p))} < \epsilon$ for all $p \in M$. Given any spray on N one can choose ϵ small enough such that exp: $O_i^\epsilon \longrightarrow S_i^\epsilon := \exp(O_i^\epsilon)$ is a diffeomorphism onto the (tubular) neighbourhood S_i^ϵ of i(M). By $P_i : S_i^\epsilon \longrightarrow i(M)$ we mean the projection along the geodesics normal to i(M), i.e. $P_i(q) := \pi_N \circ \exp^{-1}(q)$ for all $q \in S_i^\epsilon$. In the following we will use the notion of transversality: Call a smooth map $f : M \longrightarrow N$ <u>transversal</u> to a submanifold $W \subset N$ in $p \in M$, if $f(p) \in W$ and

$$Tf(p)(T_pM) + T_{f(p)}W = T_{f(p)}N.$$

Let $E^t(M,S_i^\epsilon)$ consist of all those $j \in E(M,S_i^\epsilon)$ which are transversal to the exponentiated normal space through j(p) at each $p \in M$. Then for each $j \in E^t(M,S_i^\epsilon)$

$$P_i \circ j : M \longrightarrow i(N) \subset N$$

is an embedding. Clearly

$$P_i \circ j = i \circ g$$

for some $g \in \text{Diff } M$. Let s_i^ϵ be made of those $j \in E^t(M,S_i^\epsilon)$ satisfying

$$P_i \circ j = i.$$

Obviously

$$E^t(M,S_i^\epsilon) = \beta(s_i^\epsilon \times \text{Diff } M).$$

By construction

$$\pi_U\colon s_i^\epsilon \longrightarrow U(M,N)$$

is injective and maps thus onto an open subset. (It is now easy to show that U(M,N) in fact is a Hausdorff space). Thus

$$\pi_U^{-1}\colon \pi_U(s_i^\epsilon) \longrightarrow s_i^\epsilon$$

is a local section in E(M,N). Next we choose a natural chart $(\mathscr{W},\phi)$ of i associated with the given spray on N. Then if $\bar{s}_i^\epsilon := \mathscr{W} \cap s_i^\epsilon$

$$\phi \circ \pi_U^{-1} : \pi_U(\bar{s}_i^\epsilon) \longrightarrow \{\ell \in \Gamma i^* TN \mid P_i \circ \exp \circ \bar{i} \circ \ell = i\}$$

is a chart map of $\pi_U(\bar{s}_i^\epsilon)$ onto an open subset of the Fréchet space $\Gamma i^* \nu_i$. (Here $\bar{i}$ denotes the identity map $\bar{i} : i^* TN \longrightarrow TN$).

Now given $i \in E(M,N)$, S_i^ϵ and s_i^ϵ as above, the map

$$\pi_U(\bar{s}_i^\epsilon) \times \text{Diff } M \longrightarrow \bar{s}_i^\epsilon \circ \text{Diff } M = \pi_U^{-1}(\pi_U(\bar{s}_i^\epsilon))$$
$$(W,g) \longmapsto \beta(\pi_U^{-1}(W),g)$$

is a diffeomorphism and therefore a local trivialization. From here on one easily verifies that E(M,N) is indeed a smooth principal bundle with structure group Diff M over U(M,N), the Fréchet space consisting of all submanifolds of N being diffeomorphic to M.

d) The group of volume preserving diffeomorphisms modelled over Sobolev spaces

We assume here that M is a compact and oriented smooth Riemannian manifold of dimension m with metric g and (smooth) Riemannian volume μ.

$\text{Diff}_\mu M$ denotes the subgroup of Diff M consisting of all those diffeomorphisms φ which preserve μ, i.e. $\varphi^* \mu = \mu$ or more explicitly

$$\mu(\varphi(p))(T\varphi(p)(v_1),\ldots,T\varphi(p)(v_m)) = \mu(p)(v_1,\ldots,v_m)$$

for all $v_1,\ldots,v_m \in T_pM$ and for all $p \in M$.

We will first "enlarge" Diff M to $\mathrm{Diff}^{2,s+1}M$, where $s \geq 1$ is an integer. Hereby we follow [E,M]. Each mapping $f: M \longrightarrow \mathbb{R}^{2m+1}$ can be viewed as a section $\bar{f}$ in the trivial bundle $\mathbb{E} = M \times \mathbb{R}^{2m+1}$. Let

$$H^{2,s+1}(M,\mathbb{R}^{2m+1})$$

denote the set of all mappings $f : M \longrightarrow \mathbb{R}^{2m+1}$ satisfying $\bar{f} \in H^{2,s+1}(\mathbb{E})$. Define $\mathrm{Diff}^{2,s+1}M$ as the collection of all bijections φ from M into M, such that φ and φ^{-1} belong to $H^{2,s+1}(M,M)$ (where the latter set is identified with $H^{2,s+1}(M,\mathbb{R}^{2m+1})$ via a smooth embedding of M into $\mathbb{R}^{2m+1}$). Clearly Diff M $\subset \mathrm{Diff}^{2,s+1}M$. Moreover $\mathrm{Diff}^{2,s+1}M$ is open in $H^{2,s+1}(M,M)$ and hence a Hilbert manifold. Now we define the cohomology class $[\mu]_{2,s}$ of μ as follows: Write

$$H^{2,s}\Lambda^m$$

for the Sobolev space $H^{2,s}(\Lambda^m T^* M)$ where $\Lambda^m T^* M$ is the m–th exterior power of $T^* M$.

Clearly the exterior derivative d defined for smooth forms extends to

$$d: H^{2,s+1}\Lambda^{m-1} \longrightarrow H^{2,s}\Lambda^m.$$

Then the cohomology class $[\mu]_{2,s}$ of the volume form μ of M is given by

$$[\mu]_{2,s} = \mu + d(H^{2,s+1}\Lambda^{m-1}).$$

Now we will show that $[\mu]_{2,s} \subset H^{2,s}\Lambda^m$ is an affine submanifold. To this end we use Hodge's theory. Associated with d, mapping k–forms into k+1–forms, we have d^*, given by

$$d^* = (-1)^{m(k+1)+1} * d *,$$

mapping (m–k)–forms into (m–k–1)–forms, where

$$*: \Gamma\Lambda^r T^* M \longrightarrow \Gamma\Lambda^{m-r} T^* M$$

is the so called Hodge star operator, cf. [G,H,V] or [A,M,R]. Then we define the <u>Laplacian</u> by

$$\Delta = d^* d + d d^*.$$

Extending d^* and Δ to

$$d^*: H^{2,s}\Lambda^{m+1} \longrightarrow H^{2,s}\Lambda^m$$

and

$$\Delta: H^{2,s}\Lambda^m \longrightarrow H^{2,s}\Lambda^m$$

we have by the Hodge decomposition theorem

$$H^{2,s}\Lambda^m = \ker \Delta \oplus d(H^{2,s+1}\Lambda^{m-1}).$$

Since $\ker \Delta$ is one dimensional over $H^{2,s}(M,\mathbb{R})$ we conclude that

$$d(H^{2,s+1}\Lambda^{m-1})$$

is a Hilbert space. Thus $[\mu]_{2,s}$ is an affine Hilbert manifold.

Define $D\mu^{2,s+1}(M)$ by

$$D\mu^{2,s+1}(M) := \{\varphi \in \mathrm{Diff}^{2,s+1}M \mid \varphi^*[\mu]_{2,s} = [\mu]_{2,s}\},$$

and observe that this set is the union of components of $\mathrm{Diff}^{2,s+1}M$. Hence $D_\mu^{2,s+1}(M)$ itself is a Hilbert manifold, and a differentiable group as well. Now consider the map

$$\Psi: D\mu^{2,s+1}(M) \longrightarrow [\mu]_{2,s}$$

sending each φ into $\varphi^*\mu$. This map is smooth. The tangent map $T\Psi(\varphi)$ is given by

$$T\Psi(\varphi)(Y) = \varphi^* L_{Y\circ\varphi^{-1}}\mu = \varphi^* d(i_{Y\circ\varphi^{-1}}\mu),$$

where $Y \in H^{2,s+1}(\varphi^*TM)$ corresponds to $Y\circ\varphi^{-1} \in H^{2,s+1}(TM)$ under the identifications

$$H^{2,s+1}(\varphi^*TM) \cong T_\varphi(\mathrm{Diff}^{2,s+1}M) \cong$$
$$\cong T_{id}(\mathrm{Diff}^{2,s+1}M) \cong H^{2,s+1}(TM),$$

$L : H^{2,s+1}(TM) \times H^{2,s+1}(\Lambda^m) \longrightarrow H^{2,s}(\Lambda^m)$ denotes the Lie derivative and i the interior multiplication. Since

$$\begin{aligned} \operatorname{im} T\Psi(\varphi) &= \varphi^*\{d(i_X\mu) \mid X \in H^{2,s+1}(TM)\} \\ &= \varphi^* d(H^{2,s+1}\Lambda^{m-1}) \\ &= d(H^{2,s+1}\Lambda^{m-1}) \end{aligned}$$

we conclude that φ is in fact a submersion.

Thus by the implicit function theorem (which holds for smooth maps between Hilbert spaces)

$$\mathrm{Diff}_{\mu}^{2,s+1}M := \{ \varphi \in \mathrm{Diff}^{2,s+1}M \mid \varphi^{*}\mu = \mu \} = \Psi^{-1}(\mu)$$

is a Hilbert manifold and evidently a group. The operations are in fact smooth.

Clearly $\mathrm{Diff}_{\mu}M \subset \mathrm{Diff}_{\mu}^{2,s+1}M$ for all $s \geq 1$. If s is large enough the charts of the identity do not shrink any more if one enlarges the smoothness degree further, as shown in Omori's very rich lecture notes [O]. Thus $\mathrm{Diff}_{\mu}M$ is in fact a Fréchet manifold for which the operations are smooth. Exactly in the analogous way one proceeds for the symplectic group as shown in [E,M].

5.6 Gauge groups

At this point, we include a brief discussion of an entirely different class of examples of manifolds of maps because of their importance in the applications: the (restricted) automorphisms of principal bundles known as gauge transformations since the development, in the mid–70's, of the mathematical apparatus needed for (classical) Yang–Mills field theories; more recently, "gauge groups" also have begun to play an important role in differential geometry. We begin with a special case of independent interest:

(A) Current groups

Let G be a (finite–dimensional) Lie group with Lie algebra $\mathfrak{g}$, M a compact manifold. The exponential map exp: $\mathfrak{g} \to G$ induces a diffeomorphism of some open neighbourhood V of $0 \in \mathfrak{g}$ onto an open neighbourhood U of $e \in G$, say.

We know from earlier that $C^\infty(M,G)$ is a Fréchet manifold and the goal now is to obtain more information about this manifold.

Multiplication $G \times G \longrightarrow G$ and inversion $G \longrightarrow G$ are smooth maps and hence define, by composition, smooth maps

$$C^\infty(M,G) \times C^\infty(M,G) \longrightarrow C^\infty(M,G) \text{ and } C^\infty(M,G) \longrightarrow C^\infty(M,G)$$

whose description is obvious: for $f,g \in C^\infty(M,G)$, $(fg)(x) = f(x)g(x)$ and $f^{-1}(x) = (f(x))^{-1}$, $(x \in M)$, the usual pointwise operations.

Thus, we see that $C^\infty(M,G)$ is a differentiable group modelled on a Fréchet space. However, there is more: The smooth map $\exp_* : h \longmapsto \exp \circ h$ is a diffeomorphism of the open neighbourhood $C^\infty(M,V)$ of $0 \in C^\infty(M,\mathfrak{g})$ onto the open neighbourhood $C^\infty(M,U)$ of $e \in C^\infty(M,G)$: the differentiable group $C^\infty(M,G)$ admits a classical canonical chart at e, thus is a (Fréchet) Lie group in the strict sense. In particular, any element g sufficiently close to e lies on a 1–parameter subgroup – as opposed to what generally happens e.g. for the groups Diff(M).

The tangent space $T_e(C^\infty(M,G)) = C^\infty(M,\mathfrak{g})$ is a Fréchet Lie algebra under pointwise operations: $[h,k](x) = [h(x),k(x)]$, i.e. the bracket is a continuous bilinear map. It is this function space which plays the role of the Lie algebra of $C^\infty(M,G)$.

Remark: Since the bracket is continuous, the closure of a subalgebra resp. an ideal again is a subalgebra resp. an ideal. Moreover, for each $x \in M$, the kernel of the evaluation map $h \longmapsto h(x)$ is a proper closed ideal $\mu_x \subset C^\infty(M,\mathfrak{g})$ so that, in particular, a current algebra $C^\infty(M,\mathfrak{g})$ cannot be simple. One therefore has to expect some complications in the ideal theory of $C^\infty(M,\mathfrak{g})$ even when $\mathfrak{g}$ is simple; cf. e.g. [C,M] for more details.

A far–reaching generalization of these "current groups" is indicated by the following simple observation: Let $M \times G$ be the trivial group bundle with fibre G; then it is obvious that $C^\infty(M,G)$ is isomorphic to $\Gamma(M \times G)$. The natural question thus arises whether it might be possible to treat more general groups $\Gamma(\mathbb{G})$ where $\mathbb{G} \longrightarrow M$ is a group bundle with fibre G, in a similar manner. For a large and particularly important class, namely the gauge groups introduced earlier, this actually is the case, but in order to arrive at a sufficently simple approach, we need some preliminary considerations which are of interest in their own right:

(B) The manifolds $\Gamma(\mathbf{B})$

Let $\pi : \mathbf{B} \longrightarrow M$ be a smooth fibre bundle with fibre a finite–dimensional manifold B. $\Gamma(\mathbf{B})$ is the set of smooth sections of $\mathbf{B}$ and the aim now is to make this into Fréchet manifold (which, by the way, will be a closed submanifold of $C^\infty(M,\mathbf{B})$); moreover, we wish to obtain a "concrete" description of the tangent spaces $T_s(\Gamma(\mathbf{B}))$ as mapping spaces for later use. A possible choice for the latter can easily be obtained by the following heuristic consideration: Suppose that $t \longmapsto s_t$ is a curve in $\Gamma(\mathbf{B})$, $s = s_0$, which is differentiable at least in the sense that for each $x \in M$ the curve $t \longmapsto s_t(x) \in \mathbf{B}_x = \pi^{-1}(x)$ is differentiable. Then, since the $s_t(x)$ form a differentiable curve in the manifold $\mathbf{B}_x$, $\dot{s}_0(x)$ must be a tangent vector to $\mathbf{B}$ which actually is tangent to the fibre $\mathbf{B}_x$. Such vectors are called "vertical" as before and form the vertical bundle $V(\mathbf{B})$ of $\mathbf{B}$, $V(\mathbf{B}) = \ker(T\pi)$. Thus, the brief remarks just given indicate that $\dot{s}_0$ should be interpreted as a (smooth) map $M \longrightarrow V(\mathbf{B})$ over s, i.e. as a section of $s^* V(\mathbf{B})$. A reasonable candidate for $T_s(\Gamma(\mathbf{B}))$ therefore is $\Gamma_s(\mathbf{B}) = \Gamma(s^* V(\mathbf{B}))$ which is a Fréchet space in its C^∞–topology; we shall return to this shortly.

In order to obtain the desired differentiable structure on $\Gamma(\mathbf{B})$, one may proceed as follows: Let $s \in \Gamma(\mathbf{B})$; then there exist a vector bundle $\epsilon : \mathbb{E} \longrightarrow M$, an open neighbourhood V of its "zero–section" $0(M) \subset \mathbb{E}$, an open neighbourhood N_s of $s(M) \subset \mathbf{B}$ and a fibre–preserving diffeomorphism j (covering the identity of M) of V onto N_s. Such a system of data usually is called a tubular neighbourhood of the submanifold s(M) in $\mathbf{B}$. j_* then is a homeomorphism of $\Gamma(V)$ onto $\Gamma(N_s)$ and the pair $(\Gamma(N_s, j_*^{-1})$ is defined to be a chart at s. It follows from the construction that the coordinate changes will be induced by bundle maps and so are smooth; we refer to the literature for details, e.g. to [Mi]. In this manner, one obtains a C^∞–manifold structure on $\Gamma(\mathbf{B})$. The tangent map Tj maps $V(\mathbb{E})|0(M)$ isomorphically onto $V(\mathbf{B})|s(M)$ and, using that $V(\mathbb{E}) = \mathbb{E} \oplus \mathbb{E}$ (fibred over M) and assuming (without loss of generality) that $V = \mathbb{E}$, one verifies that $Tj|(V(\mathbb{E})$ yields a linear and topological isomorphism of $\Gamma(\mathbb{E})$ onto $\Gamma(s^* V(\mathbf{B}))$ – as mentioned above.

We briefly indicate how one could show that $\Gamma(\mathbf{B})$ is a (closed) submanifold of $C^\infty(M,\mathbf{B})$: Choose a Riemannian metric on $\mathbf{B}$ such that the fibres $\mathbf{B}_x$ are totally geodesic. Use this to define a supplement $H(\mathbf{B}) = V(B)^\perp$ of $V(\mathbf{B})$ in $T(\mathbf{B})$. Then $\Gamma(s^* T(\mathbf{B}))$ splits into $\Gamma(s^* T(\mathbf{B})) = \Gamma(s^* V(\mathbf{B})) \oplus \Gamma(s^* H(\mathbf{B}))$ as Fréchet spaces. The geodesic spray of $\mathbf{B}$ then will yield "submanifold charts" at the $s \in \Gamma(\mathbf{B})$. Details are omitted.

Next, suppose that $\mathbf{B} \longrightarrow M$, $\mathbb{C} \longrightarrow M$ are smooth fibre bundles and that $f: \mathbf{B} \longrightarrow \mathbb{C}$ is a smooth strong bundle map. Then the induced map

$$f_* : \Gamma(\mathbf{B}) \longrightarrow \Gamma(\mathbb{C})$$

is smooth: This is the general form of the so called "Ω–lemma" (namely of lemma 5.3.2). The proof consists of a reduction to the latter, using tubular neighbourhoods $(\mathbb{E}_1, j_1)$ of $s(M) \subset \mathbf{B}$ and $(\mathbb{E}_2, j_2)$ of $(f_* s)(M) = f(s(M)) \subset \mathbb{C}$. Then $\tilde{f} := j_2^{-1} f j_1$ is a (locally defined) bundle map $\mathbb{E}_1 \longrightarrow \mathbb{E}_2$ and hence $\tilde{f}_*$ is smooth; cf. also

[Ha], II.2.

It is this last result which now will be used to obtain the desired Lie group structures on gauge groups.

(C) Gauge groups of principal bundles

Let $\pi : P \longrightarrow M$ be a principal G–bundle, $\mathfrak{g}$ the Lie algebra of G and, as always in this context, M compact. The gauge group $\mathcal{G} = \mathcal{G}_P$ of P was introduced in ch. 4 where we also established the following, now very convenient isomorphism:

$$\mathcal{G} \cong \Gamma(\mathbb{G})$$

where $\mathbb{G} = P \times_G G$ is the naturally associated group bundle, G acting on itself by inner automorphisms. In view of (B), thus, $\Gamma(\mathbb{G})$ now becomes a Fréchet manifold. Moreover, it is a differentiable group:

Inversion $\mathbb{G} \longrightarrow \mathbb{G}$ and multiplication $\mathbb{G} \times_M \mathbb{G} \longrightarrow \mathbb{G}$, the fibrewise group operations, where $\mathbb{G} \times_M \mathbb{G} = \bigcup_{x \in M} (G_x \times G_x)$ is the fibred product over M, are smooth bundle maps. Since clearly $\Gamma(\mathbb{G} \times_M \mathbb{G}) = \Gamma(\mathbb{G}) \times \Gamma(\mathbb{G})$ including the manifold structures, one concludes from the Ω–lemma that the group operations $\Gamma(\mathbb{G}) \times \Gamma(\mathbb{G}) \longrightarrow \Gamma(\mathbb{G})$ and $\Gamma(\mathbb{G}) \longrightarrow \Gamma(\mathbb{G})$ are smooth, establishing the claim.

Generalizing (A), however, one obtains considerably more: The Lie algebra of $\Gamma(\mathbb{G})$ is easily seen to be $\Gamma(P(\mathfrak{g}))$, $P(\mathfrak{g})$ the adjoint bundle used in ch.4: This results from the isomorphism $e^* V(\mathbb{G}) \cong P(\mathfrak{g})$ which we leave as an exercise. Furthermore, the exponential map $\exp: \mathfrak{g} \longrightarrow G$ is equivariant with respect to inner automorphisms in G and Ad in $\mathfrak{g}$ and therefore induces a smooth bundle map $\exp: P(\mathfrak{g}) \longrightarrow \mathbb{G}$ which, in turn, yields a smooth map $\exp: \Gamma(P(\mathfrak{g})) \longrightarrow \Gamma(\mathbb{G})$. With just a little work one now shows that this exponential map yields a diffeomorphism of some open neighbourhood of $0 \in \Gamma(P(\mathcal{g}))$ onto an open neighbourhood of $e \in \Gamma(\mathbb{G})$: $\Gamma(\mathbb{G})$ possesses a canonical chart at e and hence is a Lie group in a more traditional sense.

Remarks:

(i) If P is the trivial bundle $M \times G$, then so is $\mathbb{G}$ and $\Gamma(\mathbb{G})$ reduces to $C^\infty(M,G)$ considered in (A). This also is the case whenever the structure group G of P is abelian, for in this case, inner automorphism and Ad are trivial, so that $\mathbb{G} = M \times G$ and $P(\mathcal{g}) = M \times \mathcal{g}$. Since the gauge group of a principal bundle with abelian group is $C^\infty(M,G)$, it evidently does not retain any information at all about P! On the other hand, if G is e.g. compact simple, then $\Gamma(\mathbb{G})$ determines P in a very precise sense, c.f. [C,M], and so provides complete information on P.

(ii) The realization $C^\infty_G(P,G)$ of the gauge group, while often quite convenient, has the drawback that for non–compact G, the construction of the Lie group structure we wish to use becomes very awkward since P no longer will be a compact manifold. Hence the preference for $\Gamma(\mathbb{G})$, at least in this section.

(iii) In many applications one has to consider gauge groups of vector bundles in lieu of those of principal bundles. However, the former are a special case of the latter: If $\epsilon : \mathbb{E} \longrightarrow M$ is a (real or complex) vector bundle, P its full linear frame bundle (group GL(E)), the group $\mathcal{G}_{\mathbb{E}}$ of all automorphisms of $\mathbb{E}$ over the identity of M is naturally isomorphic to $\mathcal{G}_P$ and so will be a (real or complex) Fréchet Lie group. More information on this is available e.g. in [Fi].

(D) Aut(P) for compact G

We assume that the Lie group G is compact. Since we furthermore assume that M is compact, the total space P of the given principal G–bundle $\pi : P \longrightarrow M$ is compact as well and this makes it possible for us to use standard differentiable structures on the mapping spaces we have to consider. As an example, the group Diff(P) of all diffeomorphisms of P is a Fréchet group whose Lie algebra is $\mathfrak{X}(P)$, the algebra of smooth vector fields on P equipped with its C^∞–topology. As usual, the construction of natural charts on Diff(P) involves the choice of a spray on P, but the differentiable structure itself is independent of any such choice. This will be used below where we shall have to choose particular sprays which are adapted to the problems to be treated. Let $\mathrm{Aut}(P) = \mathrm{Diff}_G(P)$ be the full automorphism group of P (as a G–space), i.e. the group of all equivariant diffeomorphisms of P, and let $\mathfrak{X}_I(P) \subset \mathfrak{X}(P)$ be the obviously closed subalgebra of invariant vector fields on P. The result of the considerations to follow is:

Theorem 5.6.1 Aut(P) is a differentiable subgroup of Diff(P) with Lie algebra $\mathfrak{X}_I(P)$ and the gauge group $\mathcal{G}_P$ is a closed normal and differentiable subgroup of Aut(P) with Lie algebra $\mathfrak{X}_I^V(P)$, the algebra of invariant vertical vector fields. Moreover, the natural homomorphism ρ of Aut (P) into Diff(M) is a principal $\mathcal{G}_P$–bundle over the open subgroup $\mathrm{im}(\rho) \subset \mathrm{Diff}(M)$.

The proof will be given in several steps of which the first one consists of showing that $\mathfrak{X}_I(P)$ is a topological direct summand $\mathfrak{X}(P)$. To this end, we begin with a few general remarks: $\mathfrak{X}(P)$ is a closed subspace of the Fréchet space $C^\infty(P,TP)$ on which G acts in two different ways, both being quite obvious. There is, first of all, the "right regular" action r^* of G: $(r_g^* f)(p) = f(pg)$. It is known – and elementary to see – that $r_g^* f$

is jointly continuous in g and f, i.e. that $g \longmapsto r_g^*$ is a representation of G in $C^\infty(P,TP)$ in the usual sense. We can obtain more precise information by interpreting $(p,g) \longmapsto f(pg)$ as a smooth map $P\times G \longrightarrow TP$, i.e. as an element of $C^\infty(P\times G,TP)$. Since P,G are compact, $C^\infty(P\times G,TP) = C^\infty(G,C^\infty(P,TP))$ and this shows that $g \longmapsto r_g^* f$ is smooth for each f. We already know that $r_g^* f$ is jointly continuous and linear in f, so we may conclude that $(g,f) \longrightarrow r_g^* f$ is a <u>smooth</u> map. In particular, every $f \in C^\infty(P,TP)$ is a C^∞–vector for the representation r^* of G.

On the other hand, the operation r of G on P induces one on T(P), namely $R: g \longmapsto R_g$. Thus, G also operates on $C^\infty(P,TP)$ by $(f,g) \longmapsto R_g \circ f$. Moreover, $(u,g) \longmapsto R_g u$ is a smooth map $TP \times G \longrightarrow TP$ and therefore so is the induced map $C^\infty(P,TP)$, i.e. $(f,g) \longmapsto R_g f$.

Combining the two results, we conclude that

$$(g,f) \longmapsto g\cdot f := R_g^{-1} r_g^* f$$

is <u>smooth</u> $G \times C^\infty(P,TP) \longrightarrow C^\infty(P,TP)$. The subspace $\mathfrak{X}(P)$ is stable under this left G–action and thus the above map defines a <u>smooth representation</u> of G in $\mathfrak{X}(P)$.

It is quite immediate that $\mathfrak{X}_I(P)$ is precisely the set of <u>invariants</u> of this representation of G and this now implies that it is a topological summand: One constructs a linear map $\sigma: \mathfrak{X}(P) \longrightarrow \mathfrak{X}_I(P)$ by

$$\sigma X = \int_G g\cdot X\, dg$$

where dg denotes the normalized Haar measure of G. It is obvious that σ is a linear idempotent map with $\mathrm{im}(\sigma) = \mathfrak{X}_I(P)$ and there remains the continuity of this map: We have to show that for each continuous seminorm p on $\mathfrak{X}(P)$ there exist a continuous seminorm q and a constant $m > 0$ such that $p(\sigma X) \leq mq(X)$ for all $X \in \mathfrak{X}(P)$. To this end, note that

$$p(\sigma X) \leq \int_G p(g,X)dg \leq \sup_G(p(g\cdot X)) := \hat{p}(X).$$

Now $\hat{p}$ has finite values since G is compact and p is continuous, and also is a seminorm.

As a supremum of continuous functions, p is lower semicontinuous and hence continuous ($\mathfrak{X}(P)$ is a Fréchet space, hence barrelled!).

We thus far have shown that $\mathfrak{X}_I(P)$ is a topological direct summand of $\mathfrak{X}(P)$; furthermore, σ induces a linear and topological isomorphism $\mathfrak{X}(P) \cong \mathfrak{X}_I(P) \times \ker(\sigma)$ in the usual manner.

The remaining parts of the proof essentially follow [Cu]; in particular, we shall use <u>invariant</u> sprays on P on which we add the following remarks, referring to [Cu] for more detail: Since G acts smoothly on P, it acts on TP, cf. above, as well as on $T^2P := T(TP)$, and a spray ξ on P is called <u>invariant</u> if it is an invariant vector field on TP for these two actions. An example is provided by the geodesic spray of an invariant Riemannian structure on P, and this also proves the existence of invariant sprays (G is compact!). In such a case, $\exp_\xi$ and $\mathrm{Exp}_\xi = (\tau,\exp_\xi)$: $TP \longrightarrow P \times P$, τ the bundle projection $TP \longrightarrow P$, both are equivariant maps and the domain of $\exp_\xi$ is G–invariant in TP. One then concludes, e.g. by using an invariant metric on P, that there is an invariant open neighbourhood $\mathcal{O}$ of the zero–section in TP such that Exp_ξ: $\mathcal{O} \longrightarrow P \times P$ is a diffeomorphism onto some open neighbourhood $\mathcal{V}$ of the diagonal which then will be invariant under the diagonal action of G on $P \times P$; this is a form of an equivariant tubular neighbourhood theorem. Exp_ξ yields a chart at $e \in \mathrm{Diff}(P)$; the vector field X with values in $\mathcal{O}$ defines a diffeomorphism α of P by means of $\mathrm{Exp}(X(p)) = (p,\alpha(p))$, i.e. by $\alpha = \exp_\xi(X)$. Since Exp_ξ is a diffeomorphism of $\mathcal{O}$, one easily shows that α is equivariant iff X is invariant. Writing $\Gamma(\mathcal{O})$ for the open neighbourhood of $0 \in \mathfrak{X}(P)$ consisting of vector fields with values in $\mathcal{O}$, we thus see that $\exp_\xi(\Gamma(\mathcal{O}) \cap \mathfrak{X}_I(P)) = (\exp(\Gamma(\mathcal{O}))) \cap \mathrm{Aut}(P)$. This shows that Aut(P) is a differentiable subgroup of Diff(P) whose Lie algebra is $\mathfrak{X}_I(P)$.

There remain the properties of ρ stated in the theorem: Again following [Cu], we say that a spray ξ on P is a "G–lift" of the spray η on M if $\eta \cdot T\pi = T^2\pi \cdot \xi$, i.e. if ξ is $T\pi$–related to η. The existence of such lifts is established in [Cu] and we here merely mention some of the properties of their exponential maps: $T\pi$ maps the domain of $\exp_\xi$ into the one of $\exp_\eta$ and therefore $\exp_\eta \cdot T\pi = \pi \cdot \exp_\xi$ (pointwise!). On the other hand, it is well–known that $T\pi$ induces an algebra homomorphism $\pi_*: \mathfrak{X}_I(P) \longrightarrow \mathfrak{X}(M)$, i.e. that invariant vector fields are "projectable". Thus, restriction to $\mathfrak{X}_I(P)$ yields identity $\exp_\eta \circ \pi_* = \pi \circ \exp_\xi$, valid in some open neighbourhood of $0 \in \mathfrak{X}_I(P)$. From this and the definition of ρ one concludes that if $\alpha = \exp_\xi(X)$, $X \in \mathfrak{X}_I(P)$ sufficiently small, then $\rho(\alpha) = \mathrm{Exp}_\eta(\pi_* X)$: with respect to the charts used here, ρ is represented by π_*. By ch. 3, π_* is induced by a bundle map and hence is a continuous linear map of Fréchet spaces. Lastly, the choice of a connection on P induces a "horizontal lift" $\lambda: \mathfrak{X}(M) \longrightarrow \mathfrak{X}_I(P)$ which is a right inverse of π_*, showing that π_* is surjective and hence a homomorphism of $\mathfrak{X}_I(P)$ onto $\mathfrak{X}(M)$. It follows from this that ρ is a smooth and <u>open</u> map, in particular: that $\mathrm{im}(\rho)$ is an open, hence differentiable subgroup of Diff(M).

Evidently, $\ker(\pi_*) = \mathfrak{X}_I^v(P)$. If once more one chooses a connection on P and if HP is its horizontal bundle, then the equivariance of the projections $TP \longrightarrow VP$ and $TP \longrightarrow HP$ shows that $\mathfrak{X}_I(P) = \mathfrak{X}_I^v(P) \oplus \mathfrak{X}_I^h(P)$, $\mathfrak{X}_I^h(P)$ the space of invariant horizontal vector fields. Accordingly, $\ker(\pi_*)$ is a split subspace and π_* induces a linear and topological isomorphism $\mathfrak{X}_I^h(P) \cong \mathfrak{X}(M)$ whose inverse is λ which thus also is a <u>continuous</u> linear map. Therefore, for f sufficiently close to $\mathrm{id} \in \mathrm{Diff}(M)$, $\exp_\xi(\lambda \exp_\eta^{-1}(f))$ is well–defined and this construction defines a smooth local section of ρ because of $\pi_* \lambda = \mathrm{id}$.

Recall that $\rho(\alpha) = \exp_\eta(\pi_* X)$ if α is sufficiently close to the identity. Clearly, this will be equal to 1 iff $\pi_* X = 0$, i.e. if and only if X is vertical. Accordingly, $\exp_\xi(X) \in \mathcal{G}_P$ iff $X \in \mathfrak{X}_I^v(P)$, X small enough, e.g. in $\Gamma(\mathcal{O})$. This shows that $\exp(\Gamma(\mathcal{O}) \cap \mathfrak{X}_I^v(P)) = \exp(\Gamma(O)) \cap \mathcal{G}_P$. $\mathcal{G}_P$ is a closed <u>differentiable</u> subgroup of Aut(P). Together with the last remark on ρ this also shows that ρ admits "enough" smooth local sections defined in some fixed open neighbourhood of $\mathrm{id} \in \mathrm{Diff}(M)$: one uses right multiplication by $\mathcal{G}_P$. In conclusion, it follows that ρ is a principal $\mathcal{G}_P$–bundle over $\mathrm{im}(\rho)$ and this completes the proof of the theorem.

Remark: A straightforward argument which we omit here shows that the induced differentiable structure obtained in the theorem for the gauge group $\mathcal{G}_P$ coincides with the structure "of classical type" defined earlier on these groups.

The preceding arguments also establish the following assertion:

Corollary 5.6.2 Let $H \subset \mathrm{Diff}(M)$ be a closed differentiable subgroup. Then the "enlarged gauge group" $\rho^{-1}(H) \subset \mathrm{Aut}(P)$ is a differentiable subgroup and $\rho : \rho^{-1}(H) \longrightarrow H \cap \mathrm{im}(\rho)$ is a principal $\mathcal{G}_P$–bundle.

The easy verification of this is omitted here.

5.7 On the deformation of differentials of immersions

Here we present a formalism which describes the deformations of differentials of immersions. For further details we refer to [Bi,Pe] and [Bi].

Throughout this section we assume a scalar product $< , >$ in $\mathbb{R}^n$ to be fixed. Let us furthermore fix an immersion i of an m–dimensional compact smooth manifold into $\mathbb{R}^n$. Within its connected component $0_i \subset I(M,\mathbb{R}^n)$ any two maps can be joined by a smooth curve. Hence i is isotopic to any other $j \in 0_i$. The smooth tangential representation $\tilde{j}$ of j assignes to any $p \in M$ the tangent plane $dj(p)(T_pM)$, regarded as an element in the Grassmanian G(m,n) of all m–planes in $\mathbb{R}^n$. By dj we mean the principal part of the tangent map $Tj : TM \longrightarrow \mathbb{R}^n \times \mathbb{R}^n$.

The description of dj by di is achieved as follows: Denote by ξ and η the canonical m– resp. (n–m)–plane bundle over G(m,n). Their direct sum is the trivial bundle $G(m,n) \times \mathbb{R}^n$. Since i and j are isotopic the pull–backs $\tilde{i}^* \xi$ and $\tilde{j}^* \xi$ as well as $\tilde{i}^* \eta$ and $\tilde{j}^* \eta$ are isomorphic. ($\tilde{j}^* \xi \cong TM$ and $\tilde{j}^* \eta$ is isomorphic to the normal bundle $\bar{\nu}(j)$, regarded as a bundle over M.) Thus there is a strong smooth bundle isomorphism

$$B : \tilde{i}^*(\xi \oplus \eta) \longrightarrow \tilde{j}^*(\xi \oplus \eta)$$

preserving the direct sum. Both domain and range are canonically isomorphic to $M \times \mathbb{R}^n$ and thus B is described by a smooth map $\Psi : M \longrightarrow GL(n)$. Choose B such that Ψ satisfies $dj(p)(v_p) = \Psi(p)(di(p)(v_p))$ for all $v_p \in T_pM$ and all $p \in M$. In short we have

$$dj = \Psi \cdot di.$$

Now let $g \cdot \varphi$ be the polar decomposition of Ψ formed pointwise with respect to <,> on $\mathbb{R}^n$, where $g \in C^\infty(M,SO(n))$ and $\varphi \in C^\infty(M,GL(n))$. The latter is pointwise symmetric and positive definite. Based on the next observation the description $dj = g \cdot \varphi \cdot di$ will be defined as follows: The metrics m(i) and m(j), the pullbacks of <,> by i and j, are related by

$$m(i)(AX,Y) = m(j)(X,Y) = \langle \varphi^2 \cdot diX, diY \rangle$$

for a unique smooth strong bundle map $A : TM \longrightarrow TM$ and any two smooth vector fields X,Y on M. The map A is (pointwise) selfadjoint and positive definite with respect to m(i) and hence admits a unique positive square root $f : TM \longrightarrow TM$, again formed pointwise. Hence we have for each $v_p \in T_pM$ and each $p \in M$

$$dj(p)(v_p) = g(p)(di(f(p)(v_p))),$$

or in short

$$dj = g \cdot di \cdot f.$$

This is the relative description of dj by di, in which m(j) is reflected by the symmetric factor f. The role of the orthogonal factor g will be apparent later. In contrast to the uniqueness of the symmetric factor f the orthogonal one g is not unique: For any $q \in C^\infty(M,SO(n))$ which is the identity on $\tilde{i}^* \xi$ we have obviously

$$dj = q \cdot g \cdot di \cdot f .$$

If M is simply connected, then given f, a map $g \in C^\infty(M,SO(n))$ determines a differential of an immersion iff $\delta(g \cdot di \cdot f) = 0$, where δ is the exterior derivative. Thus we state:

Theorem 5.7.1 Let $i \in I(M,\mathbb{R}^n)$ be fixed, then for an immersion j in the connected component O_i of i there is $g \in C^\infty(M,SO(n))$ and a strong bundle map $f : TM \longrightarrow TM$, positive definite and symmetric with respect to m(i), such that

$$dj = g \cdot di \cdot f .$$

Any Riemannian metric G on M can be represented as

$$G(X,Y) = m(i)(f^2X,Y)$$

for any pair of smooth vector fields X,Y on M and a well defined strong bundle isomorphism f of TM, positive definite and symmetric with respect to m(i). In general $\delta(di \cdot f) \neq 0$. Call $g \in C^\infty(M,SO(n))$ an integrating factor for $di \cdot f$ if

$$\delta(g \cdot di \cdot f) = 0.$$

Theorem 5.5.4 immediately yields

Corollary 5.7.2 For $n \geq \frac{m}{2}(3m+11)$ any smooth strong bundle map f of TM, positive definite and symmetric with respect to m(i), admits an integrating factor $g \in C^\infty(M,SO(n))$ and

$$g \cdot di \cdot f$$

is a differential of an immersion.

Theorem 5.7.1 suggests the following interpretation for embeddings i and j: The form $di\cdot f$ reparametrizes the tangent spaces of $i(M)$ in $\mathbb{R}^n$, however, in order to envelope the integral manifold $j(M)$, they have to be displayed orthogonally by the integrating factor g to the appropriate places.

Next we revisit the notions of connections, torsions and second fundamental form.

To this end let α be a smooth $\mathbb{R}^n$–valued one form of maximal rank everywhere regarded as a section α of $L(TM, M\times\mathbb{R}^n)$. Let $P(\alpha)$ be the unique smooth section of $L(M\times\mathbb{R}^n, TM)$ such that $P(\alpha)\cdot\alpha = id_{TM}$ (pointwise formed composition) and for which at each $p \in M$ the kernel $\ker_p P(\alpha)$ is the orthogonal complement $\alpha(T_pM)^\perp$ in $\mathbb{R}^n$. The form α defines a covariant derivative $\nabla(\alpha)$ by setting

$$\nabla(\alpha)_X Y = P(\alpha)\cdot d(\alpha Y)(X)$$

for all smooth vector fields X,Y on M. If α is one of the forms di or $g\cdot di$ with $g \in C^\infty(M, SO(n))$ the associated covariant derivative is denoted by $\nabla(i)$ or $\nabla(g)$ respectively.

Given $i,j \in I(M,\mathbb{R}^n)$ the Levi–Civitá connection $\nabla(j)$ of $m(j)$, i.e. the connection associated to the form $dj = g\cdot di\cdot f$, is expressed by

$$\nabla(j)_X Y = f^{-1}\nabla(i)_X fY + f^{-1}\cdot P(i)\cdot g^{-1}\cdot dg(X)\cdot di\cdot fY =$$
$$= f^{-1}\nabla(g)_X fY.$$

The connection $\nabla(f)$ given by $f^{-1}\nabla(i)f$ leaves $m(j)$ parallel but has in general a non–vanishing torsion tensor $T(f)$ determined by

$$T(f)(X,Y) = f^{-1}(\nabla(i)_X(f)\cdot Y - \nabla(i)_Y(f)\cdot X).$$

Hence $f^{-1}\cdot P(i)\cdot g^{-1}\cdot dg(\ )\cdot di\cdot f$ kills this torsion. On the other hand $\nabla(g)$ leaves $m(i)$

parallel and has to be gauged by f in order to become the Levi–Cività connection of m(j).

Associated with the form α of maximal rank everywhere define $\tilde{\alpha} : M \longrightarrow G(m,n)$ by $\tilde{\alpha}(p) = \alpha(T_pM)$ for all $p \in M$ and let $\pi(\alpha)^{\perp}: M \longrightarrow L(M \times \mathbb{R}^n, \tilde{\alpha}^* \eta)$ be the fibrewise formed orthogonal projection from $M \times \mathbb{R}^n$ to $\tilde{\alpha}^* \eta$. The "second fundamental tensor" $S(\alpha)$ is given by

$$S(\alpha)(X,Y) = \pi(\alpha)^{\perp} d(\alpha Y)(X).$$

If $\alpha = g \cdot di \cdot f$ with $g \in C^{\infty}(M,SO(n))$ and f a smooth bundle isomorphism of TM replace $\nabla(\alpha)$ and $S(\alpha)$ by $\nabla(g,f)$ and $S(g,f)$ respectively. Then:

Theorem 5.7.3 Let M be simply connected and $i \in I(M,\mathbb{R}^n)$ be fixed. Given $g \in C^{\infty}(M,SO(n))$ and a bundle isomorphism f of TM, symmetric and positive definite with respect to m(i), the following statements are equivalent:

(1) $\partial(g \cdot di \cdot f) = 0$

(2) $g \cdot di \cdot f = dj$ for some $j \in I(M,\mathbb{R}^n)$

(3) $\nabla(g,f)$ is torsion free and $S(g,f)$ is symmetric.

In conclusion of this excursion we remark that the curvature tensor R(j) of m(j) is

$$R(j)(X,Y)Z = f^{-1}R(g)(X,Y)fZ,$$

where R(g) is the curvature tensor of $\nabla(g)$. How the Laplacian and the Dirac operator change if f changes can be found in [Bi,Pf] and [Pf].

Next we will investigate the infinitesimal situation of the deformations. To this end we begin with

$$dj = g \cdot di \cdot f$$

for some fixed $i \in I(M,\mathbb{R}^n)$ and any $j \in O_i$. Moreover let $\nu(i)$ be the normal bundle over $i(M)$. First we will replace the factor g to make it unique. We define $\bar{g} : M \longrightarrow \text{End}\,\mathbb{R}^n$ by

$$\bar{g}|\text{di TM} = g|\text{di TM} \text{ and } \bar{g}|\nu(i) = 0.$$

The factors $\bar{g}$ and f by construction vary smoothly with j. Let $j(t) \in I(M,\mathbb{R}^n)$ vary smoothly with $t \in \mathbb{R}$. Then

$$\frac{d}{dt}\,dj(t) = d\,\frac{d}{dt}\,j(t) = (\frac{d}{dt}\,\bar{g}(t))\cdot di\cdot f(t) + \bar{g}(t)\cdot di\cdot\frac{d}{dt}\,f(t).$$

In particular this formula yields

$$dh = \frac{d}{dt}\,\bar{g}(t)|_{t=0}\,di + di\cdot\frac{d}{dt}\,f(t)|_{t=0}$$

provided that $j(0) = i$ and $\frac{d}{dt}\,j(t)|_{t=0}$ is denoted by h. To simplify our notation we set

$$a := \frac{d}{dt}\,\bar{g}(t)|_{t=0} \text{ and } b := \frac{d}{dt}\,f(t)|_{t=0}\,.$$

Clearly $b : TM \longrightarrow TM$ is with respect to m(i) a selfadjoint strong bundle endomorphism which satisfies for any choice of $X,Y \in \Gamma TM$

$$\begin{aligned} Dm(i)(h)(X,Y) &= \frac{d}{dt} < dj(t)X, dj(t)Y > |_{t=0} \\ &= 2m(i)(bX,Y). \end{aligned}$$

The factor a obeys

$$<a\,diX, diY> = -<diX, a\cdot diY>\,.$$

Hence <a di...,di...> is a two form on M which can be represented via m(i) by a smooth strong bundle endomorphism

$$C : TM \longrightarrow TM$$

which is (fibrewise) skew adjoint with respect to m(i). Hence

$$a\cdot diX = di\cdot CX + r\cdot diX$$

with $r \in C^\infty(M,\text{End}\,\mathbb{R}^n)$. However since $\bar{g}(t)|\nu(i) = 0$ for all t we find $r|\nu(i) = 0$. Therefore we can replace r by a uniquely determined $c \in C^\infty(M,so(n))$ for which

$$c\cdot di = r\cdot di\,.$$

We now easily verify

<u>Proposition 5.7.4</u> Given $i \in I(M,\mathbb{R}^n)$ then the differential dh for any $h \in C^\infty(M,\mathbb{R}^n)$ is uniquely representable as

$$dh = c\cdot di + di\cdot C + di\cdot b.$$

Here $c \in C^\infty(M,so(n))$ mapping $c(diTM) \subset \nu(i)$ and hence can be viewed as a tangent field along the Gauss map $\tilde{i}$. Moreover C and b are strong bundle endomorphisms of TM which are respectively skew– and selfadjoint with respect to m(i). In addition b and C are determined by

$$Dm(i)(h)(X,Y) = 2m(i)(bX,Y)$$

and

$$m(i)(CX,Y) = -\langle diX,dhY\rangle + \frac{1}{2}Dm(i)(h)(X,Y)$$

holding both for all $X,Y \in \Gamma TM$.

The description of dh in the above proposition allows dh to be represented by

$$dh = (c+\overline{C}+\overline{b})\cdot di$$

where $\overline{C}$ and $\overline{b}$ both vanish on $\nu(i)$ and on di TM are given by

$$\overline{C}\cdot di = di\cdot C \text{ and } \overline{b}\cdot di = di\cdot b.$$

In short dh can be described by

$$dh = s\cdot di$$

with $s \in C^\infty(M,\text{End}\,\mathbb{R}^n)$. Clearly the exterior differential ∂ applied to dh vanishes. Hence

$$\partial(s\cdot di) = 0.$$

Proposition 5.7.5 Given $i \in I(M,\mathbb{R}^n)$. If the differential of $h \in C^\infty(M,\mathbb{R}^n)$ is given by

$$dh = s \cdot di$$

where $s = c+\overline{C}+\overline{b}$, then for each natural number r

$$s^r \cdot di = d\ell$$

for some $\ell \in C^\infty(M,\mathbb{R}^n)$. Hence $\{dh | h \in C^\infty(M,\mathbb{R}^n)\}$ inherits a di–dependent Lie algebra structure with the commutator of the coefficients as the bracket.

Proof: We form

$$F(s',j) = s' \cdot dj$$

where s' varies in $C^\infty(M,\text{End } \mathbb{R}^n)$ and j in O_i respectively. By proposition 5.7.4 we may assume that F depends smoothly on its variables. The total derivative of F at (s,i) in the direction of $s_1 \in C^\infty(M,\text{End } \mathbb{R}^n)$ and $k \in C^\infty(M,\mathbb{R}^n)$ is

$$DF(s,i)(s_1,k) = s_1 \cdot di + s \cdot dk = s_1 \cdot di + s \cdot s_2 \cdot di$$

where we set $dk = s_2 \cdot di$. We demand w.l.o.g. that $s_i \cdot di = dh'$ and that $DF(s,i)(s_1,k) = dk'$ for some $h',k' \in C^\infty(M,\mathbb{R}^n)$. Hence $s \cdot s_2 \cdot di = dk''$ for some $k'' \in C^\infty(M,\mathbb{R}^n)$. A simple induction on the power r completes the proof up to an obvious rest.

Now we are able to show that $m^{-1}(m(i))$ is a manifold. To do so we let for any $K \subset C^\infty(M,\mathbb{R}^n)$

$$K/\mathbb{R}^n := \{dh \mid h \in K\}$$

and define

$$\Omega_{\exp}: C^\infty(M,\mathbb{R}^n)/\mathbb{R}^n \to C^\infty(M,\mathbb{R}^n)/\mathbb{R}^n$$

to be

$$\Omega_{\exp}(dh) = (\exp \circ s) \cdot di$$

where $dh = s \cdot di$ with $s = c + \overline{C} + \overline{b}$ and c,C and b as in proposition 5.7.4. Obviously $\Omega_{\exp}$ is smooth by the Ω–lemma. Hence there is an open set $Q \subset C^\infty(M,\mathbb{R}^n)/\mathbb{R}^n$ such that

$$\Omega_{\exp}(Q) \subset I(M,\mathbb{R}^n)/\mathbb{R}^n .$$

The next lemma which shows how to integrate $s \cdot di$ for some s is an immediate consequence of 5.7.5.

Lemma 5.7.6 If $i \in I(M,\mathbb{R}^n)$ and $h \in C^\infty(M,\mathbb{R}^n)$ such that $dh = s \cdot di$ with $s = c+\overline{C}+\overline{b}$, then

$$\partial((\exp \circ s)\cdot di) = 0$$

where $\exp\colon \mathrm{End}\,\mathbb{R}^n \longrightarrow GL(n)$ is the usual exponential map. In fact there is an immersion $j \in C^\infty(M,\mathbb{R}^n)$ such that

$$dj = (\exp \circ s)\cdot di.$$

Observe that for $dh \in Q$ and $i \in I(M,\mathbb{R}^n)$

$$\Omega_{\exp}(dh) \subset m^{-1}(m(i))/\mathbb{R}^n$$

holds iff $dh = c \cdot di + di \cdot C$. Hence

$$T_i(m^{-1}(m(i))/\mathbb{R}^n) = \{dh \in C^\infty(M,\mathbb{R}^n)/\mathbb{R}^n \mid Dm(i)(h) = 0\}.$$

Then using the idea of canonical charts of second kind in the theory of Lie groups one shows that there is a neighbourhood $V \subset m^{-1}(m(i))/\mathbb{R}^n$ for which $\Omega_{\exp}^{-1}(V) \subset T_i(m^{-1}(m(i))/\mathbb{R}^n)$ is a neighbourhood of zero. Due to the fact that $\exp : \mathrm{End}\,\mathbb{R}^n \longrightarrow GL(n)$ is injective on a small enough neighbourhood of zero in $\mathrm{End}\,\mathbb{R}^n$ the set V can be chosen that small that

$$\Omega_{\exp} : \Omega_{\exp}^{-1}(V) \longrightarrow V$$

is bijective. In this case $\Omega_{\exp}$ is also a homeomorphism. Hence $m^{-1}(m(i))/\mathbb{R}^n$ is a manifold. Thus we have in turn

Theorem 5.7.7 Given $i \in E(M,\mathbb{R}^n)$ then $m^{-1}(m(i))$ is a Fréchet manifold.

References, Chapter 5

[A,M,R] Abraham,R. Marsden, J. Ratiu, T. — Manifolds, Tensor Analysis and Applications, Addison–Wesley, London, (1983)

[Bi] Binz, E. — On the levi–Cività Connection of a Gauged Levi–Cività Connection, C.R. Math. Rep. Acad. Canada, Vol. IV (1982), No. 2, pp. 117–122.

[Bi,Fi] Binz, E. Fischer, H.R. — On the manifold of embeddings of a closed manifold. Proceedings of the conference on: Differential Geometric Methods in Math. Physics, Techn. Universitaet Clausthal, (1978), LN in Math. Physics, vol. 134, (1981), pp. 310–324.

[Bi,Pe] Binz, E. Peter,Th. — On Deformation of Differentials of Immersions, Differential Geometric Methods in Math. Physics, Math. Phys. Stud. 6, pp. 225–240, R. Reidel, Dordrecht, (1984).

[Bi,Pf] Binz, E. Pferschy, R. — The Dirac Operator and the Change of the Metric, C.R. Math. Rep. Acad. Sci. Canada, Vol. V, (1983), No. 6, pp. 269–274.

[Cu] Curtis, W.D. — The Automorphism Group of a Compact Group Action, Trans. AMS 203, (1975), pp. 45–54.

[C,M] Curtis, W.D. Miller, F.R. — Gauge Groups and Classification of Bundles with Simple Structure Group, Pac. J. Math. 68, (1977), pp. 331–338.

[C,W,D] Choquet–Bruhat, Y. deWitt–Morette, C. Dillard–Bleick, M. — Analysys Manifolds and Physics, North Holland, Amsterdam, (1977).

[dR] de Rham, G. — Variétés différentiables, Hermann, Paris, (1960).

[D,I] Dieudonné, J. — Foundations of Modern Analysis, Vol. I, Academic Press, New York, (1969).

[E,M] Ebin, D.G. Marden, J. — Groups of Diffeomorphisms and the Motion of an Incompressible Fluid, Ann. of Math. 92, (1970), pp. 102–163.

[Fi]	Fischer, H.R.	Notes on Gauge Groups, preprint.
[G,H,V]	Greub, W. Halperin, S. Vanstone, R.	Connections, Curvature and Cohomology, Vol. I,II, Academic Press, New York, (1972/73).
[Go,Gui]	Golubitski, M. Guillemin, V.	Stable Mappings and their Singularities, Springer GTM 14, Berlin, (1973).
[Gu]	Gutknecht, J.	Die C^{∞}_{Γ}–Struktur auf der Diffeomorphismen–gruppe einer kompakten Mannigfaltigkeit, Diss. ETH 5879, Zuerich, (1977).
[Ha]	Hamilton, R.S.	The Inverse Function Theorem of Nash and Moser, Bull. of AMS, (New Series) 7, (1982), No. 1, pp. 65–222.
[Heu]	Heuser, H.G.	Functional Analysis, Wiley–Interscience, Chichester, (1982).
[Hi,1]	Hirsch, M.W.	Differential Topology, Springer GTM, Berlin, (1976).
[Hi,2]	Hirsch, M.W.	Immersions of Manifolds, Trans. AMS 93, (1959), pp. 243–276.
[Jac]	Jacobowitz, H.	Implicit Function Theorems and Isometric Embeddings, Ann. of Math. 95, (1972), pp. 191–225.
[Ja]	Jarchow, H.	Locally Convex Spaces, B.G. Teubner, Stuttgart, (1981).
[Ke]	Kelley, J.	General Topology, Van Nostrand, Princeton, (1955).
[Kel]	Keller, H.H.	Differential Calculus in Locally Convex Spaces, Lecture Notes in Math., vol. 417, (1974), Springer Verlag, Berlin, Heidelberg, New York.
[La]	Lang, S.	Real Analysis, Addison–Wesley, Reading, Mass., (1969).
[Mi]	Michor, P.W.	Manifolds of Differentiable Mappings, Shiva Math. Series 3, Shiva Publishing, Orpington, Kent, UK, (1980).
[N]	Nash, J.	The Embedding Problem for Riemannian Manifolds, Ann. of Math. 63, (1956), pp. 20–63.

[O]	Omori, H.	Infinite Dimensional Lie Transformation Groups, Lecture Notes in Math., 427, Springer, Berlin, (1974).
[Pa]	Palais, R.	Foundations of Non–linear Global Analysis, Benjamin, New York, (1968).
[Pe]	Peter, Th.	Zur Struktur der Immersionen einer Mannig–faltigkeit in einen Euklidischen Raum, Diss. Universitaet Zuerich, (1983).
[Pf]	Pferschy, R.	Die Abhaengigkeit des Dirac–Operators von der pseudoriemannschen Metrik, Diss. Graz, (1983).
[Schm]	Schmid, R.	Die Symplektomorphismengruppe als Fréchet–Lie–Gruppe, Diss. Universitaet Zuerich, (1978).
[Sch]	Schwarz, J.T.	Nonlinear Functional Analysis, Gordon and Breach, New York, (1969).
[Tr]	Treves, F.	Topological Vector Spaces, Distributions and Kernels, Fourth Printing, Academic Press, New York, (1973).

PART II

COVARIANT HAMILTONIAN DYNAMICS

Chapter 6

NON–RELATIVISTIC DYNAMICS

The aim of this chapter is to present dynamics of systems with a finite number of degrees of freedom in a way which could serve as a model for a subsequent formulation of dynamics of classical fields. The principle of stationary action is a starting point of the presentation adopted here. For regular time independent Lagrangians it leads in the usual way to the Hamiltonian formalism. Techniques of symplectic geometry and Poisson algebras are basic tools of the Hamiltonian formalism, and they are briefly reviewed. Since the Lagrangians appearing in important examples in field theory are not regular, degenerate Lagrangians and the corresponding constraints are studied in the spirit of the Dirac theory of constraints.

The main emphasis of the presentation adopted here is on symmetries of the theory. The standard Hamiltonian formalism describes the motion of the system under consideration relative to an arbitrarily chosen frame of reference. The choice of a frame of reference introduces an extrinsic object to the theory which obscures the symmetry group. Since our aim is to have an explicitely covariant, intrinsic, Hamiltonian formulation of dynamics we return to the principle of stationary action. An equivalent variational principle, in which the Lagrangian is replaced by the corresponding Cartan form, leads to equations of motion given by the kernel of the exterior differential of the Cartan form. For degenerate Lagrangians the Cartan form pushes forward to the Hamilton form on the primary constraint manifold. We determine conditions under which, for each direction of evolution (corresponding to a choice of the reference frame) the equations of motion split into constraint equations and Hamiltonian equations of evolution. This splitting is covariant, and the corresponding Hamiltonian is given by the evaluation of the Hamilton form on the vector field describing the direction of evolution. For the sake of simplicity we consider here only the case when the direction of evolution is given by an infinitesimal symmetry.

The constraint equations determine the constraint set of the theory which need not have a manifold structure. Using the Second Noether Theorem we relate the constraint set to the action of the group $\mathscr{H}$ of localizable symmetries. We identify the space of $\mathscr{H}$ orbits in the constraint set with the space of physical states of the system under consideration, and use the reduction of Poisson algebras to discuss its structure.

6.1 Action principle

The time is an absolute element of Newtonian dynamics. It can be geometrically described as an oriented 1–dimensional Euclidean space T. We denote by dt the volume form on T defined by the orientation and the Euclidean structure of T. It can be considered as the differential of an orientation preserving isometry from T to the real line defined by an arbitrary choice of a point $0 \in T$ representing the origin of a coordinate system in T. We denote by $\frac{\partial}{\partial t}$ the unique vector field on T such that

$$<dt, \frac{\partial}{\partial t}> = 1 \tag{1.1}$$

Configurations of a dynamical system form a locally trivial fibre bundle $\mathbb{Q}$ over T with the projection map $\pi_{T\mathbb{Q}}$ and typical fibre Q. We shall refer to $\mathbb{Q}$ as the *configuration bundle* and Q as the *configuration space* of the system. *Histories* of the system are given by sections $\sigma_{\mathbb{Q}T}: T \to \mathbb{Q}$ of $\pi_{T\mathbb{Q}}$. A trivialization of $\mathbb{Q}$ corresponds to a projection map $\pi_{Q\mathbb{Q}}$ from $\mathbb{Q}$ to the typical fibre Q. The composition $\pi_{Q\mathbb{Q}} \circ \sigma_{\mathbb{Q}T} : T \to Q$ describes the *motion*, corresponding to the history $\sigma_{\mathbb{Q}T}$, *relative to the trivialization* $\pi_{Q\mathbb{Q}}$.[1]

We denote by $\mathbb{I}$ the space of 1–jets of sections of $\mathbb{Q}$,

$$\mathbb{I} = J^1\mathbb{Q} \tag{1.2}$$

and by $\pi_{T\mathbb{I}}$ and $\pi_{\mathbb{Q}\mathbb{I}}$ the source and the target maps, respectively.[2] A trivialization $\pi_{Q\mathbb{Q}}$ of $\mathbb{Q}$ induces a trivialization $\pi_{TQ\mathbb{I}}: \mathbb{I} \to TQ$ of $\mathbb{I}$ as a bundle over T defined as follows. If $z = j^1_t(\sigma_{\mathbb{Q}T})$ is the 1–jet at $t \in T$ of a section $\sigma_{\mathbb{Q}T}$, then $\pi_{TQ\mathbb{I}}(z)$ is the image of the value at t of the vector field $\frac{\partial}{\partial t}$ under the derived mapping of $\pi_{Q\mathbb{Q}} \circ \sigma_{\mathbb{Q}T}$,

$$\pi_{TQ\mathbb{I}}(z) = T(\pi_{Q\mathbb{Q}} \circ \sigma_{\mathbb{Q}T})(\frac{\partial}{\partial t}|_t). \tag{1.3}$$

---

[1]The structure of various spaces appearing in dynamics is discussed in Synge [1960], see also Lanczos [1962]. For a differential geometric treatment see Klein [1962], Souriau [1970], Abraham and Marsden [1978], and the references quoted there.

[2]See Sec. 2.8.

We assume that the dynamics of the system under consideration is determined by a *Lagrangian* $L \in C^\infty(\mathbb{Z})$ through the Principle of Stationary Action given below. The *Lagrangian form* corresponding to a Lagrangian L is the product Λ of L and the pull–back to $\mathbb{Z}$ of the 1–form dt on T,

$$\Lambda = L\, \overset{*}{\pi}_{T\mathbb{Z}}\, dt \tag{1.4}$$

It associates to each local section $\sigma_{\mathbb{Q}T}$ of $\pi_{T\mathbb{Q}}$ with compact domain $[t_0,t_1]$ the action integral

$$A[\sigma_{\mathbb{Q}T}] = \int_{t_0}^{t_1} (j^1\sigma_{\mathbb{Q}T})^*\Lambda \tag{1.5}$$

We denote by $\mathbb{Q}|[t_0,t_1]$ the pull back of $\mathbb{Q}$ to $[t_0,t_1] \subset T$, and by $\Gamma(\mathbb{Q}|[t_0,t_1])$ the manifold of sections of $\mathbb{Q}$ with domain $[t_0,t_1]$. A variation of a section $\sigma_{\mathbb{Q}T} \in \Gamma(\mathbb{Q}|[t_0,t_1])$ is a vector $\delta\sigma_{\mathbb{Q}T} \in T_{\sigma_{\mathbb{Q}T}}\Gamma(\mathbb{Q}|[t_0,t_1])$ which can be identified with a map $\delta\sigma_{\mathbb{Q}T}$ from $[t_0,t_1]$ to Ker $T\pi_{T\mathbb{Q}}$ covering $\sigma_{\mathbb{Q}T}$.[3]

6.1.1 <u>Principle of Stationary Action</u>. A section $\sigma_{\mathbb{Q}T}$ with domain $[t_0,t_1]$ describes the restriction to $[t_0,t_1]$ of a *dynamically admissible history* of the system if and only if it is a stationary point of the action integral under the variations $\delta\sigma_{\mathbb{Q}T} \in T_{\sigma_{\mathbb{Q}T}}\Gamma(\mathbb{Q}|[t_0,t_1])$ such that

$$\delta\sigma_{\mathbb{Q}T}(t_0) = 0 \text{ and } \delta\sigma_{\mathbb{Q}T}(t_1) = 0.$$

Let $\pi_{Q\mathbb{Q}}$ be a trivialization of $\mathbb{Q}$. It induces a trivialization $\pi_{TQ\mathbb{Z}}: \mathbb{Z} \to TQ$ defined as follows. For each $t \in T$, and each section $\sigma_{\mathbb{Q}T} : T \to \mathbb{Q}$, $\pi_{TQ\mathbb{Z}}(j^1_t(\sigma_{\mathbb{Q}T}))$ is the vector in TQ at $q = \pi_{Q\mathbb{Q}}(\sigma_{\mathbb{Q}T}(t))$given by

$$\pi_{TQ\mathbb{Z}}(j^1_t(\sigma_{\mathbb{Q}T})) = T\sigma_{\mathbb{Q}T}\circ(\frac{\partial}{\partial t})(t) \tag{1.6}$$

Let (q^i) be local coordinates in Q, (v^i,q^i) the induced coordinates in TQ, and (q^i,t) and

---

[3]See Ch. 5.

(v^i,q^i,t) the corresponding coordinates in $\mathbb{Q}$ and $\mathbb{I}$, respectively. A section $\sigma_{\mathbb{Q}T}$ of $\pi_{T\mathbb{Q}}$ can be described by specifying the coordinates q^i as functions of t,

$$q^i = q^i(t) \tag{1.7}$$

This also gives the coordinate description of the motion, relative to $\pi_{Q\mathbb{Q}}$, which corresponds to $\sigma_{\mathbb{Q}T}$. Similarly, the jet extension $j^1\sigma_{\mathbb{Q}T}$ of $\sigma_{\mathbb{Q}T}$ is given by Eq. (1.7) and

$$v^i = \dot{q}^i(t), \tag{1.8}$$

where the dot denotes the differentiation with respect to t. The Lagrangian L can be expressed as a function $L(v^i,q^i,t)$. The *Lagrange equations* corresponding to L are given by

$$-\frac{d}{dt}\frac{\partial L}{\partial v^i}(\dot{q}^j(t),q^j(t),t) + \frac{\partial L}{\partial q^i}(\dot{q}^j(t),q^j(t),t) = 0. \tag{1.9}$$

Though the Lagrange equations are given here in terms of local coordinates, they have an intrinsic invariant meaning as the Euler–Lagrange equations for the action integral $A[\sigma_{\mathbb{Q}T}]$.[4]

6.1.2. Proposition. A section $\sigma_{\mathbb{Q}T} \in \Gamma(\mathbb{Q}|[t_0,t_1])$ is a stationary point of the action integral A, with respect to the variations $\delta\sigma_{\mathbb{Q}T}$ which vanish at t_0 and t_1, if and only if it satisfies the Lagrange equations.

Proof. If $\sigma_{\mathbb{Q}T}([t_0,t_1])$ is contained in the domain of a local coordinate system (q^i,t) in $\mathbb{Q}$, we can express the action integral in terms of local coordinate functions as follows

$$A[\sigma_{QT}] = \int_{t_0}^{t_1} L(\dot{q}^j(t),q^j(t),t)dt \tag{1.10}$$

Writing $\delta\sigma_{\mathbb{Q}T}$ in terms of functions $\delta q^i(t)$ we have

$$<dA[\sigma_{\mathbb{Q}T}],\delta\sigma_{\mathbb{Q}T}> = \int_{t_0}^{t_1} [\frac{\partial L}{\partial q^i}\delta q^i(t) + \frac{\partial L}{\partial v^i}\delta\dot{q}^i(t)]dt$$

[4]For an intrinsic discussion of the Lagrange equations see Godbillon [1969].

$$= \int_{t_0}^{t_1} [\frac{\partial L}{\partial q^i} - \frac{d}{dt}\frac{\partial L}{\partial v^i}]\delta q^i(t) + [\frac{\partial L}{\partial v^i}\, \delta q^i(t)] \Big|_{t_0}^{t_1} .$$

The boundary term vanishes if $\delta q^i(t_1) = 0$ and $\delta q^i(t_0) = 0$. The Fundamental Theorem in the Calculus of Variations implies that the right hand side vanishes for all $\delta q^i(t)$, such that $\delta q^i(t_1) = 0$ and $\delta q^i(t_0) = 0$, if and only if the Lagrange equations are satisfied.

If $\sigma_{QT}([t_0,t_1])$ is not contained in the domain of a coordinate system (q^i,t), we can subdivide $[t_0,t_1]$ into subintervals $[t_i,t_{i+1}]$ such that $\sigma_{QT}([t_i,t_{i+1}])$ is contained in the domain of a coordinate system for every i, and repeat the same argument in each subinterval. The boundary terms from subintervals with common end points will cancel out, and we obtain the same result. ∎

6.2 Canonical Hamiltonian formalism.[1]

A Lagrangian $L \in C^{\infty}(\mathbb{Z})$ is said to be *time independent* if there exists a trivialization $\pi_{Q\mathbb{Q}}$ of $\mathbb{Q}$ such that L is constant along the fibres of the induced trivialization $\pi_{TQ\mathbb{Q}} : \mathbb{Z} \longrightarrow TQ$ of $\mathbb{Z}$. In this case L induces a function $L \in C^{\infty}(TQ)$ such that

$$L = L \circ \pi_{TQ\mathbb{Z}} \tag{2.1}$$

Identifying T with the real line, we can represent motions relative to $\pi_{Q\mathbb{Q}}$ by curves in TQ. They satisfy the Lagrangian equations (1.9), which can be rewritten in terms of L as follows

$$-\frac{d}{dt}\frac{\partial L}{\partial v^i}(\dot{q}^j(t),q^j(t)) + \frac{\partial L}{\partial q^i}(\dot{q}^j(t),q^j(t)) = 0. \tag{2.2}$$

Lagrangian systems with time independent Lagrangians are called *autonomous.* We shall refer to $L \in C^{\infty}(TQ)$ as a Lagrangian of an autonomous system.

The *Legendre transformation* corresponding to a Lagrangian L is the fibre derivative $\mathbb{F}L : TQ \longrightarrow T^*Q$ defined by

$$<\mathbb{F}L(v),w> = <dL(v),w> \tag{2.3}$$

for every $q \in Q$ and every $v,w \in T_qQ$ (c.f. Sec. 3.4). If (q^i) are local coordinates in Q, (v^i,q^i) are the induced coordinates in TQ, and (p_i,q^i) are the induced coordinates in T^*Q, then $\mathbb{F}L$ is given by

$$p_i = \frac{\partial L}{\partial v^i}(v^j,q^j).$$

A Lagrangian L is said to be *regular* if $\mathbb{F}L$ is a submersion.

We denote here by θ_{T^*Q} the canonical 1–form on T^*Q introduced in Sec. 3.4. For each $p \in T^*Q$, and each $u \in T_p(T^*Q)$,

$$<\theta_{T^*Q},u> = <p,T\pi_{QT^*Q}(u)>, \tag{2.4}$$

---

[1]The presentation of the material in this section follows of Abraham and Marsden [1978].

where π_{QT^*Q} is the cotangent bundle projection. Similarly, we denote by ω_{T^*Q} the canonical 1–form of T^*Q, given by the negative of the exterior differential of the canonical 1–form,

$$\omega_{T^*Q} = -\,d\theta_{T^*Q}. \tag{2.5}$$

In terms of local coordinates in T^*Q introduced above we have

$$\theta_{T^*Q} = p_i dq^i , \tag{2.6}$$

where the summation over the repeated indices is assumed, and

$$\omega_{T^*Q} = -\,dp_i \wedge dq^i . \tag{2.7}$$

For a regular Lagrangian L the pull–back of ω_{T^*Q} by $\mathbb{F}L$ is a symplectic form ω_{TQ} on TQ. That is,

$$\omega_{TQ} = (\mathbb{F}L)^* \omega_{T^*Q} \tag{2.8}$$

is closed and non–degenerate. The energy function E on TQ corresponding to a Lagrangian L is given by

$$E(v) = \langle \mathbb{F}L(v), v \rangle - L(v) \tag{2.9}$$

for every $v \in TQ$. Since ω_{TQ} is non–degenerate, and TQ is finite dimensional, there exists a unique vector field ξ_E on TQ such that

$$\xi_E \,\lrcorner\, \omega_{TQ} = dE . \tag{2.10}$$

It is called the Hamiltonian vector field of E relative to the symplectic form ω_{TQ}.

6.2.1. Proposition. A curve $t \longrightarrow q(t)$ in Q satisfies the Lagrange equations corresponding to a regular Lagrangian L if and only if it is the projection to Q of an integral curve of the Hamiltonian vector field ξ_E of the energy function corresponding to L.

Proof. In terms of local coordinates (q^i) in Q and the induced coordinates (v^i, q^i) in TQ we have

$$E = \frac{\partial L}{\partial v^i} v^i - L \tag{2.11}$$

and

$$\omega_{TQ} = -\frac{\partial^2 L}{\partial v^i \partial v^j} dv^i \wedge dv^j - \frac{\partial^2 L}{\partial q^i \partial v^j} dq^i \wedge dq^j. \tag{2.12}$$

Eq. (2.10) for the Hamiltonian vector field

$$\xi_E = u^i(v^i,q^k) \frac{\partial}{\partial v^i} + w^i(v^j,q^k) \frac{\partial}{\partial q^i}$$

of E yields

$$\frac{\partial^2 L}{\partial v^i \partial v^j} (w^j - v^j) = 0 \tag{2.13}$$

and

$$\frac{\partial^2 L}{\partial v^j \partial v^i} u^j + \frac{\partial^2 L}{\partial q^j \partial v^i} w^j - \frac{\partial L}{\partial q^i} = \frac{\partial^2 L}{\partial v^j \partial q^i} (w^j - v^j) \tag{2.14}$$

Since L is a regular Lagrangian, it follows that Eq. (2.13) implies $w^j(v^i,q^k) = v^j$, and Eq. (2.14) becomes

$$\frac{\partial^2 L}{\partial v^j \partial v^i} u^j + \frac{\partial^2 L}{\partial q^j \partial v^i} w^j - \frac{\partial L}{\partial q^i} = 0 \tag{2.15}$$

If $t \longrightarrow (v(t),q(t))$ is an integral curve of ξ_E, then $v^i(t) = \dot{q}^i(t)$ and Eq. (2.15) implies Eq. (2.2). Conversely, the tangent vector of the curve $t \longrightarrow (\dot{q}^i(t),q^i(t))$ satisfying Eq. (2.2) coincides with the restriction of ξ_E to the curve. ■

Proposition 6.2.1. implies that the dynamics of a Lagrangian system with a regular Lagrangian $L \in C^\infty(TQ)$ is completely determined by the energy function E and the symplectic form ω_{TQ} corresponding to L.

A Lagrangian $L \in C^\infty(TQ)$ is said to be *hyperregular* if $\mathbb{F}L : TQ \longrightarrow T^*Q$ is a diffeomorphism. For hyperregular Lagrangians we can use $\mathbb{F}L$ to push the description of dynamics in TQ forward to T^*Q obtaining the canonical Hamiltonian formulation of dynamics. Let H be the push–forward of the energy function by the Legendre transformation

$$H = (\mathbb{F}L)_* E\,. \tag{2.16}$$

The Hamiltonian vector field of H is the unique vector field ξ_H on T^*Q such that

$$\xi_H \lrcorner\, \omega_{T^*Q} = dH. \tag{2.17}$$

Since $\xi_H = (\mathbb{F}L)_* \xi_E$ and $\mathbb{F}L : TQ \to T^*Q$ covers the identity transformation in Q we obtain the following corollary.

6.2.2. Corollary. A curve $t \to q(t)$ in Q satisfies the Lagrange equations corresponding to a hyperregular Lagrangian L if and only if it is the projection to Q of an integral curve of the Hamiltonian vector field ξ_H of the Hamiltonian H corresponding to L.

In terms of local coordinates (p_i, q^i) in T^*Q used in Eq. (2.7) we can express ξ_H in the form

$$\xi_H = -\frac{\partial H}{\partial q^i}\frac{\partial}{\partial p^i} + \frac{\partial H}{\partial p_i}\frac{\partial}{\partial q^i}\,. \tag{2.18}$$

Hence, an integral curve $t \to (p_i(t), q^i(t))$ of ξ_H satisfies the equations

$$\begin{aligned} \dot{p}_i(t) &= -\frac{\partial H}{\partial q^i}(p_j(t), q^k(t)) \\ \dot{q}^i(t) &= \frac{\partial H}{\partial p_i}(p_j(t), q^k(t)) \end{aligned} \tag{2.19}$$

called the *canonical equations* of Hamilton.

6.3 Symplectic manifolds and Poisson algebras

In the preceding section we passed, for regular Lagrangians, from the Lagrange equations to the Hamilton equations in TQ given by the Hamiltonian vector field of the energy function. For hyper–regular Lagrangians we proceeded to the canonical Hamiltonian equations in T^*Q. In either case the structure underlying the Hamiltonian formalism was a symplectic form; in T^*Q it was the canonical symplectic form ω_{T^*Q} of the cotangent bundle, and in TQ it was the pull–back of ω_{T^*Q} by the Legendre transformation. Since symplectic structures and their generalizations will play a fundamental role in the sequel, we are going to review them briefly in this section.[1]

In general, a symplectic structure on a manifold P is given by a closed, non–degenerate 2–form ω_P on P, called a *symplectic form*. A manifold P endowed with a symplectic structure is called a *symplectic manifold*.

Let (P,ω_P) be a symplectic manifold. Diffeomorphisms of P which preserve ω_P are called *canonical transformations*. *Infinitesimal canonical transformations* are given by vector fields ξ_P on P which preserve ω_P. Since ω_P is closed, it follows that ξ_P is an infinitesimal canonical transformation if and only if $\xi_P \lrcorner\, \omega_P$ is closed. Vector fields ξ_P on P, such that $\xi_P \lrcorner\, \omega_P$ is exact, are called *Hamiltonian vector fields*. A function $f \in C^\infty(P)$ admits a Hamiltonian vector field if there exists a vector field ξ_f on P such that

$$\xi_f \lrcorner\, \omega_P = df. \qquad (3.1)$$

Since ω_P is non–degenerate, ξ_f is uniquely determined by f. It is called the *Hamiltonian vector field* of f (relative to ω_P). If P is finite dimensional, then every $f \in C^\infty(P)$ admits a Hamiltonian vector field. For infinite dimensional symplectic manifolds, which appear in applications, not all functions admit Hamiltonian vector fields. In this section we restrict our considerations to finite dimensional manifolds.

The mapping associating to each $f \in C^\infty(P)$ the Hamiltonian vector field of f

[1]For a comprehensive discussion of symplectic geometry and its applications see Guillemin and Sternberg [1984] and references quoted there.

induces in $C^\infty(P)$ the structure of a Lie algebra, called the *Poisson algebra* of (P,ω_P). The *Poisson bracket* $\{f,h\}$ of $f,h \in C^\infty(P)$ is given by

$$\{f,h\} = \xi_f h = -\omega_P(\xi_f,\xi_h). \tag{3.2}$$

Since, for every $f,h \in C^\infty(P)$

$$[\xi_f,\xi_h] = \xi_{\{f,h\}}, \tag{3.3}$$

it follows that the mapping $f \longrightarrow \xi_f$ is a Lie algebra homomorphism[2]. Its kernel consists of locally constant functions; constants if P is connected.

Let G be a Lie group acting in P by canonical transformations. For each ξ in the Lie algebra $\mathfrak{g}$ of G we denote by ξ_P the vector field on P generating the action of the one parameter group corresponding to ξ. The *action* of G in P is said to be *Hamiltonian* if there exists a G–equivariant mapping $J_P^{\mathfrak{g}} : P \longrightarrow \mathfrak{g}^*$ such that, for each $\xi \in \mathfrak{g}$, ξ_P is the Hamiltonian vector field of the function J^ξ obtained by the evaluation of $J_P^{\mathfrak{g}}$ on ξ,

$$J^\xi = \langle J_P^{\mathfrak{g}},\xi\rangle. \tag{3.4}$$

$J_P^{\mathfrak{g}}$ is called a *momentum map*[3] corresponding to the Hamiltonian action of G on P. The function J^ξ is called the *momentum* associated to ξ. The mapping $\xi \longrightarrow J^\xi$ is a homomorphism of $\mathfrak{g}$ into the Poisson algebra of (P,ω_P): for every $\xi,\zeta \in \mathfrak{g}$,

$$\{J^\xi,J^\zeta\} = J^{[\xi,\zeta]}. \tag{3.5}$$

As a special case consider an exact symplectic form,

$$\omega_P = -d\theta_P, \tag{3.6}$$

where θ_P is a 1–form on P. If the action of G on P preserves θ_P then it is Hamiltonian,

---

[2]With the sign convention usually adopted in physics this mapping is a Lie algebra antihomomorphism.

[3]In literature one usually defines momentum maps without the equivariance condition. Since all the momentum maps discussed here are equivariant, we have incorporated this condition in the definition.

and the momentum map is given by the evaluation of θ_P. For each $\xi \in \mathfrak{g}$,

$$J^{\xi} = \langle \theta_P, \xi_P \rangle . \tag{3.7}$$

Let M be a submanifold of P. The pull–back of ω_P to M induces a closed 2–form ω_M on M which may be degenerate. Let N be the subset of $TP|M$ such that, for each $m \in M$, $N_m = N \cap T_mP$ is given by

$$N_m = \{v \in T_mP \mid \omega_P(v,u) = 0 \ \forall \ u \in T_mM\}. \tag{3.8}$$

We denote by K the intersection of N with TM,

$$K = N \cap TM. \tag{3.9}$$

M is an *isotropic submanifold* of (P,ω_P) if $N \supset TM$. If $N \subset TM$, M is a *co–isotropic submanifold*. A *Lagrangian submanifold* is both isotropic and co–isotropic. If $K = \{0\}$, then ω_M is non–degenerate and M is a *symplectic submanifold*.

Assume that *dim* K is constant. Then K is an involutive distribution on M. Let R denote the space of integral manifolds of K and $\pi_{RM} : M \longrightarrow R$ be the canonical projection. If R has a manifold structure such that π_{RM} is a submersion then there exists a unique symplectic form ω_R in R such that

$$\omega_M = \overset{*}{\pi}_{RM} \, \omega_R . \tag{3.10}$$

In this case (M,ω_M) is called a *presymplectic manifold* and (R,ω_R) is the corresponding *reduced symplectic manifold.*[4]

In dynamics we encounter manifolds M endowed with a closed 2–form ω_M, which are not presented as submanifolds of a symplectic manifold. It is convenient to generalize the notions of Hamiltonian vector fields and momentum maps to such manifolds. Given a function $f \in C^{\infty}(M)$, we say that it admits a Hamiltonian vector field if there exists a vector field ξ_M on M such that $\xi_M \rfloor \omega_M = df$. Clearly, f admits a

[4]This construction of the reduced symplectic manifold of a presymplectic manifold is implicit in Dirac [1950], and Bergmann and Goldberg, [1955]. The formulation given here follows Śniatycki and Tulczyjew [1971].

Hamiltonian vector field only if df annihilates the kernel K of ω_M. We shall refer to ξ_M as *a Hamiltonian vector field of* f. It is determined by f up to an arbitrary vector field on M with values in K. Let G be a Lie group acting in M by diffeomorphisms preserving ω_M. A mapping $J_M^{\mathfrak{g}} : M \longrightarrow \mathfrak{g}^*$ such that, for each $\xi \in \mathfrak{g}$, the corresponding vector field ξ_M on M is a Hamiltonian vector field of $J_M^{\xi} = \langle J_M, \xi \rangle$, will be called a *momentum map* for the action of G in (M,ω_M). The action of G in (M,ω_M) is said to be *Hamiltonian* if it admits an equivariant momentum map. As an example consider the case when ω_M is exact, $\omega_M = -\,d\theta_M$, and G acts on M by diffeomorphisms preserving θ_M. Hence, for each $\xi \in \mathfrak{g}$, $0 = £_{\xi_M} \theta_M = \xi_M \lrcorner\, d\theta_M + d\langle \theta_M, \xi_M \rangle$ which implies that ξ_M is a Hamiltonian vector field of $J_M^{\xi} = \langle \theta_M, \xi_M \rangle$. The corresponding momentum map $J_M^{\mathfrak{g}} : M \longrightarrow \mathfrak{g}^*$ is equivariant, and the action of G in (M,ω_M) is Hamiltonian.

Another example is provided by the case when M is a submanifold of a symplectic manifold (P,ω_P), and ω_M is the pull–back of ω_P to M. Suppose we have a Hamiltonian action of G in (P,ω_P) which leaves M stable. Then the induced action of G on M is Hamiltonian. If $J_P^{\mathfrak{g}} : P \longrightarrow \mathfrak{g}^*$ is an equivariant momentum map for the action of G in P, then the pull back of $J_P^{\mathfrak{g}}$ to M is an equivariant momentum map for the action of G in M.

Let $J_P^{\mathfrak{g}} : P \longrightarrow \mathfrak{g}^*$ be a momentum map for a Hamiltonian action of G in (P,ω_P) and M be the zero level of $J_P^{\mathfrak{g}}$,

$$M = (J_P^{\mathfrak{g}})^{-1}(0). \tag{3.11}$$

If 0 is a regular value of $J_P^{\mathfrak{g}}$, then M is a submanifold of P. For each $p \in M$, each $v \in T_pP$ and each $\xi \in \mathfrak{g}$,

$$\omega_P(\xi_P(p),v) = \langle dJ^{\xi}, v \rangle = \langle TJ_P^{\mathfrak{g}}(v), \xi \rangle .$$

Hence, $v \in T_pM$ if and only if $v \lrcorner\, \omega_P$ annihilates all vectors tangent to the orbit of G through p. This means that the set $K \subset TM$, defined by Eq. (3.9), consists of vectors tangent to the G–orbits in M. If G acts freely and properly in M, then the space M/G of the G orbits in M is a quotient manifold of M which coincides with the reduced symplectic manifold of (R,ω_R) of (M,ω_M),

$$R = M/G. \tag{3.12}$$

This presentation of the reduced symplectic manifold as the space of G orbits in $M = J^{-1}(0)$ is called a *Marsden–Weinstein reduction*[5].

We have seen that the space $C^\infty(P)$ of smooth functions on a symplectic manifold (P,ω_P) has the structure of a Lie algebra given by the Poisson bracket. In general, if $\mathcal{B}$ is a commutative, associative algebra, a *Poisson bracket* on $\mathcal{B}$ is an antisymmetric bilinear map $\{,\} : \mathcal{B}\times\mathcal{B} \longrightarrow \mathcal{B}$ satisfying the Jacobi identity

$$\{f,\{g,h\}\} + \{f,\{f,g\}\} + \{g,\{h,f\}\} = 0 \tag{3.13}$$

and the Leibniz identity

$$\{f,gh\} = \{f,g\}h + g\{f,h\} \tag{3.14}$$

for every $f,g,h \in \mathcal{B}$. A commutative associative algebra endowed with a Poisson brachet will be called a *Poisson algebra.* A manifold P such that the space $C^\infty(P)$ has the structure of a Poisson algebra is called a *Poisson manifold.*[6] Clearly, a symplectic manifold is a Poisson manifold. Another example of a Poisson manifold is provided by a manifold P endowed with a closed 2–form ω_P and a submersion $\pi_{TP}: P \longrightarrow T$ such that, for each $t \in T$, the pull–back of ω_P to the fibre $P_t = \pi_{TP}^{-1}(t)$ is symplectic. For every $f,h \in C^\infty(P)$, the Poisson bracket $\{f,h\}$ of f and h at $p \in P$ is given by the value at p of the Poisson bracket in $C^\infty(P_t)$, where $t = \pi_{TP}(p)$, of the restrictions of f and h to P_t,

$$\{f,h\}(p) = \{f|P_t,h|P_t\}(p)\ ; \tag{3.15}$$

here we use the vertical bar to denote the restriction, and the Poisson bracket on the right hand side is evaluated in $C^\infty(P_t)$.

[5]Marsden and Weinstein [1974].

[6]See Jost [1964], where the term canonical manifold is used.

We can extend to Poisson manifolds the notions of a Hamiltonian action of a Lie group and of a momentum map. Let G be a Lie group acting on a Poisson manifold P in such a way that the induced action in $C^\infty(P)$ preserve the Poisson bracket, and there exists a G–equivariant mapping $J_P^{\mathfrak{g}} : P \longrightarrow \mathfrak{g}^*$ such that,for each $\xi \in \mathfrak{g}$ and each of $f \in C^\infty(P)$,

$$\xi_P f = \{J^\xi, f\} \tag{3.16}$$

where $J^\xi = \langle J_P^{\mathfrak{g}}, \xi \rangle$. In this case we say that the *action of G in P is Hamiltonian* and refer to $J_P^{\mathfrak{g}} : P \longrightarrow \mathfrak{g}^*$ as a *momentum map*. Note that the map $\xi \longrightarrow J^\xi$ is a Lie algebra homomorphism of $\mathfrak{g}$ into $C^\infty(P)$. This observation enables us to further generalize the notions of a Hamiltonian action of a Lie group and of a momentum map.

Let $\mathcal{B}$ be a Poisson algebra, which need not be isomorphic to the Poisson algebra of smooth functions of a Poisson manifold, and G a Lie group acting in $\mathcal{B}$ by automorphisms. For each $f \in \mathcal{B}$ and $g \in G$, we denote by $g_{\mathcal{B}}f$ the element of $\mathcal{B}$ associated to f by g. We say that the *action of G in B is Hamiltonian* if there exists a Lie algebra homomorphism $J : \mathfrak{g} \longrightarrow \mathcal{B} : \xi \longrightarrow J^\xi$ such that, for each $\xi \in \mathfrak{g}$, the action of ξ in $\mathcal{B}$ is given by the Poisson bracket with J^ξ, that is

$$\frac{d}{dt}(\exp t\xi)_{\mathcal{B}} f \,|_{t=0} = \{J^\xi, f\}. \tag{3.17}$$

for every $f \in \mathcal{B}$. As before we shall refer to J^ξ as *the momentum associated to* ξ and call J the *momentum homomorphism*.

A Hamiltonian action of a connected Lie group G in a Poisson algebra $\mathcal{B}$ gives rise to a *reduced Poisson algebra* $\mathcal{O}$ defined as follows[7]. Let $\mathcal{J}$ be the ideal in the associative algebra structure of $\mathcal{B}$ generated by the momenta J^ξ, $\xi \in \mathfrak{g}$,

$$\mathcal{J} = \{ \sum f_\xi J^\xi \mid f_\xi \in \mathcal{B}, \xi \in \mathfrak{g}\}, \tag{3.18}$$

[7]The construction below is a generalization of the construction given in Śniatycki and Weinstein [1983].

and $\rho : \mathcal{B} \longrightarrow \mathcal{B}/\mathcal{J}$ the canonical projection. Since $J : \mathfrak{g} \longrightarrow \mathcal{B}$ is a Lie algebra homomorphism it follows that $\mathcal{J}$ is also an ideal in the Lie algebra structure of $\mathcal{B}$. The assumption that G is connected implies that $\mathcal{J}$ is stable under the action of G in $\mathcal{B}$. Hence, the action of G in $\mathcal{B}$ induces an action in $\mathcal{B}/\mathcal{J}$ such that, for each $g \in G$,

$$g_{\mathcal{B}/\mathcal{J}}\, \rho(f) = \rho(g_{\mathcal{B}} f) \, . \tag{3.19}$$

Let $\mathcal{O}$ be the set of G invariant elements of $\mathcal{B}/\mathcal{J}$,

$$\mathcal{O} = \{\rho(f) \in \mathcal{B}/\mathcal{J} \mid g_{\mathcal{B}/\mathcal{J}}\, \rho(f) = \rho(f) \ \forall\, g \in G\}. \tag{3.20}$$

Clearly, $\mathcal{O}$ inherits from $\mathcal{B}/\mathcal{J}$ the structure of an associative algebra.

6.3.1. Proposition. $\mathcal{O}$ inherits from $\mathcal{A}$ a Poisson bracket $\{,\}$ such that, for every $f,h \in \rho^{-1}(\mathcal{O}) \subset \mathcal{B}$

$$\{\rho(f),\rho(h)\} = \rho(\{f,h\}) \tag{3.21}$$

where $\{f,h\}$ is the Poisson bracket of f and h in $\mathcal{B}$.

Proof. Since G is connected, $\rho(\{f,h\})$ is in $\mathcal{O}$ if and only if $\{J^\xi,\{f,h\}\} \in \mathcal{J}$ for every $\xi \in \mathfrak{g}$. The assumption that $\rho(f)$, $\rho(h) \in \mathcal{O}$ implies that $\{J^\xi,f\} = \sum_\zeta f_\zeta\, J^\zeta$ and $\{J_\xi,h\} = \sum_\eta h_\eta\, J^\eta$ for some f_ζ, $h_\eta \in \mathcal{B}$ where ζ, $\eta \in \mathfrak{g}$. Hence,

$$\begin{aligned}
\{J^\xi,\{f,h\}\} &= \{\{J^\xi,f\}\},h\} + \{f,\{J^\xi,h\}\} \\
&= \sum_\zeta(\{\{J^\xi,f_\zeta\}J^\zeta,h\} + \{\{J^\xi,J^\zeta\}f_\zeta,h\}) \\
&\quad + \sum_\eta(\{f,\{J^\xi,h_\eta\}J^\eta\} + \{f,h_\eta\{J^\xi,J^\eta\}\}) \\
&= \sum_\zeta(\{J^\xi,f_\zeta\}\{J^\zeta,h\} + \{\{J^\xi,f_\zeta\}h\}J^\zeta + \\
&\quad + f_\zeta\{\{J^\xi,J^\zeta\},h\}\ \} + \{f_\zeta,h\}\{J^\xi,J^\zeta\}) \\
&\quad + \sum_\eta(\{f,\{J^\xi,h_\eta\}\}J^\eta + \{J^\xi,h_\eta\}\{f,J^\eta\} + \\
&\quad \{f,h_\eta\}\{J^\xi,J^\eta\} + h_\eta\{f,\{J^\xi,J^\eta\}\}) \in \mathcal{J}
\end{aligned}$$

since every term in the sum belongs to $\mathcal{J}$. Moreover, if $f \in \mathcal{J}$, $\{f,h\} \in \mathcal{J}$ so that $\{\rho(f)\rho(h)\}$ is well defined by Eq. (3.21). The Poisson bracket properties of $\{\rho(f),\rho(h)\}$ follow immediately from the corresponding properties for $\{f,h\}$. ■

Consider a special case when $\mathcal{B} = C^{\infty}(P)$ and the momentum homomorphism is given by a momentum map $J_P^{\mathfrak{g}} : P \longrightarrow \mathfrak{g}^*$ corresponding to a Hamiltonian action in P of a connected Lie group G. If $0 \in \mathfrak{g}^*$ is a regular value of $J_P^{\mathfrak{g}}$, then $M = (J_P^{\mathfrak{g}})^{-1}(0)$ is a submanifold of P. We can identify $C^{\infty}(P)/\mathcal{J}$ with $C^{\infty}(M)$ and ρ with the restriction map. If the assumptions of the Marsden–Weinstein reduction are satisfied then the reduced Poisson algebra $\mathcal{O}$ is canonically isomorphic to the Poisson algebra of the reduced symplectic manifold (R, ω_R). If 0 is not a regular value of $J_P^{\mathfrak{g}}$, the space of G orbits in $(J_P^{\mathfrak{g}})^{-1}(0)$ need not have a manifold structure. In this case the reduced Poisson algebra $\mathcal{O}$ may be used to resolve the singularities in $(J_P^{\mathfrak{g}})^{-1}(0)/G$.[8]

[8]See for example Śniatycki and Weinstein [1983], Gotay [1988], and references quoted there.

6.4 Degenerate Lagrangians and constraints.

We can now return to the discussion of dynamics of Lagrangian systems under assumptions weaker than the regularity of the Lagrangian, which we required in Sec. 6.2. A Lagrangian L is said to be *degenerate* if the Legendre transformation $\mathbb{F}L : TQ \longrightarrow T^*Q$ is not a local diffeomorphism. For degenerate Lagrangians not all vectors $v \in TQ$ admit solutions of the Lagrange equations with the initial datum v. The integrability conditions are usually referred to as constraints. Our aim is to separate the constraint and the evolution equations, and to cast the evolution equations in a Hamiltonian form.

Let P denote the image of the Legendre transformation,

$$P = \mathbb{F}L(TQ) . \tag{4.1}$$

We assume that P is a submanifold of T^*Q and that the induced mapping $\pi_{PTQ}: TQ \longrightarrow P$ is a submersion with connected fibres. We shall refer to P as the *primary constraint manifold* [1]. It is endowed with a closed 2–form ω_P given by the pull–back of the canonical symplectic form ω_{T^*Q} by the inclusion map $i_P : P \longrightarrow T^*Q$,

$$\omega_P = i_P^* \, \omega_{T^*Q} \, . \tag{4.2}$$

The energy function E on TQ pushes forward to a function H on P which is called the *Hamiltonian function* corresponding to L,

$$E = H \circ \pi_{PTQ} \, . \tag{4.3}$$

For each $p \in P$, we consider vectors $w \in T_pP$ satisfying the equation

$$w \,\lrcorner\, \omega_P = dH(p) \, . \tag{4.4}$$

6.4.1. Proposition. A curve $t \longrightarrow q(t)$ in Q satisfies the Lagrange equations if and only if the tangent vector to the curve $t \longrightarrow \pi_{PTQ}(\dot{q}(t))$, where $\dot{q}(t)$ is the tangent vector to $t \longrightarrow q(t)$, satisfies Eq. (4.4).

[1]This terminology is due to P.A.M. Dirac, [1950], who was the first to study degenerate Lagrangian systems.

Proof. Since π_{PTQ} is a submersion, a vector $w \in T_pP$ satisfies Eq. (4.4) if and only if there exists $v \in \pi_{PTQ}^{-1}(p)$ and a vector $u \in T_vTQ$ such that $w = T\pi_{PTQ}(u)$, and

$$u \lrcorner\, \omega_{TQ} = dE(v) \tag{4.5}$$

where $\omega_{TQ} = (\mathbb{F}L)^* \omega_{T^*Q}$ as is Sec. 6.2. If (q^i) are local coordinates in Q and (q^i, v^i) are the induced coordinates in TQ, a vector $u = u^i \frac{\partial}{\partial v^i} + w^i \frac{\partial}{\partial q^i} \in T_vTQ$ satisfies Eq. (4.5) if and only if Eqs. (2.13) and (2.14) are satisfied. If u is the tangent vector of $t \longrightarrow \dot{q}(t)$, then $w^j = v^j = \dot{q}^j(t)$, so that Eq. (2.13) is satisfied and Eq. (2.14) is equivalent to the Lagrange equation (2.2). ∎

According to Proposition. 6.5.1 we can split the problem of finding solutions of the Lagrange equations into two steps. In the first step we look for curves $t \longrightarrow p(t)$ in P such that their tangent vectors $\dot{p}(t)$ satisfy Eq. (4.4). That is, we look for solutions of the differential equation

$$\dot{p}(t) \lrcorner\, \omega_P = dH(p(t)). \tag{4.6}$$

In the second step we check which solutions $t \longrightarrow p(t)$ of Eq. (4.6) satisfy the condition

$$p(t) = \pi_{PTQ}(\dot{q}(t)) , \tag{4.7}$$

where q(t) is the projection of p(t) to Q. Eq. (4.7) is known as the *second order equation condition.*[2] It is rather cumbersome from the point of view of the Hamiltonian formalism. However, there is as class of Lagrangians for which a curve $t \longrightarrow q(t)$ in Q satisfies the Lagrange equations if and only if it is the projection to Q of a curve $t \longrightarrow p(t)$ in P satisfying Eq. (4.6), and we are going to restrict our considerations to such Lagrangians. In this case one does not have to be concerned with the second order equation condition and can concentrate on Eq. (4.6).

---

[2]The second order differential equation condition was first studied in Gotay and Nester [1980]. For more recent developments see Cariñena and Lopez [1987], and the references quoted there.

If ω_P is symplectic then, for every $p \in P$, there is a unique solution of Eq. (4.6) passing through p. If ω_P is not symplectic, Eq. (4.6) admits solutions only for some initial data.

6.4.2. Definition. The *constraint set* in P is the set C of points in P which lie on curves $t \longrightarrow p(t)$ satisfying Eq. (4.6).

A determination of the constraint set is one of the main problems in the study of dynamics of systems with degenerate Lagrangians. The main difficulty with the constraint set is that in many interesting cases it is not a manifold but a variety with singular points corresponding to physically important solutions.

A way of finding, or at least approximating, the constraint set is provided by the *Gotay–Nester constraint algorithm* outlined below.[3] Set $C_1 = P$ and denote by C_2 the set of all points $p \in P$ such that Eq. (4.4) admits a solution at p. Introducing

$$K = \{w \in TP \mid w \,\lrcorner\, \omega_P = 0\} \tag{4.8}$$

we can characterize C_2 as follows:

$$p \in C_2 \iff \langle dH(p), w\rangle = 0 \;\; \forall \;\; w \in K \cap T_pP\,. \tag{4.9}$$

The set C_2 is called the *secondary constraint set.* The *tertiary constraint set* is the set C_3 of all points $p \in C_2$ which admit smooth curves $t \longrightarrow p(t)$ in P such that $p(t) \in C_2$ for all t in a neighbourhood of 0, $p(0) = p$ and the tangent vector $\dot{p}(0)$ satisfies Eq. (4.4). By induction we can construct a sequence C_n of subsets $\{C_n\}$ of P such that $p \in C_{n+1}$ if and only if there exists a smooth curve $t \longrightarrow P(t)$ in P which is contained in C_n and such that $p(0) = p$ and $\dot{p}(0)$ satisfies Eq. (5.4). We denote by C_∞ the intersection of all C_n, $n = 1,2,\ldots$. Clearly, the constraint set C is contained in C_∞. In some cases the algorithm terminates in a finite number of steps, and $C = C_n$ for some n. Even if the constraint algorithm does not determine C completely, it provides an upper bound for C.

[3]Gotay and Nester [1979].

Since P is a submanifold of T^*Q and C is a subset of P, one can characterize C in terms of functions in $C^\infty(T^*Q)$ which vanish on C.[4] Such functions are called *constraints.* They form an ideal $\mathcal{J}$ in the associative algebra structure of $C^\infty(T^*Q)$,

$$\mathcal{J} = \{f \in C^\infty(T^*Q) \mid f|C = 0\}. \tag{4.10}$$

A constraint $f \in \mathcal{J}$ is said to be *first class* if, for every constraint f′, the Poisson bracket of f and f′ is a constraint. First class constraints form an ideal $\mathcal{J}^1$ contained in $\mathcal{J}$,

$$\mathcal{J}^1 = \{f \in \mathcal{J} \mid \{f,f'\} \in \mathcal{J} \ \forall \ f' \in \mathcal{J}\}. \tag{4.11}$$

The Jacobi identity, Eq. (3.15), implies that $\mathcal{J}^1$ is also an ideal of the Lie algebra structure of $C^\infty(T^*Q)$. The constraints in the complement of $\mathcal{J}^1$ are called *second class constraints.* If C is closed, then it is completely determined by the constraints.

The *Dirac constraint algorithm* starts with the ideal $\mathcal{J}_1$ of the associative algebra structure in $C^\infty(T^*Q)$ consisting of functions which vanish on P. Clearly, $\mathcal{J}_1 \subset \mathcal{J}$. The functions in $\mathcal{J}_1$ are called *primary constraints.* Let h be an extension of $H \in C^\infty(P)$ to $C^\infty(T^*Q)$, that is

$$H = h|P. \tag{4.12}$$

Clearly, h is determined by H up to an arbitrary primary constraint. If $n = \dim T^*Q - \dim P$, we can locally extend Eq. (4.6) to a system of equations for a curve $t \longrightarrow p(t)$ in T^*Q such that

$$\dot{p}(t) \,\lrcorner\, \omega_{T^*Q} = d(h + \lambda_1 f_1 + \cdots + \lambda_n f_n), \tag{4.13}$$

$$f_i(p(t)) = 0 \ , \ i = 1,2,\ldots,n \ , \tag{4.14}$$

where $f_1,\ldots,f_n$ are functionally independent primary constraints and $\lambda_1,\ldots,\lambda_n$ are Lagrange multipliers. Every constraint f vanishes along the curve $t \longrightarrow p(t)$ on account of Eq. (4.14). Differentiating $f(p(t)) = 0$ with respect to t and taking into account Eq. (4.13) and the definition of the Poisson bracket, Eq. (3.7), we obtain $\{h + \lambda_1 f_1 + \cdots + \lambda_n f_n, f\}(p(t)) = 0$. Assume that all the primary constraints are first

---

[4]Dirac [1950].

class. Then $\{\lambda_1 f_1 + \cdots + \lambda_n f_n, f\}(p(t))$ vanishes identically, and we are left with the condition $\{h,f\}(p(t)) = 0$. Since this has to hold for every solution of Eq. (4.6) it follows that $\{h,f\} \in \mathcal{J}$. Thus, commutators of constraints with h lead to constraints. Let $\mathcal{J}_2$ be the ideal generated by $\mathcal{J}_1$ and the commutators of h with primary constraints. The constraints in the complement of $\mathcal{J}_1$ in $\mathcal{J}_2$ are called the *secondary constraints.* Similarly, let $\mathcal{J}_3$ be the ideal generated by $\mathcal{J}_2$ and the commutators of h with secondary constraints. The constraints contained in the complement of $\mathcal{J}_2$ in $\mathcal{J}_3$ are called *tertiary constraints.* Following this procedure inductively we obtain a sequence of ideals $\{\mathcal{J}_k\}$ such that $\mathcal{J}_k \subset \mathcal{J}_{k+1}$. As before, we denote by $\mathcal{J}_\infty$ the union of all $\mathcal{J}_k$'s. Clearly, $\mathcal{J}_\infty \subset \mathcal{J}$. If the sequence terminates at some k such that $\mathcal{J}_k = \mathcal{J}_\infty = \mathcal{J}$, then we can obtain all constraints in a finite number of steps. Otherwise, we obtain a lower bound on the constraints which corresponds to an upper bound (in the partial order defined by inclusion of sets) on the constraint set C.

The condition of the Dirac constraint algorithm that the primary constraints should be first class is a condition on the embedding of P into T^*Q, equivalent to the condition that P should be a co–isotropic submanifold of T^*Q, and not a condition on the embedding of C into P which we want to determine. In order to eliminate this condition P.A.M. Dirac proposed a modification of the Poisson bracket. The transition from the Poisson bracket in $C^\infty(T^*Q)$ to the modified bracket, called the *Dirac bracket,* is equivalent to finding a symplectic submanifold S of T^*Q containing P, such that P is as co–isotropic submanifold of S, and replacing the Poisson algebra $C^\infty(T^*Q)$ by the Poisson algebra $C^\infty(S)$[5].

One could eliminate the dependence of the Dirac constraint algorithm on the embedding of P into T^*Q if one worked directly with $C^\infty(P)$. Let us assume that (P,ω_P) is a presymplectic manifold. We denote by (R,ω_R) the corresponding reduced

[5]Śniatycki [1974].

symplectic manifold and by $\pi_{RP} : P \longrightarrow R$ the canonical projection. We assume further that π_{RP} is a trivial fibration with typical fibre A. Choose a projection $\pi_{AP} : P \longrightarrow A$ inducing in P a product structure,

$$P = A \times R. \tag{4.15}$$

Since (R,ω_R) is a symplectic manifold, each fibre of π_{AP} inherits the structure of a symplectic manifold isomorphic to (R,ω_R). Hence, π_{AP}: $P \longrightarrow A$ is a symplectic fibration. It induces in P the structure of a Poisson manifold, and in $C^{\infty}(P)$ the structure of a Poisson algebra.

Let $t \longrightarrow a(t)$ be a curve in F and $t \longrightarrow r(t)$ a curve in R such that

$$\dot{r}(t) \,\lrcorner\, \omega_R = dH(a(t),r(t)). \tag{4.16}$$

If the curve $t \longrightarrow (a(t),r(t))$ in P, identified with $A \times R$, is contained in the secondary constraint set C_2, then it satisfies Eq. (4.6). Thus, the evolution component of Eq. (4.6) can be replaced by Hamiltonian equations in (R,ω_R), or more precisely, in the fibres of π_{AP}. For some Lagrangian systems there exists a choice of π_{AP} such that the condition $(a(t),r(t)) \in C_2$ for all t can be replaced by the condition on the initial data $(a(0),r(0)) \in C$. However, in order to determine an appropriate choice of π_{AP}, we have to discuss the symmetries of the system under consideration. This is best done in terms of the Cartan formalism independent of the choice of trivialization of the configuration bundle $\pi_{TQ} : Q \longrightarrow T$.

6.5 Cartan equations and symmetries.

In order to express the Euler–Lagrange equations for the action integral $A[\sigma_{QT}]$ is terms of exterior differential forms, it is convenient to modify appropriately the Lagrangian form Λ. Recall that the canonical 1–form of the jet bundle $J^1\mathbb{Q}$, denoted here by $\mathbb{Z}$, is a linear mapping $\omega_1 : T\mathbb{Z} \longrightarrow T\mathbb{Q}$ such that, for each local section $\sigma_{\mathbb{Q}T}$, each $z \in range\, j^1\sigma_{\mathbb{Q}T}$, and each $u \in T_z\mathbb{Z}$,

$$\omega_1(u) = T\pi_{\mathbb{Q}\mathbb{Z}}(u) - T\sigma_{\mathbb{Q}T}(T\pi_{T\mathbb{Z}}(u)), \tag{5.1}$$

c.f. Sec. 2.9. Moreover, a section $\sigma_{\mathbb{Z}T}$: $T \longrightarrow \mathbb{Z}$ of the source map $\pi_{T\mathbb{Z}}$ is the jet extension of its projection $\sigma_{\mathbb{Q}T} = \pi_{\mathbb{Q}\mathbb{Z}} \circ \sigma_{\mathbb{Z}T}$ to $\mathbb{Q}$ if and only if $\omega_1 \circ T\sigma_{\mathbb{Z}T} = 0$. Therefore, a modification of the Lagrangian form Λ by a term proportional to ω_1 does not change the value of the action integral, and it leads to an equivalent variational principle.

6.5.1. Proposition. There exists a unique 1–form Ξ on $\mathbb{Z}$ such that

$$\Xi(u) = \Lambda(u) \text{ whenever } u \in ker\, \omega_1 \tag{5.2}$$

$$d\Xi(u,v) = 0 \text{ whenever } u,v \in ker\, \omega_1 \tag{5.3}$$

Proof Eq. (5.2) implies that Ξ differs from Λ by a term proportional to ω_1,

$$\Xi = \Lambda + <\lambda,\omega_1>,$$

where λ is a section of the pull–back of $T^*\mathbb{Q}$ by the target map $\pi_{\mathbb{Q}\mathbb{Z}} : \mathbb{Z} \longrightarrow \mathbb{Q}$. In order to determine λ introduce a local trivialization $\pi_{Q\mathbb{Q}} : \mathbb{Q} \longrightarrow Q$ of $\pi_{T\mathbb{Q}} : \mathbb{Q} \longrightarrow T$, a local coordinate system (q^i) in Q, and the corresponding coordinates (t,q^i,v^i) in $\mathbb{Z}$. We have the following local expressions

$$\Lambda = L(t,q^i,v^j)dt, \tag{5.4}$$

and

$$\omega_1 = \frac{\partial}{\partial q^i} \otimes (dq^i - v^i dt), \tag{5.5}$$

Hence $\Xi = Ldt + \lambda_i(dq^i - v^i dt)$. Substituting this expressions into Eq. (5.3) we obtain

$\lambda_i = \frac{\partial L}{\partial v^i}$ so that

$$\Xi = \frac{\partial L}{\partial v^i} dq^i - (\frac{\partial L}{\partial v^i} v^i - L)dt \tag{5.6}$$

is the unique form satisfying Eqs. (5.2) and (5.3). ∎

We shall refer to the form Ξ given by Eq. (5.6) as the *Cartan form* corresponding to Λ[1]. For time independent Lagrangians the coefficient of dt in Eq. (5.6) is the pull–back by $\pi_{Q\mathbb{Q}}$ of the energy function E defined by Eq. (2.9). Denoting this coefficient by E,

$$E = \frac{\partial L}{\partial v^i} v^i - L , \tag{5.7}$$

we can rewrite the Cartan form as follows

$$\Xi = \frac{\partial L}{\partial v^i} dq^i - Edt \tag{5.8}$$

It should be noted that Ξ is defined independently of any trivialization of $\pi_{T\mathbb{Q}} : \mathbb{Q} \longrightarrow T$. The first term on the right hand side of Eq. (5.8) can be interpreted as the pull–back of Ξ to the fibres of $\pi_{T\mathbb{I}}$, and it coincides with $\mathbb{F}L$. On the other hand the function E given by Eq. (5.7) depends on the choice of a trivialization of $\pi_{T\mathbb{Q}}$.

The construction of the Cartan form ensures that the action integral $A[\sigma_{\mathbb{Q}T}]$ can be rewritten in the form

$$A[\sigma_{\mathbb{Q}T}] = \int (j^1\sigma_{\mathbb{Q}T})^* \Xi . \tag{5.9}$$

6.5.2. Proposition. A section $\sigma_{\mathbb{Q}T} : [t_0,t_1] \longrightarrow \mathbb{Q}$ is a stationary point of the action integral $A[\sigma_{\mathbb{Q}T}]$ with respect to the variations of $\sigma_{\mathbb{Q}T}$ vanishing at t_0 and t_1 if and only if

$$(j^1\sigma_{\mathbb{Q}T})^*(\xi_{\mathbb{I}} \lrcorner d\Xi) = 0 \tag{5.10}$$

for every vector field $\xi_{\mathbb{I}}$ on $\mathbb{I}$.

Proof Variations of sections of $\pi_{T\mathbb{Q}}$ are described by vector fields tangent to the fibres.

---

[1]This form was used by E. Cartan, [1922], in his studies of variational problems with one independent variable.

The corresponding variations of the jet extensions of sections are given by lifts to vector fields to $\mathbb{Z}$ which preserve the canonical form ω_1. Hence, a section $\sigma_{\mathbb{Q}T} : [t_0,t_1] \longrightarrow \mathbb{Q}$ is a stationary point of the action integral with respect to the variations of $\sigma_{\mathbb{Q}T}$ which vanish at t_0 and t_1, if and only if

$$\int (j^1\sigma_{\mathbb{Q}T})^* \pounds_{\xi_{\mathbb{Z}}} \Xi = 0$$

for every vector field $\xi_{\mathbb{Z}}$ on $\mathbb{Z}$ which preserves ω_1 and projects to a vertical vector field on $\mathbb{Q}$ vanishing on the fibres $\overset{-1}{\pi}_{T\mathbb{Q}}(t_0)$ and $\overset{-1}{\pi}_{T\mathbb{Q}}(t_1)$. Substituting

$$\pounds_{\xi_{\mathbb{Z}}} \Xi = d(\xi_{\mathbb{Z}} \lrcorner\, \Xi) + \xi_{\mathbb{Z}} \lrcorner\, d\Xi$$

and using Stokes Theorem, the boundary conditions on $\xi_{\mathbb{Z}}$ and Eq. (5.2) we obtain

$$\int (j^1\sigma_{\mathbb{Q}T})^* (\xi_{\mathbb{Z}} \lrcorner\, d\Xi) = 0.$$

Eq. (5.3) ensures that the vanishing of this integral for all $\xi_{\mathbb{Z}}$ preserving ω_1 and projecting to vector fields in *ker* $T\pi_{T\mathbb{Q}}$ implies that the integral vanishes for all vector fields on $\mathbb{Z}$.[2] Using the Fundamental Theorem in the Calculus of variation we obtain Eq. (5.10). ■

By a *symmetry* of a Lagrangian theory we mean here a triplet $g = (g_T, g_{\mathbb{Q}}, g_{\mathbb{Z}})$ of diffeomorphisms of $T, \mathbb{Q}$, and $\mathbb{Z}$, respectively, such that

$$\overset{*}{g}_{\mathbb{Z}}\Lambda = \Lambda \tag{5.11}$$

$$\omega_1 \circ Tg_{\mathbb{Z}} = Tg_{\mathbb{Q}} \circ \omega_1 \tag{5.12}$$

and the following diagram commutes

[2]For details see Śniatycki (1970a).

$$\begin{array}{ccc} \mathbb{Z} & \xrightarrow{g_{\mathbb{Z}}} & \mathbb{Z} \\ {\scriptstyle \pi_{\mathbb{Q}\mathbb{Z}}}\downarrow & & \downarrow{\scriptstyle \pi_{\mathbb{Q}\mathbb{Z}}} \\ \mathbb{Q} & \xrightarrow{g_{\mathbb{Z}}} & \mathbb{Q} \\ {\scriptstyle \pi_{\mathrm{T}\mathbb{Q}}}\downarrow & & \downarrow{\scriptstyle \pi_{\mathrm{T}\mathbb{Q}}} \\ \mathrm{T} & \xrightarrow{g_{\mathrm{T}}} & \mathrm{T} \end{array}$$

This is a very restrictive notion of a symmetry, but it suffices for our purposes. Symmetries form a Lie group $\mathcal{G}$. The Lie algebra $\mathfrak{g}$ of $\mathcal{G}$ consists of triplets $\xi = (\xi_{\mathrm{T}}, \xi_{\mathbb{Q}}, \xi_{\mathbb{Z}})$ of vector fields on T,$\mathbb{Q}$ and $\mathbb{Z}$, respectively which are related by the corresponding projection maps and preserve the Lagrangian form Λ and the canonical form ω_1.

Since the Cartan form Ξ is constructed in terms of Λ and ω_1, it follows that it is preserved by symmetries. In particular, for every $\xi \in \mathfrak{g}$,

$$\pounds_{\xi_{\mathbb{Z}}} \Xi = 0 . \tag{5.13}$$

This is equivalent to

$$\xi_{\mathbb{Z}} \lrcorner (-d\Xi) = d\langle \Xi, \xi_{\mathbb{Z}} \rangle . \tag{5.14}$$

If $-d\Xi$ were a symplectic form on Z, Eq. (5.14) would imply that the action of the symmetry group on $\mathbb{Z}$ is Hamiltonian with the momentum map given by the evaluation of Ξ on $\xi_{\mathbb{Z}}$, c.f. Sec. 6.3. However, $d\Xi$ is degenerate, and Eq. (5.14) corresponds to a generalization of the notion of a Hamiltonian action of a Lie group. We shall refer to the value of Ξ on $\xi_{\mathbb{Z}}$ as the *momentum* corresponding to an infinitesimal symmetry ξ.

6.5.3. Proposition. Let ξ be an infinitesimal symmetry of the theory. For each section $\sigma_{\mathbb{Z}\mathrm{T}}$ of the source map $\pi_{\mathrm{T}\mathbb{Z}} : \mathbb{Z} \longrightarrow \mathrm{T}$ such that

$$\sigma^*_{\mathbb{Z}\mathrm{T}}(\zeta_{\mathbb{Z}} \lrcorner d\Xi) = 0 \tag{5.15}$$

for all vector fields $\zeta_{\mathbb{Z}}$ on $\mathbb{Z}$, the momentum $\langle \Xi, \xi_{\mathbb{Z}} \rangle$ is constant on the image of $\sigma_{\mathbb{Z}\mathrm{T}}$,

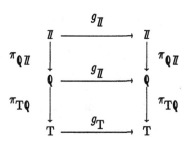

$$d[<\Xi,\xi_{\mathbb{Z}}> \circ \sigma_{\mathbb{Z}T}] = 0\,. \tag{5.16}$$

Proof Pull back Eq. (5.14) by $\sigma_{\mathbb{Z}T}$ and use Eq. (5.15) to obtain Eq. (5.16). ∎

Substituting $\sigma_{\mathbb{Z}T} = j^1\sigma_{QT}$ in Eq. (5.15) we obtain a special case of the First Noether Theorem.[3]

6.5.4. Corollary. The momenta corresponding to infinitesimal symmetries are constants of motion.

We shall refer to Eq. (5.15) as the *Cartan equation* for the Lagrangian system under consideration. It will play an important role in the Hamiltonian formulation of dynamics discussed in the next section.

In order to formulate an analogue of the Second Noether Theorem we need a notion of localizable infinitesimal symmetries. We say that an *infinitesimal symmetry* ξ *is localizable* if, for each pair U_1,U_2 of open sets in T with disjoint closures, there exist an infinitesimal symmetry ξ' such that

$$\xi'_{\mathbb{Z}}(z) = \xi_{\mathbb{Z}}(z) \;\; \forall \;\; z \in \pi^{-1}_{T\mathbb{Z}}(U_1)\,, \tag{5.17}$$

and

$$\xi'_{\mathbb{Z}}(z) = 0 \;\; \forall \;\; z \in \pi^{-1}_{T\mathbb{Z}}(U_2). \tag{5.18}$$

Localizable infinitesimal symmetries form an ideal $\mathscr{h}$ of the Lie algebra $\mathscr{g}$ of infinitesimal symmetries. We denote by $\mathscr{H}$ the connected subgroup of $\mathscr{G}$ with the Lie algebra $\mathscr{h}$ and refer to it as the group of *localizable symmetries.*

6.5.5. Proposition. The momenta corresponding to infinitesimal localizable symmetries vanish on solutions of the Cartan equations.

Proof Let ξ be an infinitesimal localizable symmetry and $\sigma_{\mathbb{Z}T}$ a solution of the Cartan equations. For each $t \in$ *domain* $\sigma_{\mathbb{Z}T}$, there exist two open sets U_1 and U_2 with disjoint

[3] Noether, [1918].

closures, which are contained in the domain of $\sigma_{\mathbb{I}T}$, and such that $t \in U_2$. Since ξ is localizable, there exists an infinitesimal symmetry ξ' satisfying Eqs. (5.17) and (5.18). Thus $<\Xi,\xi_{\mathbb{I}}> \circ \sigma_{\mathbb{I}T}$ and $\Xi,\xi'_{\mathbb{I}}> \circ \sigma_{\mathbb{I}T}$ agree in U_1, and $<\Xi,\xi'_{\mathbb{I}}>(\sigma_{\mathbb{I}T}(t)) = 0$. Since the momenta corresponding to infinitesimal symmetries are constant along $\sigma_{\mathbb{I}T}$ it follows that $<\Xi,\xi_{\mathbb{I}}>(\sigma_{\mathbb{I}T}(t)) = 0$. Hence,

$$<\Xi,\xi_{\mathbb{I}}> \circ \sigma_{\mathbb{I}T} = 0 \tag{5.19}$$

for every localizable infinitesimal symmetry ξ and every solution $\sigma_{\mathbb{I}T}$ of the Cartan equations. ∎

Setting in Eq. (5.19) $\sigma_{\mathbb{I}T} = j^1\sigma_{\mathbb{Q}T}$ we obtain a special case of the Second Noether Theorem. We shall refer to Propositions 6.5.3 and 6.5.5 as the First and the Second Noether Theorem, respectively.

6.6 Generalized Hamiltonian dynamics

Our aim in this section is to pass from the Cartan formalism discussed in Sec. 6.5 to a generalization of the Hamiltonian formalism in analogy with the discussion of Sec. 6.4[1]. This is possible for theories satisfying some additional hypotheses which will be formulated in the sequel. In order to obtain a formalism which is applicable to general relativity we shall develop the theory without assuming the relationship between the Cartan form and the Lagrangian, given by Prop. 6.5.1[2]. Thus, we are lead to the study of a dynamical system with dynamics given by the equations of motion in the form

$$(j^1\sigma_{QT})^*(\xi_{\mathbb{Z}} \lrcorner\, d\Xi) = 0, \tag{6.1}$$

where Ξ is a 1–form on the first jet bundle $\mathbb{Z}$ of the configuration bundle $\mathbb{Q}$ of the system, which need not satisfy Eq. (5.6). By a *symmetry* of such a system we understand as before a triplet $g = (g_T, g_{\mathbb{Q}}, g_{\mathbb{Z}})$ of diffeomorphisms of T, $\mathbb{Q}$ and $\mathbb{Z}$, respectively, which are intertwined by the projection maps $\pi_{\mathbb{Q}\mathbb{Z}}$, $\pi_{Y\mathbb{Q}}$, and $\pi_{T\mathbb{Z}}$, preserve the canonical 1–form ω_1 of the first jet bundle, c.f. Eq. (5.12), and such that $g_{\mathbb{Z}}$ preserves the Cartan form Ξ,

$$\overset{*}{g_{\mathbb{Z}}}\,\Xi = \Xi\,. \tag{6.2}$$

We denote by $\mathcal{G}$ the symmetry group of the system under consideration.

Substituting

$$\sigma_{\mathbb{Z}T} = j^1\sigma_{\mathbb{Q}T} \tag{6.3}$$

into Eq. (6.1) we obtain

$$\overset{*}{\sigma}_{\mathbb{Z}T}(\xi_{\mathbb{Z}} \lrcorner\, d\Xi) = 0 \tag{6.4}$$

for every vector field $\xi_{\mathbb{Z}}$ on $\mathbb{Z}$. As in Sec. 6.5, we shall refer to Eq. (6.4) for sections $\sigma_{\mathbb{Z}T}$

---

[1]The presentation given here is based on an adaptation to dynamics of the ideas developed for field theory in Gotay, Isenberg, Marsden, Montgomery, Śniatycki, and Yasskin [1988], hereafter referred to as GIMMSY.

[2]The Hilbert Lagrangian in General Relativity depends on the second derivatives of the metric, and the arguments used in Prop. 6.5.1 do not apply. However, one can define intrinsically a form Ξ such that the dynamics of the theory is given by the Cartan equations, c.f. Chapter 9.

of the source map $\pi_{T\mathbb{Z}} : \mathbb{Z} \longrightarrow T$ as the *Cartan equation.* If Ξ is the Cartan form corresponding to a regular Lagrangian, then a section $\sigma_{\mathbb{Z}T}$ satisfying the Cartan equation satisfies also Eq. (6.3) with $\sigma_{\mathbb{Q}T} = \pi_{\mathbb{Q}\mathbb{Z}} \circ \sigma_{\mathbb{Z}T}$. For degenerate Lagrangians Eq. (6.3) has to be imposed in addition to the Cartan equation. It corresponds to the second order differential equation condition discussed in Sec. 6.4. We shall circumvent this condition by adopting the following hypothesis.

6.6.1. Hypothesis. For each section $\sigma_{\mathbb{Z}T}$ satisfying the Cartan equation, the projection $\sigma_{\mathbb{Q}T}$ of $\sigma_{\mathbb{Z}T}$ to $\mathbb{Q}$ satisfies equations of motion (6.1).

This hypothesis guarantees that all solutions of equations of motion of the system can be obtained by projecting solutions of the Cartan equation to the configuration bundle. If the Cartan form Ξ corresponds to a Lagrangian, we obtain in this way all solutions of the Lagrange equations.

Let $\mathbb{N}$ denote the set of vectors in $T\mathbb{Z}$ tangent to the fibres of $\pi_{\mathbb{Q}\mathbb{Z}}$, which annihilate $d\Xi$ and Ξ,

$$\mathbb{N} = \{w \in Ker\, T\pi_{\mathbb{Q}\mathbb{Z}} \mid w \,\lrcorner\, d\Xi = 0 \text{ and } \langle \Xi, w\rangle = 0\} \qquad (6.5)$$

6.6.2. Hypothesis. $\mathbb{N}$ is an involutive distribution on $\mathbb{Z}$ and the space $\mathbb{P}$ of integral manifolds of $\mathbb{N}$ has a manifold structure such that the canonical projection $\pi_{\mathbb{P}\mathbb{Z}} : \mathbb{Z} \longrightarrow \mathbb{P}$ is a submersion. The projection $\pi_{T\mathbb{Z}} : \mathbb{Z} \longrightarrow T$ induces a locally trivial fibration $\pi_{T\mathbb{P}} : \mathbb{P} \longrightarrow T$ such that $\pi_{T\mathbb{Z}} = \pi_{T\mathbb{P}} \circ \pi_{\mathbb{P}\mathbb{Z}}$.

6.6.3. Proposition There exists a unique 1–form Θ on $\mathbb{P}$ such that

$$\Xi = \overset{*}{\pi}_{\mathbb{P}\mathbb{Z}} \Theta . \qquad (6.6)$$

A section $\sigma_{\mathbb{Z}T}$ of $\pi_{T\mathbb{Z}}$ satisfies the Cartan equation if and only if its projection $\sigma_{\mathbb{P}T} = \pi_{\mathbb{P}\mathbb{Z}} \circ \sigma_{\mathbb{Z}T}$ to $\mathbb{P}$ satisfies the equation

$$\overset{*}{\sigma}_{\mathbb{P}T} (\xi_{\mathbb{P}} \,\lrcorner\, d\Theta) = 0 \qquad (6.7)$$

for every vector field $\xi_{\mathbb{P}}$ on $\mathbb{P}$.

The symmetry group $\mathcal{G}$ acts in $\mathbb{P}$ by diffeomorphisms preserving Θ.

Proof follows directly from the definition of $\mathbb{P}$ and the assumption that $\pi_{\mathbb{P}\mathbb{I}}$ is a submersion. ∎

The bundle $\mathbb{P}$ over T will be called the *primary constraint bundle* of the theory, and Θ the *Hamilton form* corresponding to the Cartan form Ξ. We shall refer to Eq. (6.7) as the *Hamilton–Cartan equation.*

In order to relate the construction given here to the discussion of Sec. 6.4, consider the Cartan form given by Eq. (5.8), where L is a time independent Lagrangian. In this case

$$\begin{aligned} d\Xi = {} & \frac{\partial^2 L}{\partial v^i \partial v^j} dv^i \wedge dq^j + \frac{\partial^2 L}{\partial q^i \partial v^j} dq^i \wedge dq^j \\ & - \frac{\partial E}{\partial v^i} dv^i \wedge dt - \frac{\partial E}{\partial q^i} dq^i \wedge dt \,, \end{aligned} \tag{6.8}$$

and a vector $w = w^i \frac{\partial}{\partial v^i} \in \mathbb{N}$ if and only if

$$w^i \frac{\partial^2 L}{\partial v^i \partial v^j} = 0 \tag{6.9}$$

and

$$w^i \frac{\partial E}{\partial v^i} = 0 \,. \tag{6.10}$$

Taking into account Eq. (5.7) we see that Eq. (6.10) is a consequence of Eq. (6.9). On the other hand, Eq. (6.9) is equivalent to the condition that w is tangent to the fibres of the map $(\mathbb{F}L \times id_T) : TQ \times T \longrightarrow T^*Q \times T$, where we have identified $\mathbb{I}$ with $TQ \times T$ by means of the trivialization $\pi_{Q\mathbb{Q}}$ of $\pi_{T\mathbb{Q}} : \mathbb{Q} \longrightarrow T$ used here. The assumption made in Sec. 6.2 that the fibres of the Legendre transformation $\mathbb{F}L$ are connected implies that they can be identified with the integral manifolds of $\mathbb{N}$. Hence, the primary constraint bundle can be identified with the product of the image P of the Legendre transformation and the time axis T,

$$\mathbb{P} = P \times T \,. \tag{6.11}$$

For each $t \in T$, the fibre $P_t = \pi_{T\mathbb{P}}^{-1}(t)$ of $\pi_{T\mathbb{P}}$ over t can be interpreted as the *primary constraint manifold at time* t. The Hamilton form Θ is given by

$$\Theta = p_i dq^i - H\,dt, \qquad (6.12)$$

where H is the Hamiltonian on P corresponding to L, c.f. Eq. (4.3), and p_i are the restrictions to P of the coordinates on the fibres of T^*Q used in the proof of Prop. 6.4.1. Comparing Eq. (6.8) to Eq. (2.12) we see that a section $\sigma_{\mathbb{P}T}$ satisfies the Hamilton–Cartan equation if and only if vectors tangent to the its image satisfy Eq. (4.4).

Solutions of the initial value problem for the Hamilton–Cartan equation (6.7) need not exist, and if they exist they need not be unique. The lack of uniqueness is characterized by a subset $\mathbb{K}$of $T\mathbb{P}$ consisting of vectors tangent to the fibres of $\pi_{T\mathbb{P}}$ which annihilate the restrictions of $d\Theta$ to the fibres of $\pi_{T\mathbb{P}}$,

$$\mathbb{K} = \{w \in Ker\, T\pi_{T\mathbb{P}} \mid \langle w \lrcorner, v\rangle = 0 \;\forall\, v \in Ker\, T\pi_{T\mathbb{P}}\} \qquad (6.13)$$

6.6.4. Hypothesis. $\mathbb{K}$ is an involutive distribution on $\mathbb{P}$, and the space $\mathbb{R}$ of integral manifolds of $\mathbb{K}$ has a manifold structure such that the canonical projection $\pi_{\mathbb{R}\mathbb{P}} : \mathbb{P} \to \mathbb{R}$ is a submersion. The projection $\pi_{T\mathbb{P}} : \mathbb{P} \to T$ induces a locally trivial fibration $\pi_{T\mathbb{R}} : \mathbb{R} \to T$ such that $\pi_{T\mathbb{P}} = \pi_{T\mathbb{R}} \circ \pi_{\mathbb{R}\mathbb{P}}$.

6.6.5. Proposition. For each $t \in T$, the fibre $R_t = \pi_{T\mathbb{R}}^{-1}(t)$ has a unique symplectic form ω_{R_t} such that its pull–back ω_{P_t} to $P_t = \pi_{TP}^{-1}(t)$ coincides with the negative of the pull–back of $d\Theta$ to P_t. The symmetry group $\mathcal{G}$ acts in $\mathbb{R}$ by diffeomorphisms which induce symplectomorphisms of the fibres of $\pi_{T\mathbb{R}}$.

Proof. It is the direct consequence of the definition of $\mathbb{R}$, and assumptions that $\pi_{\mathbb{R}\mathbb{P}}$ is a submersion and $\pi_{T\mathbb{R}}$ is a locally trivial fibration. ∎

We shall refer to $\mathbb{R}$ as the *primary reduced phase bundle.* For each $t \in T$, (R_t, ω_{R_t}) is the reduced symplectic manifold of the presymplectic manifold (P_t, ω_{P_t}). If the Cartan form Ξ of the theory is given by Eq. (5.6), where L is a time independent Lagrangian, then $\mathbb{R} = R \times T$, where R is the reduced symplectic manifold of the primary constraint manifold (P, ω_P) introduced in Sec. 6.4.

In the following we shall restrict our considerations to systems which satisfy the hypotheses made above. Such systems will be called quasi–regular. Thus, we are lead to the following definition.

6.6.6. Definition. A dynamical system given by a Cartan form Ξ on $\mathbb{Z}$ is said to be *quasi–regular* if Hypotheses 6.6.1, 6.6.2 and 6.6.4 are satisfied.

According to the discussion at the end of Sec. 6.4, an appropriately chosen product structure in R may enable one to present the evolution component of the generalized Hamilton equations as the Hamilton equations with respect to the symplectic form ω_R. We are now in position to state a hypothesis which ensures the existence of such a product structure.

6.6.7. Atlas Hypothesis. There exists a locally trivial fibration $\pi_{T\mathbb{A}}: \mathbb{A} \rightarrow T$ and a mapping $\pi_{\mathbb{A}\mathbb{P}} : \mathbb{P} \rightarrow \mathbb{A}$ such that the following conditions are satisfied.

(i) The diagram

$$\begin{array}{ccc} \mathbb{P} & \xrightarrow{\pi_{\mathbb{R}\mathbb{P}}} & \mathbb{R} \\ {\scriptstyle \pi_{\mathbb{A}\mathbb{P}}} \downarrow & & \downarrow {\scriptstyle \pi_{T\mathbb{R}}} \\ \mathbb{A} & \xrightarrow[\pi_{T\mathbb{A}}]{} & T \end{array}$$

commutes, and it defines in $\mathbb{P}$ a structure of the product of bundles over T,

$$\mathbb{P} = \mathbb{A} \times_T \mathbb{R}\,, \qquad (6.14)$$

which is stable under the action of the group $\mathscr{G}$ of symmetries of the theory.

(ii) For each $t \in T$, and each pair (a,a') of elements of $A_t = \pi_{T\mathbb{A}}^{-1}(t)$, there exists a localizable symmetry $g \in \mathscr{H}$ such that

$$g_{\mathbb{P}}(a,r) = (a',r) \qquad (6.15)$$

for every $r \in R_t$, where $g_{\mathbb{P}} : \mathbb{P} \longrightarrow \mathbb{P}$ is the action of g in $\mathbb{P}$.

Since the fibres of $\pi_{T\mathbb{R}} : \mathbb{R} \longrightarrow T$ are symplectic, and $\mathbb{P}$ is the fibre product of $\mathbb{A}$ and $\mathbb{R}$, it follows that the fibres of $\pi_{\mathbb{A}\mathbb{P}} : \mathbb{P} \longrightarrow \mathbb{A}$ are symplectic. For each $a \in \mathbb{A}$, the fibre $P_a = \pi_{\mathbb{A}\mathbb{P}}^{-1}(a)$ equals to $\{a\} \times R_t$, where $t = \pi_{T\mathbb{A}}(a)$, and the symplectic form ω_{R_t} in R_t induces a symplectic form ω_{P_a} in P_a. Clearly, ω_{P_a} coincides with the negative of the pull–back of $d\Theta$ to P_a. Let

$$hor\, T\mathbb{P} = \{w \in T\mathbb{P} \mid \langle w \lrcorner d\Theta, v\rangle = 0 \ \forall\ v \in ker\, T\pi_{\mathbb{A}\mathbb{P}}\}. \qquad (6.16)$$

6.6.8. Proposition. *hor* $T\mathbb{P}$ is a distribution on $\mathbb{P}$ transverse to the fibres of $\pi_{\mathbb{A}\mathbb{P}}$.

Proof Since $d\Theta$ pulls–back to symplectic forms on the fibres of $\pi_{\mathbb{A}\mathbb{P}}$ it follows that *hor* $T\mathbb{P} \cap ker\, T\pi_{\mathbb{A}\mathbb{P}} = 0$. Moreover, for each $p \in \mathbb{P}$, and each $u \in T_a\mathbb{A}$, where $a = \pi_{\mathbb{A}\mathbb{P}}(p)$, there is a unique $w \in T_p\mathbb{R}$ such that $T\pi_{\mathbb{A}\mathbb{P}}(w) = u$ and $\langle w \lrcorner d\theta, v\rangle = 0$ for each $v \in ker\, T\pi_{\mathbb{A}\mathbb{P}}$. Hence, *hor* $T\mathbb{P}$ is a distribution on $\mathbb{P}$ transverse to the fibres of $\pi_{\mathbb{A}\mathbb{P}}$. ∎

The distribution *hor* $T\mathbb{P}$ can be thought of as a connection in the bundle $\pi_{\mathbb{A}\mathbb{P}} : \mathbb{P} \longrightarrow \mathbb{K}$ For each vector field $\xi_{\mathbb{P}}$ on $\mathbb{P}$, we denote by *hor* $\xi_{\mathbb{P}}$ the horizontal component of $\xi_{\mathbb{P}}$, and by *ver* $\xi_{\mathbb{P}}$ its vertical component.

Let $\sigma_{\mathbb{P}T}$ be a section of $\pi_{T\mathbb{P}}: \mathbb{P} \longrightarrow T$ satisfying the Hamilton–Cartan equations, $\overset{*}{\sigma}_{\mathbb{P}T}(\xi_{\mathbb{P}} \lrcorner d\theta) = 0$ for every vector field $\xi_{\mathbb{P}}$ on $\mathbb{P}$. Restricting this equation to vector fields $\xi_{\mathbb{P}}$ tangent to the fibres of $\pi_{\mathbb{A}\mathbb{P}}$, we see that the tangent bundle space of the image of $\sigma_{\mathbb{P}T}$ is contained in *hor* $T\mathbb{P}$. Moreover, it follows from the construction of $\mathbb{R}$ and the identification of the fibres of $\pi_{T\mathbb{A}}$ with the corresponding fibres of $\pi_{\mathbb{R}\mathbb{P}}$ that the time evolution of the variables in the fibres of $\pi_{T\mathbb{P}}$ is not determined by the Hamilton–Cartan equation. Following Isenberg we shall refer to the variables in $\mathbb{A}$ as *atlas variables* and to $\mathbb{A}$ as the *atlas bundle* of the theory.[3]

6.6.9. Definition. The *constraint set* in $\mathbb{P}$ is the set $\mathbb{C}$ of all points $p \in \mathbb{P}$ which lie on local sections $\sigma_{\mathbb{P}T}$ satisfying the Hamilton–Cartan equation.

6.6.10. Theorem. Let ξ be an infinitesimal symmetry of a quasi–regular system satisfying Atlas Hypothesis, and

$$\xi_{\mathbb{P}} = ver\ \xi_{\mathbb{P}} + hor\ \xi_{\mathbb{P}} \tag{6.17}$$

the decomposition of $\xi_{\mathbb{P}}$ into its vertical and horizontal components given by the distribution *hor* $T\mathbb{P}$ on $\mathbb{P}$. The following hold.

(i) For each $a \in \mathbb{A}$, the restriction $ver\ \xi_{\mathbb{P}} \mid P_a$ of $ver\ \xi_{\mathbb{P}}$ to $P_a = \pi_{\mathbb{A}\mathbb{P}}^{-1}(a)$ is the Hamiltonian vector field of $<\Theta,\xi_{\mathbb{P}}>$ restricted to P_a,

$$(\text{ver}\ \xi_{\mathbb{P}} \mid P_a) \lrcorner\ \omega_{P_a} = d(<\Theta,\xi_{\mathbb{P}}> \mid P_a). \tag{6.18}$$

(ii) If $\xi_{\mathbb{P}}$ is transverse to the fibres of $\pi_{T\mathbb{P}}$ then, for each p in the constraint set $\mathbb{C}$, the integral curve of $hor\ \xi_{\mathbb{P}}$ through p gives rise to a local section $\sigma_{\mathbb{P}T}$ of $\pi_{T\mathbb{P}}$ which satisfies the Hamilton–Cartan equation.

Proof. (i). For every vector field $\zeta_{\mathbb{P}}$ on $\mathbb{P}$ tangent to the fibres of $\pi_{\mathbb{A}\mathbb{P}}$, the definition of *hor* $T\mathbb{P}$ yields

$$-<ver\ \xi_{\mathbb{P}} \lrcorner\ d\Theta,\ \zeta_{\mathbb{P}}> = -<\xi_{\mathbb{P}} \lrcorner\ d\Theta,\ \zeta_{\mathbb{P}}> = \xi_{\mathbb{P}} \lrcorner\ d<\Theta,\xi_{\mathbb{P}}>.$$

[3]J. Isenberg, [1980], introduced the term "atlas field" on the basis of the analogy with an atlas of charts on a manifold. However, it seems that the term "atlas variable", or "atlas field" in field theory, is suitable also for another reason: each atlas variable $a \in \mathbb{A}$ supports a symplectic "world" (P_a, ω_{P_a}).

Restricting this equation to P_a, and noting that ω_{P_a} is the negative of the pull–back of $d\theta$ to P_a, we obtain Eq. (6.18).

(ii) Let $c_{\mathbb{P}}$ be an integral curve of *hor* $\xi_{\mathbb{P}}$ through a point $p \in \mathbb{C}$, and $\sigma_{\mathbb{P}T}$ the corresponding local section of $\pi_{T\mathbb{P}}$. For each $t \in$ *domain* $c_{\mathbb{P}}$, $c_{\mathbb{P}}(t) = \sigma_{\mathbb{P}T}(\pi_{T\mathbb{P}}(c_{\mathbb{P}}(t))$. Since $p \in \mathbb{C}$, there exists a local section $\sigma'_{\mathbb{P}T}$ of $\pi_{T\mathbb{P}}$ passing through p and satisfying the Hamilton–Cartan equation. We denote by $c'_{\mathbb{P}}$ the curve in $\mathbb{P}$ defined by $\sigma'_{\mathbb{P}T}$ and the parametrization of $c_{\mathbb{P}}$,

$$c'_{\mathbb{P}}(t) = \sigma'_{\mathbb{P}T}(\pi_{T\mathbb{P}}(c_{\mathbb{P}}(t)))$$

for every $t \in$ *domain* $c_{\mathbb{P}}$. Since $\sigma'_{\mathbb{P}T}$ satisfies the Hamilton–Cartan equation, it follows that the tangent vector $\dot{c}'_{\mathbb{P}}$ of $c'_{\mathbb{P}}$ annihilates $d\theta$ and, therefore, $\dot{c}'_{\mathbb{P}}$ is contained in *hor* $T\mathbb{P}$. Hence $c_{\mathbb{P}}$ and $c'_{\mathbb{P}}$ are horizontal curves in $\mathbb{P}$ (over $\mathbb{A}$) projecting to the same curve c_T in T. This implies that their projections to $\mathbb{R}$ coincide. The second part of Atlas Hypothesis, 6.6.6 (ii), ensures that, for each $t \in$ *domain* $c_{\mathbb{P}}$, there exists a symmetry g such that $g_{\mathbb{P}}(c'_{\mathbb{P}}(t)) = c_{\mathbb{P}}(t)$. Since $\dot{c}'_{\mathbb{P}}(t)$ annihilates $d\Theta$ and g is a symmetry, it follows that $Tg_{\mathbb{P}}(\dot{c}'_{\mathbb{P}}(t)) \lrcorner\, d\Theta = 0$. Moreover, $\dot{c}_{\mathbb{P}} - Tg_{\mathbb{P}}(\dot{c}'_{\mathbb{P}}(t))$ is tangent to the fibres of $\pi_{\mathbb{R}\mathbb{P}}$. The definition of the primary reduced phase bundle $\mathbb{R}$ implies that, for every $v \in$ *ker* $T\pi_{T\mathbb{P}}$, $<(\dot{c}_{\mathbb{P}}(t) - Tg_{\mathbb{P}}(\dot{c}'_{\mathbb{P}}(t))) \lrcorner\, d\Theta, v> = 0$. Hence $(\dot{c}_{\mathbb{P}}(t) \lrcorner\, d\Theta, v) = 0$ for all $v \in$ *ker* $T\pi_{T\mathbb{P}}$ at $c_{\mathbb{P}}(t)$. Since the tangent space to $\mathbb{P}$ at $c_{\mathbb{P}}(t)$ is spanned by $\dot{c}_{\mathbb{P}}(t)$ and the vectors in *ker* $T\pi_{T\mathbb{P}}$, it follows that $\dot{c}_{\mathbb{P}}(t) \lrcorner\, d\Theta = 0$. This argument can be repeated for every $t \in$ *domain* $c_{\mathbb{P}}$. Hence, for every vector field $\zeta_{\mathbb{P}}$ on $\mathbb{P}$, $c^*_{\mathbb{P}}(\zeta_{\mathbb{P}} \lrcorner\, d\Theta) = 0$, which is equivalent to the Hamilton–Cartan equation for $\sigma_{\mathbb{P}T}$. This completes the proof. ∎

Theorem 6.6.10 ensures that, for a quasi–regular system which satisfies Atlas Hypothesis and admits an infinitesimal symmetry ξ such that $\xi_{\mathbb{P}}$ is transverse to the fibres of $\pi_{T\mathbb{P}}$, all solutions of equations of motion (6.1) can be obtained as follows. For each p in the constraint set $\mathbb{C}$, find the integral curve $c_{\mathbb{A}}$ through $\pi_{\mathbb{A}\mathbb{P}}(p)$ of the vector field $\xi_{\mathbb{A}}$ in $\mathbb{A}$ corresponding to the action of ξ in the atlas bundle $\mathbb{A}$. For each $a \in$ *range* $c_{\mathbb{A}}$, find in (P_a, ω_{P_a}) the Hamiltonian vector field of $<\Theta, \xi_{\mathbb{P}}> \mid P_a$. In this way one obtains *ver* $\xi_{\mathbb{P}}$ restricted to $\pi^{-1}_{\mathbb{A}\mathbb{P}}(\textit{range}\ c_{\mathbb{A}})$. The section $\sigma_{\mathbb{P}T}$ of $\pi_{T\mathbb{P}}$ corresponding to the integral curve of *hor* $\xi_{\mathbb{P}} = \xi_{\mathbb{P}} - \textit{ver}\ \xi_{\mathbb{P}}$ satisfies the Hamilton–Cartan equation. Since $\pi_{\mathbb{P}\mathbb{Z}} : \mathbb{Z} \longrightarrow \mathbb{P}$ is a submersion, $\sigma_{\mathbb{P}T}$ lifts to a local section $\sigma_{\mathbb{Z}T}$ of $\pi_{T\mathbb{Z}}$. The construction of $\mathbb{P}$ implies that $\sigma_{\mathbb{Z}T}$ satisfies the Cartan equations. Hypothesis 6.6.1 guarantees that the projection $\sigma_{\mathbb{Q}T}$ of $\sigma_{\mathbb{Z}T}$ to $\mathbb{Q}$ satisfies the equations of motion. Since $\pi_{\mathbb{Q}\mathbb{Z}} = \pi_{\mathbb{Q}\mathbb{P}} \circ \pi_{\mathbb{P}Z}$, it follows that the solution $\sigma_{\mathbb{Q}T}$ of the equations of motion is given by projecting $\sigma_{\mathbb{P}T}$ to $\mathbb{Q}$, $\sigma_{\mathbb{Q}T} = \pi_{\mathbb{Q}\mathbb{P}} \circ \sigma_{\mathbb{P}T}$.

In this way we have split the Cartan equations for a quasi–regular Cartan system satisfying Atlas Hypothesis into constraints and evolution equations corresponding to an infinitesimal symmetry ξ such that $\xi_{\mathbb{P}}$ is transverse to the fibres of $\pi_{T\mathbb{P}}$. Constraints determine the constraint set $\mathbb{C}$ in $\mathbb{P}$. Evolution equations are given by the Hamiltonian equations in the fibres of $\pi_{\mathbb{A}\mathbb{P}}$. For each $a \in \mathbb{A}$, the Hamiltonian function in P_a is given by the restriction of $<\Theta, \xi_{\mathbb{P}}>$ to P_a. If should be noted that the evolution equations are independent of the constraint equations. However, only the solutions of the evolution equations, which pass through points in the constraint set, determine solutions of the Cartan equations. If our system corresponds to a Lagrangian system, then the obtained solutions of the Cartan equations satisfy the corresponding Lagrange equations.

It should be noted that the assumption that there exists an infinitesimal symmetry ξ such that $\xi_{\mathbb{P}}$ is transverse to the fibres of $\pi_{T\mathbb{P}}$ is very restrictive. For a Lagrangian theory, it corresponds to an assumption that there exist local trivializations of the configuration bundle for which the Lagrangian function is independent of time. In the usual Hamiltonian treatment of dynamics of regular Lagrangian systems, one can obtain Hamiltonian equations of evolution also for time dependent Lagrangians. One could generalize these results to quasi–regular systems, but at the cost of a rather complicated formalism which, at this stage of understanding of the theory, would tend to obscure the role played by symmetries. Atlas Hypothesis shows that the infinitesimal symmetry ξ such that $\xi_{\mathbb{P}}$ is transverse to the fibres of $\pi_{\mathbb{TP}}$ need not be unique if it exists. The arbitrariness of $\xi_{\mathbb{P}}$ is related to the action of the symmetry group on the fibres of $\pi_{\mathbb{RP}}$.

6.7 Constraints and reduction.

Points in the primary constraint bundle $\mathbb{P}$ describe classical states of the system. Evolution of the system is given by the Hamilton–Cartan equations. Only the points in the constraint set $\mathbb{C}$ lie on solutions of these equations. Hence, the points in the complement of $\mathbb{C}$ cannot represent states of the physical system described by the theory outlined here. These points are introduced into the description of the system in order to obtain equations of motion in a Hamiltonian form. Symmetries of the theory preserve the constraint set $\mathbb{C}$. It is usually assumed that points in $\mathbb{C}$ which are on the same orbit of the group $\mathscr{H}$ of localizable symmetries represent the same physical space of the system. Hence, the space S of the $\mathscr{H}$ orbits in the constraint set $\mathbb{C}$,

$$\mathrm{S} = \mathbb{C}/\mathscr{H}, \tag{7.1}$$

provides a complete description of the physical states of the system. We shall refer to S as the *space of physical states.*

The aim of this section is to describe the structure of S. This requires a knowledge of the structure of the constraint set $\mathbb{C}$. We saw in Sec. 6.4 that, for time independent Lagrangians, $\mathbb{C} = \mathrm{C} \times \mathrm{T}$ and the constraint set C can be determined, or at least approximated, by the constraint algorithm. An alternative way of determining $\mathbb{C}$ is provided by the Second Noether Theorem, Prop. 6.5.4, which asserts that, for every localizable infinitesimal symmetry ξ,

$$<\Xi,\xi_{\mathbb{Z}}> \circ \sigma_{\mathbb{Z}\mathrm{T}} = 0 \tag{7.2}$$

for every solution $\sigma_{\mathbb{Z}\mathrm{T}}$ of the Cartan equation. Since Ξ pushes forward to Θ and solutions of the Cartan equation project to solutions of the Hamilton–Cartan equation we obtain the following proposition.

6.7.1. Proposition. For every infinitesimal localizable symmetry ξ, the restriction of $<\Theta,\xi_{\mathbb{P}}>$ to the constraint set vanishes,

$$<\Theta,\xi_{\mathbb{P}}> \mid \mathbb{C} = 0\,. \tag{7.3}$$

Proof Pushing forward Eq. (7.2) by $\pi_{\mathbb{P}\mathbb{Z}} : \mathbb{Z} \longrightarrow \mathbb{P}$ we obtain

$$<\Theta,\xi_{\mathbb{P}}> \circ \sigma_{\mathbb{P}T} = 0 \tag{7.4}$$

for every solution $\sigma_{\mathbb{P}T}$ of the Hamilton–Cartan equation. This implies Eq. (7.3). ∎

The evaluation of the Hamilton form Θ on vector fields corresponding to infinitesimal localizable symmetries gives rise to a map J from $\mathbb{P}$ to the dual $\mathscr{h}^*$ of the Lie algebra $\mathscr{h}$ of $\mathscr{H}$ defined as follows. For each $p \in \mathbb{P}$ and each $\xi \in \mathscr{h}$,

$$<J(p),\xi> = <\Theta,\xi_{\mathbb{P}}(p)> . \tag{7.5}$$

Following the terminology introduced in Sec. 6.3, we shall refer to $J : \mathbb{P} \longrightarrow \mathscr{h}^*$ as the momentum map corresponding to the Hamiltonian action of $\mathscr{H}$ in $(\mathbb{P},-d\Theta)$. Prop. 6.7.1 implies that the constraint set $\mathbb{C}$ is contained in the zero level of $J : \mathbb{P} \longrightarrow \mathscr{h}^*$,

$$\mathbb{C} \subset J^{-1}(0) \tag{7.6}$$

For many systems of physical interest the constraint set $\mathbb{C}$ is equal to the zero level of J,

$$\mathbb{C} = J^{-1}(0), \tag{7.7}$$

and we restrict our considerations to such systems. Eq. (7.7) implies that the space S of physical states coinsices with the space of $\mathscr{H}$ orbits in the zero level of the momentum map for the action of $\mathscr{H}$ in $\mathbb{P}$,

$$S = J^{-1}(0)/\mathscr{H}. \tag{7.8}$$

This enables one to study the structure of S in terms of the action of $\mathscr{H}$ in $\mathbb{P}$.

Let $\pi_{S\mathbb{C}} : \mathbb{C} \longrightarrow S$ be the canonical projection associating to each $p \in \mathbb{C}$ the $\mathscr{H}$ orbit through p. The projection map $\pi_{T\mathbb{P}} : \mathbb{P} \longrightarrow T$ induces a map $\pi_{T\mathbb{C}} : \mathbb{C} \longrightarrow T$. Whether $\pi_{T\mathbb{C}}$ induces a map from S to T or not depends on the action of $\mathscr{H}$ in T. We assume here for simplicity that $\mathscr{H}$ acts in T by the identity transformation. This implies that, for each $t \in T$, the fibre $C_t = \pi_{T\mathbb{C}}^{-1}(t)$ is stable under the action of $\mathscr{H}$. Let S_t be the space of $\mathscr{H}$ orbits in C_t,

$$S_t = C_t/\mathscr{H}. \tag{7.9}$$

The space S of $\mathscr{H}$ orbits in $\mathbb{C}$ is the union of the spaces S_t, $t \in T$. Hence, we have a projection map $\pi_{TS} : S \longrightarrow T$ such that $\pi_{T\mathbb{C}} = \pi_{TS} \circ \pi_{S\mathbb{C}}$.

Atlas Hypothesis 6.6.7 ensures that $\mathscr{H}$ acts transitively on the fibres of $\pi_{T\mathbb{A}}: \mathbb{A} \longrightarrow T$. For each $a \in \mathbb{A}$ we denote by C_a the intersection of $\mathbb{C}$ with $P_a = \pi_{\mathbb{A}\mathbb{P}}^{-1}(a)$,

$$C_a = \mathbb{C} \cap P_a, \tag{7.10}$$

and by $\mathscr{H}_a$ the isotropy group of a,

$$\mathscr{H}_a = \{g \in \mathscr{H} \mid g_{\mathbb{A}}(a) = a\}. \tag{7.11}$$

Let $\mathscr{H}_a'$ denote the normal subgroup of $\mathscr{H}_a$ consisting of localizable symmetries which act in P_a by the identity transformation,

$$\mathscr{H}_a' = \{g \in \mathscr{H}_a \mid g_{\mathbb{P}}(p) = p \ \forall\, p \in P_a\}, \tag{7.12}$$

and $\mathscr{H}_a''$ be the quotient group

$$\mathscr{H}_a'' = \mathscr{H}_a \,/\, \mathscr{H}_a' . \tag{7.13}$$

The action of $\mathscr{H}_a$ in P_a induces an action of $\mathscr{H}_a''$. For each $g \in \mathscr{H}_a$, the action in P_a of the corresponding element $[g] \in \mathscr{H}_a''$ is given by the diffeomorphism g_{P_a} of P_a obtained by restricting $g_{\mathbb{P}}$ to P_a. Similarly, for each $[\xi]$ in the Lie algebra $\mathscr{h}_a''$ of $\mathscr{H}_a''$, where $\xi \in \mathscr{h}_a$, the corresponding vector field on P_a is given by the restriction ξ_{P_a} of $\xi_{\mathbb{P}}$ to P_a. For each $\xi \in \mathscr{h}_a$, the restriction of $\xi_{\mathbb{P}} \,\lrcorner\, d\Theta + d\langle\Theta,\xi_{\mathbb{P}}\rangle = 0$ to P_a yields

$$\xi_{P_a} \,\lrcorner\, \omega_{P_a} = d(\langle\Theta,\xi_{\mathbb{P}}\rangle \mid P_a). \tag{7.14}$$

This shows that the action of $\mathscr{H}_a''$ in the symplectic manifold (P_a, ω_{P_a}) is Hamiltonian with a momentum map $J_a : P_a \longrightarrow \mathscr{h}_a''^{*}$ given by

$$\langle J_a(p),[\xi]\rangle = \langle\Theta,\xi(p)\rangle . \tag{7.15}$$

Clearly, the zero level of J_a contains the intersection C_a of $J^{-1}(0)$ with P_a,

$$J_a^{-1}(0) \supset C_a = J^{-1}(0) \cap P_a . \tag{7.16}$$

Assume further that, for each $a \in \mathbb{A}$,

$$C_a = J_a^{-1}(0) . \tag{7.17}$$

Since $\mathscr{H}$ acts transitively on the fibres of $\pi_{T\mathbb{A}}$, it follows that, for each $t \in T$, the space S_t of the $\mathscr{H}$ orbits in C_t can be identified with the space of $\mathscr{H}_a$ orbits in C_a for any

$a \in A_t$. On the other hand, in C_a the $\mathscr{H}_a$ orbits are the same as the $\mathscr{H}_a''$ orbits. Taking into account Eq. (7.17) we obtain an identification of S_t with the space of $\mathscr{H}_a''$ orbits in $J_a^{-1}(0)$,

$$S_t = J_a^{-1}(0)/\mathscr{H}_a'' . \tag{7.18}$$

Consider first the most regular situation. Suppose that $\mathbb{C}$ is a submanifold of $\mathbb{P}$, $\mathbb{R}'$ is a quotient manifold of $\mathbb{C}$ and $\pi_{\mathbb{T}\mathbb{R}'}$ is a submersion. If, for each $a \in \mathbb{A}$ the assumptions of the Marsden–Weinstein reduction are satisfied, that is $0 \in \mathscr{h}_a''^{*}$ is a regular value of J_a, and $\mathscr{H}_a''$ acts freely and properly on $J_a^{-1}(0)$, then the fibres of $\pi_{\mathbb{T}\mathbb{R}'}$ are symplectic manifolds. For each $t \in T$ and each $a \in A_t$, the symplectic form on S_t pulls back to a form on $J_a^{-1}(0)$ which coincides with the pull back of ω_{P_a} by the inclusion map of $J_a^{-1}(0)$ into P_a. Thus S has the structure of a Poisson manifold and the space $C^\infty(S)$ has the structure of a Poisson algebra.

If the assumptions of the Marsden–Weinstein reduction are not satisfied S may have singularities. Nevertheless, for each $t \in T$, we can associate to S_t the reduced Poisson algebra $\mathcal{O}_t$ in the manner described in Proposition 6.3.1. The family $\mathcal{O} = \{\mathcal{O}_t\}_{t \in T}$ has also structure of a Poisson algebra. In the regular case, $\mathcal{O}$ is canonically isomorphic to the Poisson algebra $C^\infty(S)$. In a singular case it can be used to resolve the singularities of S. A more detailed discussion of the reduced Poisson algebra $\mathcal{O}$ is given in Section 7.5 in the context of field theory.

References

ABRAHAM, R. and MARSDEN, J.E. [1978], *Foundations of Mechanics*, 2nd. ed. , Addison–Wesley, Reading, Mass.

BERGMANN, P.G. and GOLDBERG, I. [1954], "Dirac Bracket Transformations in Phase Space", *Phys. Rev.*, 98, 531–538.

CARINENA, J. and LOPEZ, C. [1987], "The Time Evolution for Singular Lagrangians", 14, 203–209.

CARTAN, E. [1922], *Leçons sur les invariants integraux*, Hermann, Paris.

DIRAC, P.A.M. [1950], "Generalized Hamiltonian Dynamics", *Can. J. Math.*, 2, 129–148.

GIMMSY [1988], see Gotay, Isenberg, Marsden, Montgomery, Śniatycki and Yasskin [1988]

GODBILLON, C. [1969], *Géométrie differentielle et mécanique analytique*, Hermann, Paris.

GOTAY, M., [1988], "Reduction of Homogeneous Yang–Mills Fields", preprint, Mathematics Department, United States Naval Academy, Annapolis, MD.

GOTAY, M., ISENBERG, J., MARSDEN, J., MONTGOMERY, R., ŚNIATYCKI, J. an YASSKIN, Ph. [1988], "Momentum Maps and the Hamiltonian Treatment of Classical Field Theories with Constraints", in preparation.

GOTAY, M. and NESTER, J. [1979], "Presymplectic Lagrangian Systems I: the Constraint Algorithm and Equivalence Theorem", *Ann. Inst. H. Poincaré*, A30, 129–142.

GOTAY, M. and NESTER, J. [1980], "Presymplectic Lagrangian Systems II: the Second Order Equation Problem", *Ann.Inst. H. Poincaré*, A32, 1–13.

GUILLEMIN, V. and STERNBERG, S. [1984], *Symplectic Techniques in Physics*, Cambridge Univ. Press, Cambridge.

ISENBERG, J.A. [1980], "The Construction from Initial Data of Spacetimes with Nontrivial Spatial and Bundle Topology", *Ann. Phys.* 129, 223–248.

JOST, R. [1964], "Poisson Brackets (an unpedagogical lecture)", *Rev. Mod. Phys.*, 36, 572–579.

KLEIN, J. [1962], "Espaces variationnels et mécanique", *Ann. Inst. Fourier* (*Grenoble*), 12, 1–124.

LANCZOS, C. [1962], *The Variational Principles* of *Mechanics*, 2nd. ed., Univ. of Toronto Press.

MARSDEN, J. and WEINSTEIN, A. [1974], "Reduction of Symplectic Manifolds with Symmetry", *Rep. Math. Phys.*, 5, 121–130.

NOETHER, E. [1918], "Invariante Variationsprobleme", *Nachr. Ges. Wiss. Göttingen*, 235–257.

ŚNIATYCKI, J. and TULCZYJEW, W.M. [1971], "Canonical Dynamics of Relativistic Charged Particles", *Ann. Inst. H. Poincaré*, 15, 177–187.

ŚNIATYCKI, J. and WEINSTEIN, A. [1983], "Reduction and Quantization for Singular Momentum Mappings", *Lett. Math. Phys.*, 7, 155–161.

SOURIAU, J.–M. [1970], *Structure des systèmes dynamiques*, Dunod, Paris.

SYNGE, J.L. [1960], "Classical Dynamics", *Handbuch. d. Physik*, 3, 1–225.

Chapter 7

DYNAMICS OF CLASSICAL FIELDS

Dynamics of classical fields is usually formulated directly in terms of an action principle. The aim of this chapter is to follow the development of the preceding chapter and to obtain a covariant Hamiltonian formulation of field theory. The usual transition from the Lagrangian to the Hamiltonian formalism requires an a priori splitting of space–time into space and time. This introduces an external element and obscures the symmetries of a relativistic theory. In order to avoid it, we pass to an equivalent action principle in which the Lagrangian is replaced by the corresponding DeDonder form. In this way one obtains the field equations expressed in terms of the exterior differential of the DeDonder form. This form of field equations enables one to make a covariant passage to equations of motion for Cauchy data, provided we consider a space of Cauchy surfaces stable under the action of the symmetry group. Integration of the DeDonder form over Cauchy data gives rise to a 1–form in the space of Cauchy data which is an analogue of the Cartan form in non–relativistic dynamics. The transision from the obtained Cartan formalism to a covariant Hamiltonian formalism follows the steps described in the preceding chapter.

Since dynamics of classical fields is governed by partial differential equations, one has to consider boundary conditions satisfied by the Cauchy data. One finds that the evolution equation for Cauchy data can be cast into a Hamiltonian form only under appropriate boundary conditions. Here we consider a class of boundary conditions which includes the Dirichlet conditions and the Neumann conditions. The choice of boundary conditions determines the symmetry group of the theory.

Another essential difference between field theory and systems with a finite number of degrees of freedom is that in field theory we have to deal with manifolds of maps. Most existence results in field theory are obtained in appropriate Sobolev spaces. However, in manifolds of maps modelled on Sobolev spaces many interesting geometric objects of the theory are only densely defined. This obscures the geometric interpretation of the theory. For this reason we restrict our considerations to the category of smooth maps. In this category we have no implicit function theorem. Hence, the results in the theory of finite dimensional manifolds which depend on the implicit function theorem need not extend. Moreover, all symplectic forms appearing in applications are weak. This implies that not all smooth functions admit Hamiltonian vector fields.

Throughout the chapter we illustrate the constructions discussed here by the Klein–Gordon dynamics of a scalar field. More complicated and interesting examples will be discussed in subsequent chapters.

7.1 Action principle and field equations.

Let X be an oriented manifold with boundary ∂X, $n = \dim X$, and $\pi_{XY}: Y \longrightarrow X$ a fibre bundle over X. We denote by Z the space of first jets of sections of π_{XY},

$$Z = J^1 Y,$$

by $\pi_{YZ}: Z \longrightarrow Y$ the target map, and by $\pi_{XZ}: Z \longrightarrow X$ the source map. Clearly, $\pi_{XZ} = \pi_{XY} \circ \pi_{YZ}$. We denote by ∂Y and ∂Z the boundaries of Y and Z respectively. The fibration $\pi_{XY}: Y \longrightarrow X$ induces a fibration of ∂Y over ∂X denoted by $\pi_{\partial X \partial Y}$. Similarly, ∂Z is fibred over ∂Y adn ∂X with the projection maps $\pi_{\partial Y \partial Z}$ and $\pi_{\partial X \partial Z}$ obtained by the restriction to ∂Z of the target and the source map respectively.

A *Lagrangian form* in an n–form Λ on Z annihilating vectors tangent to the source map $\pi_{XZ}: Z \longrightarrow X$,

$$v \,\lrcorner\, \Lambda = 0 \text{ whenever } T\pi_{XZ}(v) = 0. \tag{1.1}$$

It defines a function A on the space of local sections of π_{XY} with relatively compact domains given by

$$A[\sigma_{YX}] = \int_U (j^1\sigma_{YX})^* \Lambda \tag{1.2}$$

for every local section σ_{YX} of π_{XY} with domain U. The functional A is called the *action integral* corresponding to Λ.

A section $\sigma_{YX}: U \longrightarrow Y$ of π_{XY} is a stationary point of the action integral A if, for each one parameter group g_t of diffeomorphisms of Y such that $\pi_{XY} \circ g_t = \pi_{XY}$ and $g_t(y) = y$ for all y in the boundary of $\sigma_{YX}(U)$, the derivative of $A[g_t \circ \sigma_{YX}]$ with respect to t vanishes at $t = 0$,

$$\frac{d}{dt} A[g_t \circ \sigma_{YX}] \,|_{t=0} = 0 \tag{1.3}$$

Taking into account Eq. (1.2) and differentiating with respect to t under the integral sign we see that a section σ_{YX} of π_{XY} with domain U is a stationary point of the action integral A if and only if, for each vector field ξ_Y on Y such that $T\pi_{XY} \circ \xi_Y = 0$ and $\xi_Y(y) = 0$ for all y in the boundary of $\sigma_{YX}(U)$,

$$\int_U (j^1\sigma_{YX})^* \pounds_{\xi_Z} \Lambda = 0, \tag{1.4}$$

where ξ_Z is the extension of ξ_Y to the vector field on Z preserving the canonical form ω_1 of the first jet bundle, and $\pounds$ denotes the Lie derivative. In field theory dynamically admissible fields are usually given by stationary sections of the action integral corresponding to an appropriately chosen Lagrangian form. This method of selection of dynamically admissible sections is called a *principle of stationary action.*

If X is endowed with a non–degenerate metric, then a Lagrangian form Λ can be expressed as the product of a function L on Z, called a *Lagrangian*, and the pull back to Z of the volume form vol_X on X defined by the metric and the orientation of X,

$$\Lambda = L\, \pi_{XZ}^*(vol_X) \tag{1.5}$$

Moreover, a Lagrangian is usually specified by associating to each section σ_{YX} of π_{XY} a function $L[\sigma_{YX}]$ on X given by the pull back of L by the jet extension of σ_{YX},

$$L[\sigma_{YX}] = (j^1\sigma_{YX})^* L\, .$$

Let (x^μ, y^A, z^A_μ) be a coordinate system in Z adapted to the structure of a jet bundle. That is (x^μ) are local coordinates in X, (y^A) are coordinates in a typical fibre of π_{XY}, and (z^A_μ) are coordinates in a typical fibre of π_{XZ} such that the canonical form ω_1 of $Z = J^1Y$ is given by

$$\omega_1 = (dy^A - z^A_\mu dx^\mu) \otimes \frac{\partial}{\partial y^A}\, .$$

A local expression for Λ is given by

$$\Lambda = \mathcal{L}\, d_n x$$

where $\mathcal{L}$ is a function of the coordinates (x^μ, y^A, z^A_μ) and

$$d_n x = dx^1 \wedge dx^2 \wedge \cdots \wedge dx^n.$$

Note that, if Λ is given by Eq. (1.5), then

$$\mathcal{L} = L\sqrt{|\, det\ (g_{\mu\nu})|}$$

where $(g_{\mu\nu})$ are the components of the metric g_X with respect to the coordinates (x^μ), and *det* denotes the determinant. The corresponding local expression for a section σ_{YX} is given by specifying the coordinates y_A as functions of the coordinates x^μ,

$$y^A = \sigma^A(x^\mu), \tag{1.6}$$

and the jet extension $j^1\sigma_{YX}$ of σ_{YX} is given by the equations

$$y^A = \sigma^A(x^\nu), \quad z^A_\mu = \sigma^A_{,\mu}(x^\nu), \tag{1.7}$$

where $\sigma^A_{,\mu}$ denotes the derivative of σ^A with respect to x^μ. A local expression for a vector field ξ_Y on Y tangent to the fibres of π_{XY} is given by

$$\xi_Y = \xi^A(x^\nu, y^B)\frac{\partial}{\partial y^A},$$

where the summation convention of summing over the repeated indices is adopted. The extension ξ_Z of ξ_Y to Z has a local representation

$$\xi_Z = \xi^A(x^\nu, y^B)\frac{\partial}{\partial y^A} + \xi^A_{,\mu}(x^\nu, y^B)\frac{\partial}{\partial z^A_\mu}$$

substituting these local expressions into Eq. (1.4) we obtain

$$\int_U \left[\xi^A \frac{\partial L}{\partial y^A} + \xi^A_{,\mu}\frac{\partial L}{\partial z^A_\mu}\right] d_n x = 0,$$

where all the functions under the integral sign are evaluated on the range of $j\sigma_{YX}$ given by Eq. (1.7). The condition that ξ_Y should vanish on the boundary of $\sigma_{YX}(U)$ implies that the integration by parts of the second term in the integrand does not give any boundary term, so that we obtain

$$\int_U \xi^A\left[\frac{\partial L}{\partial y^A} - \frac{\partial}{\partial x^\mu}\frac{\partial L}{\partial z^A_\mu}\right] d_n x = 0\,. \tag{1.8}$$

Since ξ_A are arbitrary inside U, the Fundamental Theorem of the Calculus of Variations implies that the integrand in Eq. (1.8) has to vanish. Therefore, a section σ_{YX} of π_{XY}, given by Eq. (1.6), is a stationary point of the action integral if and only if

$$\frac{\partial L}{\partial y^A}(x,\sigma^B(x),\ \sigma^C_\nu(x)) - \frac{\partial}{\partial x^\mu}\left[\frac{\partial L}{\partial z^A_\mu}(x,\sigma^B(x),\sigma^C_\nu(x))\right] = 0 \tag{1.9}$$

at each point x of its domain. Eq. (1.9) is the *Euler–Lagrange equation* corresponding to the action integral A given by Eq. (1.2). We have given it here in terms of local coordinates rather than in terms of an intrinsic geometric construction because the latter is somewhat involved, and in the following we shall not use Euler–Lagrange equations but an equivalent system of equations in exterior differential forms.

Since ω_1 annihilates vectors tangent to jet extensions of sections of π_{XY}, it follows that a modification of Λ off $\wedge^n ker\ \omega_1$ does not change the action integral. An appropriate modification of Λ enables one to replace integration by parts in the derivation of the Euler–Lagrange equation by an application of the Stokes' theorem to an intrinsically defined differential form.

7.1.1 Proposition. There exists a unique form Ω on Z satisfying the following conditions

(i) $\Lambda(u_1,\ldots,u_n) = \Omega(u_1,\ldots,u_n)$
whenever $u_1,\ldots,u_n \in ker\ \omega_1$,

(ii) $u \lrcorner \Omega = 0$ for all $u \in ker\ T\pi_{YZ}$

(iii) $d\Omega(u_1,\ldots,u_{n+1}) = 0$
whenever $u_1,\ldots,u_{n+1} \in ker\ \omega_1$

(iv) $(u_1 \wedge u_2) \lrcorner \Omega = 0$ for all $u_1, u_2 \in ker\ T\pi_{XZ}$.

Proof. Using a local coordinate description of Λ we see that the conditions (i) through (iv) specify Ω uniquely. A local description of Ω is given by

$$\Omega = L d_n x + \frac{\partial L}{\partial z^A_\mu}(dy^A - z^A_\nu\, dx^\nu)\wedge d_{n-1}x_\mu, \tag{1.10}$$

where

$$d_{n-1}x_\mu = \frac{\partial}{\partial x^\mu} \lrcorner\, d_n x\,. \qquad \blacksquare$$

We shall refer to Ω as the *DeDonder form* corresponding to Λ.[1]

Replacing in the action integral the Lagrangian form Λ by the DeDonder fotm Ω, and taking into account the condition (i) of Proposition 7.1.1, we obtain

$$A[\sigma_{YX}] = \int_U (j^1\sigma_{YX})^* \Omega . \tag{1.11}$$

7.1.2 Proposition. A section σ_{YX} is a stationary point of action integral if and only if

$$(j^1\sigma_{YX})^*(\xi_Z \rfloor d\Omega) = 0 \tag{1.12}$$

for every vector field ξ_Z on Z.

Proof Eq. (1.11) implies that we can replace Λ by Ω in Eq. (1.4),

$$\int_U (j^1\sigma_{YX})^*(\pounds_{\xi_Z}\Omega) = 0 ,$$

where ξ_Z is the extension to Z of a vector field ξ_Y tangent to the fibres of π_{XY}. Taking into account the boundary conditions on ξ_Z we can rewrite this equation in the form

$$\int_U (j^1\sigma_{YX})^*(\xi_Z \rfloor d\Omega) = 0 . \tag{1.13}$$

It can be easily shown that, for each $z \in Z$, every vector in T_zZ can be expressed as sum of a vector in *ker* ω_1 and the value at z of the extension to Z of a vector field on Y tangent to the fibres od π_{XY}. Hence, the condition (ii) of Proposition 7.1.1 implies that Eq. (1.13) is valid for all vector fields on Z. Taking into account the Fundamental Theorem in the Calculus of Variations we obtain the thesis of the proposition.[2] ■

---

[1]This form appears in the literature under various names. Tulczyjew, [1968], uses the term "multisymplectic structure", García and Pérez–Rendón, [1969], call it "Poincaré–Cartan form", and Goldschmidt and Sternberg, [1973], use the term "Hamilton–Cartan". In his [1933] paper Cartan attributed this form to DeDonder, [1929], and we use the term "DeDonder form" to distinguish it from the Cartan and the Hamilton forms in mechanics, and their generalizations to field theory introduced in the next section.

[2]Propositions 7.1.1 and 7.1.2 are due to Lepage, [1936], who constructed a family of

We shall refer to Eq. (1.12) as the *field equations* of the theory. They are equivalent to the Euler–Lagrange equation (1.9).

It is convenient to extend Eq. (1.12) to sections σ_{ZX} of the source map π_{XZ} which need not be jet extensions of sections of π_{XY}. That is, we consider sections σ_{ZX} of π_{XZ} such that

$$\sigma_{ZX}{}^{*}(\xi_Z \lrcorner \, d\,\Omega) = 0 \tag{1.14}$$

for every vector field ξ_Z on Z. We shall refer to Eq. (1.14) as the *DeDonder equations* of the theory. For some special choices of Lagrangians, if a section σ_{ZX} satisfies the DeDonder equations, then it is the jet extension of its projection $\sigma_{YX} = \pi_{YZ} \circ \sigma_{ZX}$ to Y, $\sigma_{ZX} = j^1 \sigma_{YX}$, and σ_{YX} is a stationary point of the action integral. However, many Lagrangians used in field theory do not belong to this special class. On the other hand, in all applications considered here the following hypothesis is satisfied.

7.1.3 Hypothesis. A section σ_{YX} satisfies the field equations if and only if there exists a section σ_{ZX} satisfying the DeDonder equations and projecting to σ_{YX}, that is $\sigma_{YX} = \pi_{YZ} \circ \sigma_{ZX}$.

It is a field theory analogue of Hypothesis 6.6.1 imposed on dynamical systems in order to circumvent the second order equation problem.

As an example we consider a scalar field on a four–dimensional space–time X endowed with a Lorentian metric g_X of signature $(-,+,+,+)$. In this case Y is the product of X and the real line, and a section σ_{YX} of $\pi_{XY} : Y \longrightarrow X$ corresponds to a function $\varphi \in C^\infty(X)$. The Klein–Gordon Lagrangian is given by

$$L[\varphi] = \tfrac{1}{2}\{\langle d\varphi | d\varphi \rangle + m^2 \varphi^2\}, \tag{1.15}$$

---

forms satisfying conditions (i) through (iii) leading to a family of theories in the calculus of variations. The presentation given here follows Śniatycki [1970a].

where $<d\varphi|d\varphi>$ is the square of the norm of 1–form $d\varphi$ defined in terms of the metric g_X, and m is a positive parameter. The corresponding Lagrangian form Λ on Z is given by

$$\Lambda = \tfrac{1}{2}\{g^{\mu\nu}z_\mu z_\nu + m^2 y^2\}\sqrt{-det\ (g_{\alpha\beta})}d_4x \tag{1.16}$$

where $(g^{\mu\nu})$ is the inverse of $(g_{\mu\nu})$. The DeDonder form is

$$\Omega=\sqrt{-det\ (g_{\alpha\beta})}\{g^{\mu\nu}z_\mu dy\wedge d_3x_\nu-\tfrac{1}{2}(g^{\mu\nu}z_\mu z_\nu-m^2y^2)d_4x\} \tag{1.17}$$

The Lagrange–Euler equation corresponding to the Lagrangian $L[\varphi]$ is the Klein–Gordon equation

$$g^{\mu\nu}D_\mu D_\nu\varphi - m^2\varphi = 0 \tag{1.18}$$

where D_μ denotes the covariant differentiation with respect to the Levi–Civita connection of g_X. A section σ_{ZX} of the source map $\pi_{XZ} : Z \rightarrow X$ can be locally expressed as

$$y = \varphi(x), \quad z_\mu = \psi_\mu(x). \tag{1.19}$$

It follows from the expression for Ω, and the identity

$$dx^\nu \wedge d_3x_\mu = \delta^\nu_\mu\, d_4x \ , \tag{1.20}$$

that the DeDonder equations yield

$$\psi_\mu - \varphi,_\mu = 0 \ , \tag{1.21}$$

and

$$D_\mu(g^{\mu\nu}\psi_\nu) - m^2\varphi = 0. \tag{1.22}$$

Eq. (1.20) implies that σ_{ZX} is the jet extension of its projection to Y. Substituting it into Eq. (1.21) we obtain the Klein–Gordon equation (1.18). Hence, the Klein–Gordon Lagrangian for a scalar field satisfies Hypothesis 7.1.3.

7.2 Boundary conditions, symmetries and conservation laws.

If the boundary ∂X of X is not empty, then the field equations can be supplemented by boundary conditions. A boundary condition is given by specifying a subbundle B of $\pi_{\partial X \partial Z} : \partial Z \longrightarrow \partial X$, and considering only sections σ_{ZX} of π_{XZ} such that

$$\sigma_{ZX}(\partial X) \subset B. \tag{2.1}$$

For example, *Dirichlet boundary conditions* consist of specifying a section $\sigma_{\partial Y \partial X}$ of $\pi_{\partial X \partial Y}$ and considering only sections σ_{ZX} such that $\pi_{YZ}(\sigma_{ZX}(\partial X)) = \sigma_{\partial Y \partial X}(\partial X)$. In this case

$$B = \pi_{YZ}^{-1}(\sigma_{\partial Y \partial X}(\partial X)) \tag{2.2}$$

7.2.1. Proposition. For a Dirichlet boundary condition B, the pull back of Ω to B vanishes:

$$i_B^* \Omega = 0 \tag{2.3}$$

where $i_B : B \longrightarrow Z$ is the inclusion map.

Proof. Since Ω is annihilated by vectors in *ker* $T\pi_{YZ}$ it follows that, for every $z \in B$ and $u_1,\ldots,u_n \in T_zB$, $\Omega(u_1,\ldots,u_n)$ depends only on the projections $T\pi_{YZ}(u_1)$ $T\pi_{YZ}(u_1)\ldots$, $T\pi_{YZ}(u_n)$. Eq. (2.2) implies that $dim\ T\pi_{YZ}(T_ZB) = dim\ \partial X = n - 1$. Hence, $\Omega(u_1,\ldots,u_n) = 0$. ∎

In applications, e.g. general relativity, we need more general boundary conditions than those of Dirichlet type. Such boundary conditions need not satisfy Eq. (2.3). However, all boundary conditions B considered here satisfy the following condition.[1]

7.2.2. Hypothesis. There exists a form Π on B such that

$$i_B^* \Omega = d\Pi. \tag{2.4}$$

[1]c.f. Binz and Śniatycki [1986].

Field theories corresponding to the same Lagrangian, but satisfying different boundary conditions are different. We shall consider here a boundary condition B, and a corresponding (n–1)–form Π on B as essential elements of a theory.

There are various notions of symmetries of field theory, depending on which part of the structure of the theory is to be preserved under a symmetry transformation. Here, we use the most restrictive notion of symmetry, requiring that a symmetry should preserve all the structure of the Lagrangian theory of the system under consideration. In physical applications most symmetries of importance are of this type, in particular the symmetries related to the relativistic invariance and the gauge invariance in Yang–Mills theory and general relativity satisfy this condition.

7.2.3. Definition: A *symmetry* of a theory if quadruplet $g = (g_X, g_Y, g_Z, g_B)$ of diffeomorphisms of X, Y, Z and B, respectively, intertwined by the projection maps π_{XY}, π_{XZ} and the inclusion map i_B, such that

$$g_Z^* \Lambda = \Lambda \tag{2.5}$$

$$g_B^* \Pi = \Pi \tag{2.6}$$

and

$$g_Z \circ j^1\sigma_{YX} \circ g_X^{-1} = j^1(g_Z \circ \sigma_{YX} \circ g_X^{-1}) \tag{2.7}$$

for every section σ_{YX} of π_{XY}.

We denote by $\mathcal{G}$ the group of symmetries of the theory. It is a Lie group, possibly infinite dimensional, with the Lie algebra $\mathfrak{g}$ consisting of quadruplets $\xi = (\xi_X, \xi_Y, \xi_Z, \xi_B)$ of vector fields on X, Y, Z and B, related by the projection maps π_{XY}, π_{XZ}, and π_{YZ}, and the inclusion map i_B, and satisfying the infinitesimal version of the conditions (2.5) and (2.6), and (2.7). In particular, Eqs. (2.5) and (2.6) yield

$$\pounds_{\xi_Z} \Lambda = 0 \tag{2.8}$$

and

$$\pounds_{\xi_B} \Pi = 0. \tag{2.9}$$

Since symmetries preserve all the structure of the theory, and the DeDonder from Ω is intrinsically defined, it follows that symmetries preserve Ω,

$$g_Z^{\ *}\Omega = \Omega \tag{2.10}$$

for every $g \in \mathcal{G}$. Similarly, for each infinitesimal symmetry $\xi \in \mathfrak{g}$, the Lie derivative of Ω with respect to ξ_Z vanishes

$$\pounds_{\xi_Z} \Omega = 0. \tag{2.11}$$

Let M be a compact oriented (n–1) submanifold of X with boundary ∂M contained in ∂X. Each infinitesimal symmetry $\xi \in \mathfrak{g}$ and each section σ_{ZX} satisfying the boundary condition B give rise to an integral

$$J^{\xi}(\sigma_{ZX}|M) = \int_M \sigma_{ZX}^{\ *}(\xi_Z \,\lrcorner\, \Omega) + \int_{\partial M} \sigma_{ZX}^{\ *}(\xi_B \,\lrcorner\, \Pi). \tag{2.12}$$

It is the value at $\sigma_{ZX}|M$ of a function J^{ξ} on the space of Cauchy data for DeDonder equations, which is called a *momentum* associated to ξ.

7.2.4. Proposition. If σ_{ZX} satisfies DeDonder equations then $J^{\xi}(\sigma_{ZX}|M)$ depends only on the orientation of M and the equivalence class of M under the equivalence relation $M_0 \sim M_1$ if and only if M_0, M_1 and an open submanifold of ∂X bound a domain in X.

Proof. Consider a domain U in X such that its boundary ∂U can be expressed in the form

$$\partial U = M - M' + V$$

where M and M′ are oriented (n–1)–dimensional submanifolds of X, and V is an open oriented submanifold of ∂X. For every infinitesimal symmetry $\xi \in \mathfrak{g}$ and every solution

σ_{ZX} of the DeDonder equations, we have

$$\sigma_{ZX}{}^{*}d(\xi_Z \lrcorner \Omega) = \sigma_{ZX}{}^{*}\pounds_{\xi_Z}\Omega - \sigma_{ZX}{}^{*}(\xi_Z \lrcorner d\Omega) = 0.$$

By Stokes' Theorem, the integral of $\sigma_{ZX}{}^{*}(\xi_Z \lrcorner \Omega)$ over ∂U vanishes so that

$$\int_{M'} \sigma_{ZX}{}^{*}(\xi_Z \lrcorner \Omega) - \int_{M} \sigma_{ZX}{}^{*}(\xi_Z \lrcorner \Omega) = \int_{V} \sigma_{ZX}{}^{*}(\xi_Z \lrcorner \Omega)$$

However, ξ_Z is tangent to V and $\sigma_{ZX}(V) \subset B$. Taking into account Eqs. (2.4) and (2.9) we obtain

$$\sigma_{ZX}(\xi_Z \lrcorner \Omega)|_V = \sigma_{ZX}{}^{*}i_B{}^{*}(\xi_B \lrcorner d\Pi) = -\sigma_{ZX}{}^{*}i_B{}^{*}d(\xi_B \lrcorner \Pi).$$

Since $\partial M - \partial M' + \partial V = \partial\partial U = 0$, it follows that the integral over V can be expressed as the difference of the integrals over $\partial M'$ and ∂M, so that

$$\int_{M'} \sigma_{ZX}{}^{*}(\xi_Z \lrcorner \Omega) + \int_{\partial M'} \sigma_{ZX}{}^{*}i_B^{*}(\xi_B \lrcorner \Pi) =$$

$$= \int_{M} \sigma_{ZX}{}^{*}(\xi_Z \lrcorner \Omega) + \int_{\partial M} \sigma_{ZX}{}^{*}i_B^{*}(\xi_B \lrcorner \Pi),$$

which completes the proof. ∎

Prop. 7.2.4 is a version of the First Noether Theorem[2] modified by an explicit inclusion of boundary conditions. It is a conservation law stating that, for each solution of the DeDonder equations satisfying the boundary condition B and each $\xi \in \mathfrak{g}$, the momentum associated to ξ does not depend on a Cauchy surface on which it is defined.

7.2.5. <u>Definition</u>. An infinitesimal symmetry $\xi \in \mathfrak{g}$ is *localizable* if ξ_Z vanishes on ∂Z and, for every pair of open sets U, U′ in X with disjoint closures, there exists $\xi' \in \mathfrak{g}$ satisfying

$$\xi'_Z(z) = \xi_Z(z) \ \forall \ z \in \pi_{XZ}^{-1}(U), \tag{2.13}$$

[2]Noether [1918].

and

$$\xi'_Z(z) = 0 \ \forall \ z \in \pi_{XZ}^{-1}(U') \cup \partial Z. \tag{2.14}$$

Localizable infinitesimal symmetries form a Lie subalgebra $\mathscr{h}$ of the Lie subalgebra of infinitesimal symmetries. We denote by $\mathscr{H}$ the connected subgroup of $\mathscr{G}$ with the Lie algebra $\mathscr{h}$ and refer to it as the group of *localizable symmetries.* It should be noticed that the notion of localizable symmetries has a somewhat different meaning than in non–relativistic dynamics. In field theory localizability means localizability in space–time, while in non–relativistic dynamics it means localizability in time.

7.2.6. Proposition. (Second Noether Theorem) For every solution σ_{ZX} of the DeDonder equations, satisfying the boundary condition B, and every infinitesimal localizable symmetry $\xi \in \mathscr{h}$, the momentum associated to ξ vanishes,

$$J^{\xi}(\sigma_{ZX} \mid M) = 0, \tag{2.15}$$

for all Cauchy surfaces M.

Proof. Since $\mathscr{h} \subset \mathscr{g}$, the First Noether Theorem ensures that $|J^{\xi}(\sigma_{ZX} \mid M)|$ is independent of M. Using a tubular neighbourhood of M in X we can construct a submanifold M′ of X, such that there exist open sets U and U′ with disjoint closures containing M and M′ respectively. Let ξ' be an infinitesimal symmetry satisfying Eqs. (2.13) and (2.14). Then, $|J^{\xi}(\sigma_{ZX}|M)|$ is independent of M, and

$$J^{\xi}(\sigma_{ZX} \mid M) = J^{\xi'}(\sigma_{ZX} \mid M) = J^{\xi'}(\sigma_{ZX} \mid M') = 0$$

which completes the proof. ■

The Second Noether Theorem implies that a necessary condition for a section $\gamma : M \longrightarrow Z$ of π_{XZ} over a Cauchy surface M to extend to a solution of the DeDonder equations is the vanishing of the momentum $J^{\xi}(\gamma)$ associated to every infinitesimal localizable $\xi \in \mathcal{k}$,

$$J^{\xi}(\gamma) = 0 \quad \forall \quad \xi \in \mathcal{k} \tag{2.16}$$

This is one of the main sources of constraints on the Cauchy data of solutions of field equations which appear in field theory[3]. For many theories of physical interest Eq. (2.16) is also a sufficient condition for an existence of a local section σ_{XZ} of π_{XZ}, defined in a neighbourhood of M, which satisfies the DeDonder equations and agrees with γ on M.

For the Klein–Gordon theory of a scalar field introduced at the end of the preceding section, and the homogeneous Dirichlet boundary conditions,

$$\sigma_{YX} \mid \partial X = 0, \tag{2.17}$$

the symmetry group of the theory is isomorphic to the group of isometries of (X,g_X). For each $\xi \in \mathfrak{g}$, ξ_X is a Killing vector field in (X,g_X), that is

$$\pounds_{\xi_X} g_X = 0 . \tag{2.18}$$

The Killing equation, (2.18), has a finite dimensional space of solutions. Hence, there are no non–zero localizable infinitesimal symmetries, and $\mathcal{k} = \{0\}$.

Let (x^{μ}) be local coordinates in X such that $x^0 = 0$ determines a Cauchy surface M, and $\frac{\partial}{\partial x^0}$ is normal to M. Let $\xi_X = \xi^{\mu} \frac{\partial}{\partial x^{\mu}}$ be a Killing vector field of g_X and σ_{ZX} a section of π_{XZ} given by Eq. (1.19). Eq. (1.17) yields

$$\begin{aligned} \xi_Z \rfloor \Omega = \sqrt{-\det(g_{\alpha\beta})}\, \{ & g^{\mu\nu} z_{\mu} \xi^{\lambda} dy \wedge d_2x_{\lambda\nu} + \\ & - \tfrac{1}{2}[g^{\mu\nu} z_{\mu} z_{\nu} - m^2 y^2] \xi^{\lambda} d_3x_{\lambda}\} , \end{aligned} \tag{2.19}$$

where

$$d_2x_{\mu\nu} = \frac{\partial}{\partial x^{\mu}} \rfloor d_3x_{\nu} . \tag{2.20}$$

---

[3]The observation that the Second Noether Theorem leads to vanishing of the momenta corresponding to infinitesimal gauge transformations can be found in Bergmann and Goldberg [1955]. The formulation of the constraint condition in terms of the momentum map is formulated in special cases in Fischer, Marsden and Moncrief [1980], and Arms [1981]. For a general discussion see GIMMSY [1988].

Taking into account the identity

$$dx^{\lambda} \wedge d_2x_{\mu\nu} = \delta^{\lambda}_{\nu}\, d_3x_{\mu} - \delta^{\lambda}_{\mu}\, d_3x_{\nu} \quad , \tag{2.21}$$

and Eqs. (2.12) and (2.17), we obtain

$$J^{\xi}(\sigma_{ZX} \mid M) = -\int_M \sqrt{|\det(g_{\alpha\beta})|}\, \{g^{00}\psi_0\varphi_{,k}\, \xi^k + \tfrac{1}{2}[g^{00}\psi_{00} - g^{mn}(2\varphi_{,m} - \psi_m)\psi_n - m^2\varphi^2]\}d_3x_0. \tag{2.22}$$

If σ_{ZX} satisfy the DeDonder equations. then $\psi_\mu = \varphi_{,\mu}$ and we obtain

$$J^{\xi}(\sigma_{ZX} \mid M) = -\int_M \sqrt{|\det(g_{\alpha\beta})|}\, \{g^{00}\varphi_{,0}\varphi_{,m}\xi^m + \tfrac{1}{2}[g^{00}\varphi_{,0}\varphi_{,0} - g^{mn}\varphi_{,m}\varphi_n - m^2\varphi^2]\}d_3x_0\,. \tag{2.23}$$

Consider a case when (X,g_X) is a Minkowski space, and M is a space–like hyperplane in X, that is the metric g_M on M induced by g_X is positive definite. This case is not covered by our general considerations which require that Cauchy surfaces should be compact manifolds with boundary, but the homogeneous boundary condition (2.17) can be reinterpreted as an asymptotic condition and all the results can be rederived in this setting. In this case the group of symmetries is isomorphic to the Poincaré group. If ξ_X is tangent to M then Eq. (2.20) yields the usual expression for the momentum associated to ξ, that is a linear momentum if ξ is an infinitesimal translation and an angular momentum if ξ is an infinitesimal rotation,

$$J^{\xi}(\sigma_{ZX}|M) = \int_M \varphi_{,0}\varphi_{,m}\xi^m d_3x_0\,. \tag{2.24}$$

This justifies the term "the momentum associated to ξ" used here. If $\xi_X = \frac{\partial}{\partial x^0}$, then

$$J^{\xi}(\sigma_{ZX}|M) = \frac{1}{2}\int_M \{\varphi_{,0}\varphi_{,0} + g^{mn}\varphi_{,m}\varphi_{,n} + m^2\varphi^2\}d_3x_0 \tag{2.25}$$

is the energy of the field φ on the Cauchy surface M.

The parameters describing irreducible representations of the Poincaré group have the physical interpretation of mass and spin of the system. The action of the Poincaré group on scalar fields satisfying Klein–Gordon equation, Eq. (1.18) corresponds to the irreducible representation with spin zero and mass given by the parameter m appearing in the equation. Hence, m is interpreted as the mass of the field. This interpretation is retained even in the case of interacting fields, satisfying non–linear equations, which do not belong to an inreducible representation of the Poincaré group.

7.3 Cartan formalism in the space of Cauchy data

An analogy between field theory and dynamics of systems with a finite number of degrees of freedom emerges when one studies the Cauchy problem for DeDonder equations. The absolute time of Newtonian dynamics is replaced by a space $\mathcal{T}$ of Cauchy surfaces, a configuration bundle $\mathbb{Q}$ is replaced by a space $\mathcal{Q}$ of Dirichlet data, and the velocity bundle $\mathbb{U} = J^1\mathbb{Q}$ is replaced by a space $\mathcal{Z}$ of Cauchy data which satisfy the boundary condition B. A field theory analogue of the Cartan form Ξ is defined in terms of the DeDonder form Ω of the theory[1]. We shall see that, as in the case of systems with a finite number of degrees of freedom, a "time" evolution of Cauchy data of the field is given by curves in $\mathcal{Z}$ such that their tangent vectors annihilate $d\Xi$.

We assume that Cauchy surfaces are compact oriented (n–1) dimensional submanifolds of X with interiors contained in the interior of X and boundaries contained in ∂X. Let M be a typical Cauchy surface with boundary ∂M. We denote by $\mathcal{T}$ a manifold of embeddings $\tau : M \longrightarrow X$, which map the interior of M into the interior of X and ∂M into ∂X, describing the space of parametrized Cauchy surfaces under consideration. Two embeddings in $\mathcal{T}$ represent the same non–parametrized Cauchy surface if they differ by an orientation preserving diffeomorphism of M. It seems to be more convenient to work with parametrized Cauchy surfaces rather than non–parametrized ones. In the following we omit the word "parametrized" and use the term *Cauchy surfaces* for embeddings $\tau \in \mathcal{T}$. The choice of the space $\mathcal{T}$ of Cauchy surfaces depends on the field theory under consideration. In all the cases we choose $\mathcal{T}$ in such a way that it is a submanifold of the space of embeddings of M into X, and it is stable under the action in X of the symmetry group of the theory.

---

[1]The construction of the Cartan form and its role in the canonical formulation of field theory were analyzed in Tulczyjew's seminars in Warsaw, 1968 (unpublished). The ideas discussed there lead to several papers written by the participants, e.g. Śniatycki [1970b], Gawędzki [1972], Kijowski and Szczyrba [1975], and Kijowski and Tulczyjew [1979].

A *space of Cauchy data* of the theory is a manifold $\mathscr{Z}$ of embeddings $\gamma : M \longrightarrow Z$ such that $\gamma(\partial M) \subset B$, $\tau = \pi_{XZ} \circ \gamma \in \mathscr{T}$, and there exists a section σ_{YX} of π_{XY} satisfying

$$\gamma = (j^1 \sigma_{YX}) \circ \tau. \tag{3.1}$$

We denote by $\pi_{\mathscr{T}\mathscr{Z}}$ the map from $\mathscr{Z}$ to $\mathscr{T}$ assciating to each $\gamma \in \mathscr{Z}$ the Cauchy surface $\pi_{\mathscr{T}\mathscr{Z}}(\gamma) = \pi_{XZ} \circ \gamma$ on which the data γ are defined, and assume that $\pi_{\mathscr{T}\mathscr{Z}}$ is a locally trivial fibration. Moreover, we assume that $\mathscr{Z}$ is stable under the action of the group $\mathscr{G}$ of symmetries of the theory, and that $\pi_{\mathscr{T}\mathscr{Z}}$ intertwines the actions of $\mathscr{G}$ in $\mathscr{Z}$ and $\mathscr{T}$.

It is convenient to introduce also a *space of Dirichlet data* of the theory. It is the manifold $\mathscr{D}$ of all embeddings $\delta : M \longrightarrow Y$ of the form $\delta = \pi_{YZ} \circ \gamma$, for $\gamma \in \mathscr{Z}$. We denote by $\pi_{\mathscr{D}\mathscr{Z}}: \mathscr{Z} \longrightarrow \mathscr{D}$ the projection mapping defined by $\pi_{\mathscr{D}\mathscr{Z}}(\gamma) = \pi_{YZ} \circ \gamma$, for every $\gamma \in \mathscr{Z}$, and by $\pi_{\mathscr{T}\mathscr{D}}$ the unique mapping from $\mathscr{D}$ to $\mathscr{T}$ satisfying $\pi_{\mathscr{T}\mathscr{Z}} = \pi_{\mathscr{T}\mathscr{D}} \circ \pi_{\mathscr{D}\mathscr{Z}}$.

The DeDonder form Ω on Z gives rise to the Cartan form Ξ on $\mathscr{Z}$ defined as follows. For each $\gamma \in \mathscr{Z}$ and each $\xi \in T_\gamma \mathscr{Z}$,

$$\Xi(\xi) = \int_M \gamma^*(\xi \lrcorner \Omega) + \int_{\partial M} \gamma^*(\xi \lrcorner \Pi). \tag{3.2}$$

Here we identify vectors in $T_\gamma \mathscr{Z}$, i.e. maps from M to TZ covering γ, with the corresponding vector fields on Z defined on $\gamma(M)$.

7.3.1. Proposition. For every $\gamma \in \mathscr{Z}$ and every pair of vectors ξ, ζ and every pair of vectors ξ, ζ in $T_\gamma \mathscr{Z}$,

$$d\Xi(\xi,\zeta) = -\int_M \gamma^*(\zeta \lrcorner \xi \lrcorner d\Omega). \tag{3.3}$$

Proof. We have

$$-d\Xi(\xi,\zeta) = \{\xi(\Xi(\zeta)) - \zeta(\Xi(\xi)) - \Xi([\xi,\zeta])\} =$$

$$= \int_M \gamma^*\{\pounds_\xi(\zeta \rfloor \Omega) - \pounds_\zeta(\xi \rfloor \Omega) - [\xi,\zeta] \rfloor \Omega\} +$$

$$+ \int_{\partial M} \gamma^*\{\pounds_\xi(\zeta \rfloor \Pi) - \pounds_\zeta(\xi \rfloor \Pi) - [\xi,\zeta] \rfloor \Pi\}$$

$$= \int_M \gamma^*\{d(\xi \rfloor \zeta \rfloor \Omega) - \xi \rfloor \zeta \rfloor d\Omega\} +$$

$$+ \int_{\partial M} \gamma^*\{d(\xi \rfloor \zeta \rfloor \Pi) - \xi \rfloor \zeta \rfloor d\Pi\} +$$

$$+ \int_M \gamma^*(\zeta \rfloor \xi \rfloor d\Omega) + \tfrac{1}{2}\int_{\partial M} \gamma^*\{\xi \rfloor \zeta \rfloor (\Omega - d\Pi)\}$$

$$= \int_M \gamma^*(\zeta \rfloor \xi \rfloor d\Omega)$$

since $\gamma(\partial M) \subset B$ and $i_{ZB}^{\ *}\Omega = d\Pi$. ∎

Let $c_{\mathcal{T}}$ be a curve in $\mathcal{T}$. For each $t \in$ *domain* $c_{\mathcal{T}}$ and each $m \in M$, we denote by $(c_{\mathcal{T}}(t))(m)$ the point of X associated to m by the embedding of M into X induced by $c_{\mathcal{T}}(t)$. We say that $c_{\mathcal{T}}$ *splits* X if the mapping from $M \times$ *domain* $c_{\mathcal{T}}$ into X, associating to each $(m,t) \in M \times$ *domain* $c_{\mathcal{T}}$ the point $(c_{\mathcal{T}}(t))(m)$, is an embedding. If $c_{\mathcal{Z}}$ is curve in $\mathcal{Z}$ such that its projection $c_{\mathcal{T}}$ to $\mathcal{T}$ splits X, then it defines a local section σ_{ZX} of π_{XZ} satisfying

$$\sigma_{ZX}((c_{\mathcal{T}}(t))(m)) = (c_{\mathcal{Z}}(t))(m) \tag{3.4}$$

7.3.2. <u>Theorem</u>. Let σ_{ZX} be a section of π_{XZ}, and $c_{\mathcal{T}}$ a curve in $\mathcal{T}$ such that, for each $t \in$ *domain* $c_{\mathcal{T}}$, the image of M under $c_{\mathcal{T}}(t)$ is contained in domain of σ_{ZX}. Let $c_{\mathcal{Z}}$ be the lift of $c_{\mathcal{T}}$ to $\mathcal{Z}$ defined by Eq. (3.4). If σ_{ZX} satisfies the DeDonder equation then the tangent vector $\dot{c}_{\mathcal{Z}}$ of $c_{\mathcal{Z}}$ satisfies the equation.

$$\dot{c}_{\mathcal{Z}} \rfloor d\Xi = 0\,. \tag{3.5}$$

Conversely, if a curve $c_{\mathcal{Z}}$ in $\mathcal{Z}$ satisfies Eq. (3.5) and its projection $c_{\mathcal{T}}$ to $\mathcal{T}$ splits X then the section σ_{ZX} defined by Eq. (3.4) satisfies the DeDonder equations.

Proof. Suppose that σ_{ZX} satisfies the DeDonder equations, that is, for each vector field ζ_Z on Z, $\overset{*}{\sigma}_{ZX}(\zeta_Z \lrcorner\, d\Omega) = 0$. Then, for each $t \in$ *domain* $c_{\mathscr{Z}}$

$$(c_{\mathscr{Z}}(t))^{*}[\dot{c}_{\mathscr{Z}}(t) \lrcorner\, (\zeta_Z \lrcorner\, d\Omega)] = 0, \tag{3.6}$$

where we identify $\dot{c}_{\mathscr{Z}}(t)$ with the corresponding vector field on Z defined on $c_{\mathscr{Z}}(t)(M)$. Taking into account Eq. (3.3) we see that Eq. (3.6) implies Eq. (3.5). Conversely, if Eq. (3.5) is satisfied then, for every vector field ζ_Z on Z, the integral over M of the left hand side of Eq. (3.6) vanishes, which is equivalent to

$$\int_M (c_{\mathscr{Y}}(t))^{*}[\dot{c}_{\mathscr{Y}}(t) \lrcorner\, \overset{*}{\sigma}_{ZX}(\zeta_Z \lrcorner\, d\Omega)] = 0, \tag{3.7}$$

where $\dot{c}_{\mathscr{Y}}(t)$ is identified with the corresponding vector field on X defined along $c_{\mathscr{Y}}(t)(M)$. Since ζ_Z is arbitrary, the Fundamental Theorem of the Calculus of Variation implies that the integrand of the left hand side of Eq. (3.7) vanishes,

$$(c_{\mathscr{Y}}(t))^{*}[\dot{c}_{\mathscr{Y}}(t) \lrcorner\, \overset{*}{\sigma}_{ZX}(\zeta_Z \lrcorner\, d\Omega)] = 0 \tag{3.8}$$

for every vector field ζ_Z on Z. If $c_{\mathscr{Y}}$ splits X, then $\dot{c}_{\mathscr{Y}}(t)$ is transverse to $c_{\mathscr{Y}}(t)(M)$ and Eq. (3.8) implies the DeDonder equations. ∎

Eq. (3.5) is the field theory analogue of the Cartan equations in dynamics of systems with a finite number of degrees of freedom, and we shall refer to it as the *Cartan equation* of the theory. Note that the assumption that $c_{\mathscr{Y}}$ splits X is needed only to reconstruct σ_{ZX} from $c_{\mathscr{Z}}$ according to Eq. (3.4). If σ_{ZX} is given, then it defines a section $\sigma_{\mathscr{Z}\mathscr{Y}}$ of $\pi_{\mathscr{Y}\mathscr{Z}}$ such that, for each $m \in M$ and each $\tau \in \mathscr{Y}$,

$$\sigma_{\mathscr{Z}\mathscr{Y}}(\tau)(m) = \sigma_{ZX}(\tau(m)). \tag{3.9}$$

The DeDonder equations for σ_{ZX} are equivalent to

$$\overset{*}{\sigma}_{\mathscr{Z}\mathscr{Y}}(\xi_{\mathscr{Z}} \lrcorner\, d\Xi) = 0 \tag{3.10}$$

for every vector field $\xi_{\mathscr{Z}}$ on $\mathscr{Z}$.

The group $\mathscr{G}$ of symmetries of the theory acts in $\mathscr{Y}$, $\mathscr{Q}$ and $\mathscr{Z}$. For each $g \in \mathscr{G}$, the corresponding diffeomorphisms of $\mathscr{Y}$, $\mathscr{Q}$ and $\mathscr{Z}$ are given by

$$g_{\mathscr{T}}(\tau) = g_X \circ \tau \quad \forall \quad \tau \in \mathscr{T},$$
$$g_{\mathscr{Q}}(\sigma) = g_Y \circ \delta \quad \forall \quad \delta \in \mathscr{Q},$$
$$g_{\mathscr{Z}}(\gamma) = g_Z \circ \gamma \quad \forall \quad \gamma \in \mathscr{Z},$$

respectively. Similarly, to each ξ in the Lie algebra $\mathfrak{g}$ of $\mathscr{G}$, there correspond vector fields $\xi_{\mathscr{T}}$, $\xi_{\mathscr{Q}}$ and $\xi_{\mathscr{Z}}$ on $\mathscr{T}$, $\mathscr{Q}$ and $\mathscr{Z}$, respectively, given by

$$\xi_{\mathscr{T}}(\tau) = \xi_X \circ \tau \quad \forall \quad \tau \in \mathscr{T},$$
$$\xi_{\mathscr{Q}}(\delta) = \xi_Y \circ \delta \quad \forall \quad \delta \in \mathscr{Q},$$
$$\xi_{\mathscr{Z}}(\gamma) = \xi_Z \circ \gamma \quad \forall \quad \gamma \in \mathscr{Z}.$$

The action of $\mathscr{G}$ in $\mathscr{Z}$ preserves the Cartan form Ξ,

$$g_{\mathscr{Z}}^{*}\Xi = \Xi. \tag{3.11}$$

Hence, for each $\xi \in \mathfrak{g}$ the Lie derivative of Ξ with respect to $\xi_{\mathscr{Z}}$ vanishes,

$$\pounds_{\xi_{\mathscr{Z}}}\Xi = 0. \tag{3.12}$$

Eq. (3.11) is equivalent to

$$\xi_{\mathscr{Z}} \lrcorner (-\mathrm{d}\Xi) = \mathrm{d}\langle\Xi,\xi_{\mathscr{Z}}\rangle\,. \tag{3.13}$$

Comparing Eqs. (2.12) and (3.2) we see that, for every section σ_{ZX} satisfying the boundary conditions, and every $\tau \in \mathscr{T}$,

$$J^{\xi}(\sigma_{ZX} | \tau(M)) = \langle\Xi,\xi_{\mathscr{Z}}\rangle(\sigma_{ZX}\circ\tau)\,. \tag{3.14}$$

We denote by $J^{\xi}_{\mathscr{Z}}$ the function on $\mathscr{Z}$ given by the evaluation of Ξ on $\xi_{\mathscr{Z}}$,

$$J^{\xi}_{\mathscr{Z}} = \langle\Xi,\xi_{\mathscr{Z}}\rangle \tag{3.15}$$

and by $J^{\mathscr{G}}_{\mathscr{Z}}$ the map from $\mathscr{Z}$ to $\mathfrak{g}$ given by

$$\langle J^{\mathscr{G}}_{\mathscr{Z}},\xi\rangle = J^{\xi}_{\mathscr{Z}} \tag{3.16}$$

for every $\xi \in \mathfrak{g}$ As before we refer we refer to $J^{\xi}_{\mathscr{Z}}$ as the *momentum* corresponding to ξ and to $J^{\mathscr{G}}_{\mathscr{Z}}$ as the *momentum map* corresponding to the action of $\mathscr{G}$ in $\mathscr{Z}$.

7.3.3. Proposition. For each curve $c_{\mathscr{Z}}$ in $\mathscr{Z}$ satisfying the Cartan equation, Eq (3.5), and each infinitesimal symmetry $\xi \in \mathfrak{g}$, the momentum $J^{\xi}_{\mathscr{Z}}$ associated to ξ is constant along $c_{\mathscr{Z}}$.

Proof. Denoting by $\dot{c}_{\mathscr{Z}}(t)$ the tangent vector to $c_{\mathscr{Z}}$ at $c_{\mathscr{Z}}(t)$, we obtain

$$\frac{d}{dt} J^{\xi}_{\mathscr{Z}}(c_{\mathscr{Z}}(t)) = \dot{c}_{\mathscr{Z}}(t) \lrcorner\, dJ^{\xi}_{\mathscr{Z}} = -\dot{c}_{\mathscr{Z}}(t) \lrcorner\, \xi_{\mathscr{Z}} \lrcorner\, d\Xi = 0. \quad ■$$

Proposition 7.3.3 is a reformulation of the First Noether Theorem, c.f. Proposition 7.2.4, in terms of the Cartan formalism. It leads to conserved quantities associated to infinitesimal symmetries of the theory. However, in the Cartan formalism in the space of Cauchy data one can adopt a weaker notion of symmetry, analogous to that used in the case of systems with a finite number of degrees of freedom. In this case one obtains a larger class of conserved quantities.

7.3.4. Proposition. Let $\xi_{\mathscr{Z}}$ be a vector field on $\mathscr{Z}$ preserving the Cartan form Ξ. For each curve $c_{\mathscr{Z}}$ in $\mathscr{Z}$ satisfying the Cartan equation, the function $<\Xi, \xi_{\mathscr{Z}}>$ is constant along $c_{\mathscr{Z}}$.

Proof of this proposition is identical to that of Proposition 7.3.3. ■

The Second Noether Theorem can also be rephrased in terms of the Cartan formalism.

7.3.5. Proposition. For every infinitesimal localizable symmetry $\xi \in \mathscr{k}$ and every curve $c_{\mathscr{Z}}$ in $\mathscr{Z}$ satisfying the Cartan equations (3.5) and such that its projection to $\mathscr{T}$ splits X, the momentum $J^{\xi}_{\mathscr{Z}}$ associated to ξ vanishes along $c_{\mathscr{Z}}$,

$$J^{\xi}_{\mathscr{Z}} \circ c_{\mathscr{Z}} = 0 \qquad (3.17)$$

Proof. Since the projection of $c_{\mathscr{Z}}$ to $\mathscr{T}$ splits X, it follows that $c_{\mathscr{Z}}$ defines a section σ_{ZX} of π_{XZ} satisfying Eq. (3.4). The assumption that the tangent vector of $c_{\mathscr{Z}}$ satisfies Eq. (3.5) implies that σ_{ZX} satisfies the DeDonder equations, c.f. Theorem 7.3.2. Hence, the hypotheses of Proposition 7.2.5 are satisfied and Eq. (3.17) holds as a consequence of Eq. (2.15). ■

The Second Noether Theorem implies that the Cauchy data $\gamma \in \mathcal{Z}$ extend to solutions of the DeDonder equations only if $J^{\xi}_{\mathcal{Z}}(\gamma) = 0$ for all $\xi \in \mathfrak{h}$. As we have mentioned before, this is a main source of constraints in field theory. Replacing $\xi \in \mathfrak{h}$ by vector fields $\xi_{\mathcal{Z}}$ which preserve the Cartan form Ξ, project to vector fields $\xi_{\mathcal{T}}$ on $\mathcal{T}$ and can be localizable in the sense of the definition adopted in Sec. 6.5, see Eqs. (6.5.17) and (6.5.18)[2] with $\mathbb{I}$ replaced by $\mathcal{Z}$ and T replaced by $\mathcal{T}$, one can obtain a generalization of the second Noether Theorem analogous to Prop. 6.5.4, which may lead to additional constraints. However, in examples discussed here all the constraints can be obtained from the vanishing of the momenta $J^{\xi}_{\mathcal{Z}}$ for $\xi \in \mathfrak{h}$.

For the Klein–Gordon dynamics of a scalar field on a Lorentzian manifold (X,g_X) the space $\mathcal{T}$ of parametrized Cauchy surfaces consists of embeddings $\tau : M \longrightarrow X$ such that $\tau^* g_X$ is a Riemannian metric on M. Since $\mathcal{T}$ is assumed to be stable under the action of the group of symmetries $\mathcal{G}$ of the theory, that is the group of isometries of g_X, we can take $\mathcal{T}$ to be an orbit of the action $\mathcal{G}$ in the space of embeddings of M into X. In this case $\mathcal{T}$ is a finite–dimensional manifold, since the dimension of the group of isometries of a non–degenerate metric is finite. We assume here that the metric g_X admits time–like translations, and the Cauchy surfaces in $\mathcal{T}$ fill up X.

Given a Cauchy surface $\tau \in \mathcal{T}$, we denote by η_τ the future pointing unit vector field of $\tau(M)$ in X. Cauchy data on $\tau(M)$ are 1–jet extensions of sections σ_{YX} restricted to $\tau(M)$, c.f. Eq. (3.1). Since σ_{YX} is given by a function φ on X, its Cauchy data are given by the restriction φ_τ of φ to $\tau(M)$ and the normal derivative $\nu_\tau = \eta_\tau \varphi$ of φ on $\tau(M)$. We shall identify elements $\gamma \in \mathcal{Z}_\tau = \pi^{-1}_{\mathcal{T}\mathcal{Z}}(\tau)$ with pairs (φ_τ, ν_τ) of functions on $\tau(M)$,

$$\gamma = (\varphi_\tau, \nu_\tau) . \tag{3.18}$$

---

[2]Throughout Part II we refer to equations from other chapters by including the number of the chapter. Thus, Eq. (6.5.17) refers to Eq. (5.17) in Chapter 6.

Vectors $\zeta \in T_\gamma \mathscr{Z}_\tau$ can be decomposed into their components in the directions of φ_τ and ν_τ, which will be denoted by $\dot{\varphi}_\tau$ and $\dot{\nu}_\tau$, respectively. We shall write

$$\zeta = (\dot{\varphi}_\tau, \dot{\nu}_\tau) \tag{3.19}$$

suppressing the point γ at which ζ is attached. Eqs. (1.17), (2.17) and (3.2) yield

$$<\Xi(\varphi_\tau, \nu_\tau), (\dot{\varphi}_\tau, \dot{\nu}_\tau)> = \int \nu_\tau \, \dot{\varphi}_\tau \, vol\,[g_\tau] \tag{3.20}$$

where g_τ is the Riemannian metric on $\tau(M)$ obtained by the pull back of g_X by the inclusion map $\tau(M) \longrightarrow X$. In the following, in order to simplify the notation, we often identify $\tau(M)$ with M. Thus, ν_τ and $\dot{\varphi}_\tau$ can be considered as functions on M and the integral on the right hand side of Eq. (3.20) as an integral over M. This will not lead to confusion. Let ξ_X be a vector field on X. Along $\tau(M)$ we can decompose ξ_X into its component ξ_τ tangential to $\tau(M)$ and the component normal to $\tau(M)$,

$$\xi_X \mid \tau(M) = \xi_\tau - g_X(\eta_\tau, \xi_X)\eta_\tau\,, \tag{3.21}$$

where the negative sign in front of g_X is the consequence of the assumption that η_τ is time–like, i.e. $g_X(\eta_\tau, \eta_\tau) = -1$. Denoting by ξ_Z the natural extension of ξ_X to the jet bundle $Z = J^1Y$, and by $\xi_{\mathscr{Z}}$ the induced vector field in the space $\mathscr{Z}$ of the Cauchy data we obtain from Eq. (2.23),

$$\begin{aligned} <\Xi, \xi_{\mathscr{Z}}(\varphi_\tau, \nu_\tau) &= \int_{\tau(M)} \{\nu_\tau <d\varphi_\tau, \xi_\tau> - \\ &-\tfrac{1}{2}[\nu_\tau^2 + |d\varphi_\tau|^2 + m^2\varphi_\tau^2]\, g_X(\eta_\tau, \xi_X)\} vol[g_\tau], \end{aligned} \tag{3.22}$$

where $|d\varphi_\tau|^2$ is the square of the norm of 1–form $d\varphi_\tau$ defined in terms of the Riemannian metric g_τ. Since every vector $\zeta \in T_\gamma \mathscr{Z}$ can be expressed as a sum of a vector in $T_\gamma \mathscr{Z}_\tau$ and the value at γ of a vector field $\xi_{\mathscr{Z}}$ obtained from a vector field ξ_X on X, Eqs. (3.21) and (3.22) determine the Cartan form Ξ completely.

7.4 Hamiltonian formulation

In the preceding section we formulated dynamics of classical fields in terms of a generalization of a Cartan system introduced in non–relativistic dynamics. In this section we follow the steps made in Sec. 6.6 to develop a covariant Hamiltonian formulation of field theory[1].

As in Sec. 6.6 we denote by $\mathscr{N}$ the set of vectors in T $\mathscr{Z}$ tangent to the fibres of $\pi_{\mathscr{Q}\mathscr{P}}$ which annihilate $d\Xi$ and Ξ,

$$\mathscr{N} = \{\xi \in \mathit{ker}\, T\pi_{\mathscr{Q}\mathscr{Z}} \mid \xi \lrcorner d\Xi \text{ and } \langle\Xi,\xi\rangle = 0\}, \tag{4.1}$$

and assume the following hypothesis.

7.4.1. Hypothesis. $\mathscr{N}$ is an integrable distribution on $\mathscr{Z}$ and the space $\mathscr{P}$ of integral manifolds of $\mathscr{N}$ has a manifold structure such that the canonical projection $\pi_{\mathscr{P}\mathscr{Z}}: \mathscr{Z} \to \mathscr{P}$ is a submersion. The projection $\pi_{\mathscr{T}\mathscr{Z}}: \mathscr{Z} \to \mathscr{T}$ induces a locally trivial fibration $\pi_{\mathscr{T}\mathscr{P}}: \mathscr{P} \to \mathscr{T}$ such that $\pi_{\mathscr{T}\mathscr{Z}} = \pi_{\mathscr{T}\mathscr{P}} \circ \pi_{\mathscr{P}\mathscr{Z}}$.

7.4.2. Proposition. There exists a unique 1–form Θ on $\mathscr{P}$ such that

$$\Xi = \pi^{*}_{\mathscr{P}\mathscr{Z}}\Theta\ . \tag{4.2}$$

A curve $c_{\mathscr{Z}}$ in $\mathscr{Z}$ satisfies the equation $\dot{c}_{\mathscr{Z}} \lrcorner d\Xi = 0$ if and only if its projection $c_{\mathscr{P}} = \pi_{\mathscr{P}\mathscr{Z}} \circ c_{\mathscr{Z}}$ to $\mathscr{P}$ satisfies the equation

$$\dot{c}_{\mathscr{P}} \lrcorner d\Theta = 0\ . \tag{4.3}$$

The symmetry group $\mathscr{G}$ acts in $\mathscr{P}$ by diffeomorphisms preserving Θ .

Proof. It is an immediate consequence of the definition of $\mathscr{N}$ and Hypothesis 7.4.1. ■

Following the terminology introduced in dynamics we call $\mathscr{P}$ the *primary constraint*

---

[1]This section is based on ideas in GIMMSY [1988].

bundle of the theory, Θ the *Hamilton form*, and we shall refer to Eq. (4.3) as the *Hamilton–Cartan equations.*

7.4.3. Proposition. Let $c_{\mathscr{P}}$ be a curve in $\mathscr{P}$ such that its projection $c_{\mathscr{T}}$ to $\mathscr{T}$ splits X, and let σ_{YX} be the section such that, for each $t \in$ *domain* $c_{\mathscr{T}}$,

$$\pi_{\mathscr{Q}\mathscr{P}}(c_{\mathscr{P}}(t)) = \sigma_{YX} \circ c_{\mathscr{T}}(t). \tag{4.4}$$

The section σ_{YX} satisfies the field equations if $c_{\mathscr{P}}$ satisfies the Hamilton–Cartan equations.

Proof. Let $c_{\mathscr{Z}}$ be a lift of $c_{\mathscr{P}}$ to $\mathscr{Z}$ and σ_{ZX} the section of π_{XZ} defined by Eq. (3.4). Clearly $\sigma_{YX} = \pi_{YZ} \circ \sigma_{ZX}$. Proposition 7.4.2 implies that Eq. (4.3) is equivalent to Eq. (3.7) which, by Proposition 7.3.2, is equivalent to the DeDonder equations for σ_{ZX}. Hypothesis 7.1.4 ensures that σ_{YX} satisfies the field equations if σ_{ZX} satisfies the DeDonder equations. ■

Solutions of the initial value problem for the Hamilton–Cartan equations need not exist. As in non–relativistic dynamics we adopt the following definition.

7.4.4. Definition. The *constraint set* in $\mathscr{P}$ is the set $\mathscr{C}$ of all points in $\mathscr{P}$ which lie on curves $c_{\mathscr{P}}$ satisfying the Hamilton–Cartan equations and projecting to curves $c_{\mathscr{T}} = \pi_{\mathscr{T}\mathscr{P}} \circ c_{\mathscr{P}}$ in $\mathscr{T}$ which split X.

Note that the condition that $c_{\mathscr{T}}$ splits X is necessary in order to obtain a section σ_{YX} satisfying Eq. (4.4). If $c_{\mathscr{Z}}$ is a lift of $c_{\mathscr{P}}$ to $\mathscr{Z}$, then Eq. (4.3) implies, that the section σ_{ZX} given by Eq. (3.3) satisfies the DeDonder equations. Therefore the section $\sigma_{\mathscr{Z}\mathscr{T}}$ defined by Eq. (3.8) satisfies Eq. (3.9). Projecting $\sigma_{\mathscr{Z}\mathscr{T}}$ to $\mathscr{P}$ we obtain a section $\sigma_{\mathscr{P}\mathscr{T}}$ of $\pi_{\mathscr{T}\mathscr{P}}$ satisfying the equation

$$\overset{*}{\sigma}_{\mathcal{P}\mathcal{T}}(\xi_{\mathcal{P}} \rfloor \mathrm{d}\Theta) = 0 \qquad (4.5)$$

for every vector field $\xi_{\mathcal{P}}$ on $\mathcal{P}$. If $\tilde{c}_{\mathcal{T}}$ is any curve in the domain of $\sigma_{\mathcal{P}\mathcal{T}}$, Eq. (4.5) implies that the curve $\tilde{c}_{\mathcal{P}} = \sigma_{\mathcal{P}\mathcal{T}} \circ \tilde{c}_{\mathcal{T}}$ satisfies the Hamilton–Cartan equations. Hence, points in $\mathcal{C}$ admit solutions of the initial value problem for the Hamilton–Cartan equations also for curves $c_{\mathcal{P}}$ in $\mathcal{P}$ which project to curves $c_{\mathcal{T}}$ in $\mathcal{T}$ which do not split X.

The lack of uniqueness of solutions of Hamilton–Cartan equations is characterized by a subset $\mathcal{K}$ of $T\mathcal{P}$ consisting of vectors tangent to the fibres of $\pi_{\mathcal{T}\mathcal{P}}$ which annihilate the restrictions of $\mathrm{d}\Theta$ to the fibres,

$$\mathcal{K} = \{\xi \in \mathit{ker}\, T\pi_{\mathcal{T}\mathcal{P}} \mid \langle \xi \rfloor \mathrm{d}\Theta, \zeta \rangle = 0 \ \forall\ \zeta \in \mathrm{Ker}\, T\pi_{\mathcal{T}\mathcal{P}}\}\,. \qquad (4.6)$$

7.4.5. <u>Hypothesis</u>. $\mathcal{K}$ is an integrable distribution on $\mathcal{P}$, and the space $\mathcal{R}$ of integral manifolds of $\mathcal{K}$ has a manifold structure such that the canonical projection $\pi_{\mathcal{R}\mathcal{P}}: \mathcal{P} \longrightarrow \mathcal{R}$ is a submersion. The projection $\pi_{\mathcal{T}\mathcal{P}}: \mathcal{P} \longrightarrow \mathcal{T}$ introduces a locally trivial fibration $\pi_{\mathcal{T}\mathcal{R}} = \pi_{\mathcal{T}\mathcal{R}} \circ \pi_{\mathcal{R}\mathcal{P}}$.

7.4.6. <u>Proposition</u>. For each $\tau \in \mathcal{T}$, the fibre $\mathcal{R}_{\tau} = \overset{-1}{\pi}_{\mathcal{T}\mathcal{R}}(\tau)$ has a unique symplectic form $\omega_{\mathcal{R}_{\tau}}$ such that its pull back $\omega_{\mathcal{P}_{\tau}}$ to $\mathcal{P}_{\tau} = \overset{-1}{\pi}_{\mathcal{T}\mathcal{P}}(\tau)$ coincides with the negative of the pull–back of $\mathrm{d}\Theta$ to $\mathcal{P}_{\tau}$.

The symmetry group $\mathcal{G}$ acts in $\mathcal{R}$ by diffeomorphisms which induce symplectomorphisms of the fibres of $\pi_{\mathcal{T}\mathcal{R}}$.

<u>Proof</u>. It is a direct consequence of the definition of $\mathcal{R}$ and the assumptions that $\pi_{\mathcal{T}\mathcal{R}}$ is a locally trivial fibration and $\pi_{\mathcal{R}\mathcal{P}}$ is a submersion. ∎

As in non–relativistic dynamics we refer to $\mathcal{R}$ as the *primary reduced phase bundle.*

In the following we shall restrict our considerations to field theories which satisfy the hypotheses made above.

7.4.7. Definition. A field theory is *quasi–regular* if it satisfies Hypotheses 7.1.3, 7.4.1, and 7.4.4.

In order to split Hamilton–Cartan equations into constraint equations and Hamiltonian evolution equations we need a field theoretical version of Atlas Hypothesis.

7.4.8. Atlas Hypothesis. There exists a bundle $\mathcal{A}$ over $\mathcal{T}$ with the projection map $\pi_{\mathcal{T}\mathcal{A}} : \mathcal{A} \to \mathcal{T}$ and a mapping $\pi_{\mathcal{A}\mathcal{P}} : \mathcal{P} \to \mathcal{A}$ such that the following conditions are satisfied.

(i) The diagram

$$\begin{array}{ccc} \mathcal{P} & \xrightarrow{\pi_{\mathcal{R}\mathcal{P}}} & \mathcal{R} \\ {\scriptstyle \pi_{\mathcal{A}\mathcal{P}}} \downarrow & & \downarrow {\scriptstyle \pi_{\mathcal{T}\mathcal{R}}} \\ \mathcal{A} & \xrightarrow[\pi_{\mathcal{T}\mathcal{A}}]{} & \mathcal{T} \end{array}$$

commutes, and it defines in $\mathcal{P}$ a structure of the product of the bundles $\mathcal{A}$ and $\mathcal{R}$ over $\mathcal{T}$,

$$\mathcal{P} = \mathcal{A} \times_{\mathcal{T}} \mathcal{R}, \tag{4.7}$$

which is stable under the action of the group $\mathcal{G}$ of symmetries of the theory.

(ii) For each $\tau \in \mathcal{T}$ and each pair (α,α') of elements of $\mathcal{A}_\tau$ there exists a localizable symmetry $g \in \mathcal{H}$ such that

$$g_{\mathrm{P}}(\alpha,\rho) = (\alpha',\rho) \tag{4.8}$$

for every $\rho \in \mathcal{R}_\tau$.

(iii) The subset $hor\,T\mathscr{P}$ of $T\mathscr{P}$, defined by,

$$hor\,T\mathscr{P} = \{\zeta \in T\mathscr{P} \mid (\zeta \lrcorner\, d\Theta)(\eta) = 0 \;\; \forall \;\; \eta \in Ker\,T\pi_{\mathscr{A}\mathscr{P}}\}. \qquad (4.9)$$

is a distribution on $\mathscr{P}$ transverse to the fibres of $\pi_{\mathscr{A}\mathscr{P}}$.

Since the fibres of $\pi_{\mathscr{T}\mathscr{R}}: \mathscr{R} \longrightarrow \mathscr{T}$ are symplectic, and $\mathscr{P}$ is the fibre product of $\mathscr{A}$ and $\mathscr{R}$, it follows that the fibres of $\pi_{\mathscr{A}\mathscr{P}}: \mathscr{P} \longrightarrow \mathscr{A}$ are symplectic. For each $\alpha \in \mathscr{A}$, the fibre $\mathscr{P}_\alpha$ over α equals to $\{\alpha\} \times \mathscr{R}_\tau$, where $\tau = \pi_{\mathscr{T}\mathscr{A}}(\alpha)$, and the symplectic form $\omega_{\mathscr{R}_\tau}$ in $\mathscr{R}_\tau$ induces a symplectic form in $\mathscr{P}_\alpha$ which will be denoted by $\omega_{\mathscr{P}_\alpha}$. The closed form $d\Theta$ extends the family of symplectic forms on the fibres of $\pi_{\mathscr{A}\mathscr{P}}$. The distribution $hor\,T\mathscr{P}$ can be thought of as a connection in $\mathscr{P}$ considered as a bundle over $\mathscr{A}$. The bundle $\mathscr{A}$ is called an *atlas bundle* of the theory, and its elements are called *atlas fields*. Atlas fields describe the components of the Cauchy data for which the time evolution is not determined by the field equations[2].

7.4.9. <u>Theorem</u>. Let ξ be an infinitesimal symmetry of a quasi–regular theory satisfying Atlas Hypothesis, and

$$\xi_{\mathscr{P}} = ver\,\xi_{\mathscr{P}} + hor\,\xi_{\mathscr{P}} \qquad (4.10)$$

the decomposition of $\xi_{\mathscr{P}}$ into its vertical and horizontal components relative to the connection $hor\,T\mathscr{P}$ on $\mathscr{P}$. Then, the following hold.

(i) For each $\alpha \in \mathscr{A}$, the restriction $ver\,\xi_{\mathscr{P}} \mid \mathscr{P}_\alpha$ of $ver\,\xi_{\mathscr{P}}$ to $\mathscr{P}_\alpha$ is the Hamiltonian vector field of $\Theta(\xi_{\mathscr{P}})$ restricted to $\mathscr{P}_\alpha$,

$$(ver\,\xi_{\mathscr{P}} \mid \mathscr{P}_\alpha) \lrcorner\, \omega_\alpha = d(\Theta(\xi_{\mathscr{P}}) \mid \mathscr{P}_\alpha). \qquad (4.11)$$

(ii) For each $\pi \in \mathscr{P}$, such that the integral curve of $\xi_{\mathscr{T}}$ through $\pi_{\mathscr{T}\mathscr{P}}(\pi)$ splits X, the section σ_{YX}, defined by Eq. (4.4), satisfies the field equations.

---

[2]The term "atlas fields" is used by GIMMSY in a somewhat different way.

Proof of this proposition is analogous to the proof of Theorem 6.6.10.

(i) For every vector field $\zeta_{\mathcal{P}}$ on $\mathcal{P}$ such that $T\pi_{\mathcal{A}\mathcal{P}} \circ \zeta_{\mathcal{P}} = 0$, the definition of *hor* $T\mathcal{P}$ yields $(ver\ \xi_{\mathcal{P}} \rfloor d\Theta)(\zeta_{\mathcal{P}}) = (\xi_{\mathcal{P}} \rfloor d\Theta)(\zeta_{\mathcal{P}}) = -\zeta_{\mathcal{P}} \rfloor d[\Theta(\xi_{\mathcal{P}})]$. Restricting this equation to $\mathcal{P}_\alpha$ we obtain Eq. (4.11).

(ii) Let $c_{\mathcal{P}}$ be an integral curve of *hor* $\xi_{\mathcal{P}}$ through a point $\pi \in \mathcal{P}$ such that its projection $c_{\mathcal{T}}$ to $\mathcal{T}$ splits X. Since $\pi \in \mathcal{P}$, there exists a curve $c'_{\mathcal{P}}$ through π which projects to $c_{\mathcal{T}}$ and satisfies the Hamilton–Cartan equations. This implies that the tangent vector $\dot{c}'_{\mathcal{P}}$ of $c'_{\mathcal{P}}$ is contained in *hor* $T\mathcal{P}$. Thus, $c_{\mathcal{P}}$ and $c'_{\mathcal{P}}$ are horizontal curves in $\mathcal{P}$ passing though π and projecting to the same curve $c_{\mathcal{T}}$ in $\mathcal{T}$. Hence the projections of $c_{\mathcal{P}}$ and $c'_{\mathcal{P}}$ to $\mathcal{R}$ coincide giving a curve $c_{\mathcal{R}}$.

The second part of Atlas Hypothesis, 7.4.6.(ii), ensures that, for each $t \in$ *domain* $c_{\mathcal{P}}$, there exists a symmetry g such that $g_{\mathcal{P}}(c'_{\mathcal{P}}(t)) = c_{\mathcal{P}}(t)$. Since $\dot{c}'_{\mathcal{P}}(t)$ annihilates $d\Theta$ and g is a symmetry, it follows that $Tg_{\mathcal{P}}(\dot{c}'_{\mathcal{P}}(t)) \rfloor d\Theta = 0$. Moreover, $\dot{c}_{\mathcal{P}}(t) - Tg_{\mathcal{P}}(\dot{c}'_{\mathcal{P}}(t))$ is tangent to the fibres of $\pi_{\mathcal{R}\mathcal{P}}$ and the definition of the reduced phase bundle $\mathcal{R}$ implies that

$$<d\Theta,(\dot{c}_{\mathcal{P}}(t) - Tg_{\mathcal{P}}(\dot{c}'_{\mathcal{P}}(t)),\zeta> = 0$$

for every ζ in $T_{c_{\mathcal{P}}(t)}\mathcal{P}$ tangent to the fibres of $\pi_{\mathcal{T}\mathcal{P}}$. Hence,

$$<d\Theta,(\dot{c}_{\mathcal{P}}(t),\zeta> = <d\Theta,(Tg_{\mathcal{P}}(\dot{c}'_{\mathcal{P}}(t)),\zeta> + <d\Theta,(\dot{c}_{\mathcal{P}}(t) - Tg_{\mathcal{P}}(\dot{c}'_{\mathcal{P}}(t)),\zeta> = 0.$$

This argument can be repeated for every $t \in$ *domain* $c_{\mathcal{P}}$. Therefore,

$$\overset{*}{c}_{\mathcal{P}}(\zeta_{\mathcal{P}} \rfloor d\Theta) = 0$$

for every vector field $\zeta_{\mathcal{P}}$ tangent to the fibres of $\pi_{\mathcal{T}\mathcal{P}}$. This implies that a lift $c_{\mathcal{Z}}$ of $c_{\mathcal{P}}$ satisfies the equation

$$\overset{*}{c}_{\mathcal{Z}}(\zeta_{\mathcal{Z}} \rfloor d\Xi) = 0$$

for every vector field $\zeta_{\mathcal{Z}}$ tangent to the fibres of $\pi_{\mathcal{T}\mathcal{Z}}$. Integrating over the domain of $c_{\mathcal{Z}}$ and taking into account Eq. (3.3) we obtain

$$\int \overset{*}{\sigma}_{ZX}(\zeta_Z \rfloor d\Omega) = 0$$

for every vector field ζ_Z tangent to the fibres of the source map π_{XZ} and preserving the boundary condition. The fundamental Theorem of the Calculus of Variations guarantees that σ_{ZX} satisfies the DeDonder equations at each point of the interior of its domain. Hence, by Hypothesis 7.1.4, the section σ_{YX} satisfies the field equations. This completes the proof of the theorem. ∎

As a consequence of Theorem 7.4.9 we see that, for quasi–regular field theories which satisfy Atlas Hypothesis, equations of evolution in direction of an infinitesimal symmetry can be cast in a Hamiltonian form with the Hamiltonian given by the corresponding conserved momentum. Because of a simple explicit relation between the direction of evolution and the corresponding Hamiltonian, given by the momentum map, we refer to this approach to evolution equations as a *covariant Hamiltonian dynamics*. For future reference, we summarize it in the following corollary.

7.4.10. Corollary In order to find solutions of the field equations, we need to choose an infinitesimal symmetry ξ such that an integral curve $c_{\mathcal{G}}$ of $\xi_{\mathcal{G}}$ splits X, lift $c_{\mathcal{G}}$ to an integral curve $c_{\mathcal{M}}$ of $\xi_{\mathcal{M}}$ and, for each $\alpha \in$ *range* $c_{\mathcal{M}}$, find the Hamiltonian vector field in $(\mathcal{P}_\alpha, \omega_\alpha)$ of $\Theta(\xi_{\mathcal{P}}) | \mathcal{P}_\alpha$, obtaining in this way $ver\, \xi_{\mathcal{P}}$ restricted to $\pi^{-1}_{\mathcal{MP}}(range\ c_{\mathcal{M}})$. The integral curves of $hor\, \xi_{\mathcal{P}} = \xi_{\mathcal{P}} - ver\, \xi_{\mathcal{P}}$ passing through points in $\mathcal{C}$ give rise to solutions of the field equations.

It should be noted that the assumption made here, that the direction of evolution corresponds to an infinitesimal symmetry of the theory, is somewhat restrictive. For theories with too few symmetries our approach need not lead to Hamiltonian equations of motion. In the usual formulation of dynamics, or (3+1)–approach to field theory, the direction of evolution corresponds to an infinitesimal canonical transformation which need not be a symmetry. One could reproduce these results in a covariant form presented here, but at the cost of a very complicated formalism which would tend to obscure the role played by symmetries.

For the Klein–Gordon dynamics of a scalar field the constructions discussed in this section are trivial. Eqs. (3.20) and (4.1) show that $\mathscr{N} = 0$ so that $\mathscr{P} = \mathscr{Z}$. Similarly, we obtain from Eq. (4.6) that $\mathscr{K} = 0$ and, therefore, $\mathscr{R} = \mathscr{P}$. This ensures that the Klein–Gordon theory is quasi–regular in the sense of Definition 7.4.7. The symmetry group $\mathscr{G}$ is finite dimensional and the group $\mathscr{H}$ of localizable symmetries consists of the identity element. Atlas Hypothesis is satisfied with $\mathscr{A} = \mathscr{T}$. There are no constraints, that is $\mathscr{C} = \mathscr{P}$. The evolution of Cauchy data on Cauchy surfaces normal to the direction of evolution, given by a unit Killing vector field of g_X, is given by the Hamiltonian equations of motion with the Hamiltonian equal to the energy function.

7.5 Constraints and Reduction.

Atlas Hypothesis asserts that the primary constraint bundle $\mathscr{P}$ has the structure of a symplectic fibration over the atlas bundle $\mathscr{A}$. For each $\alpha \in \mathscr{A}$, the symplectic form $\omega_{\mathscr{P}_\alpha}$ in $\mathscr{P}_\alpha = \pi^{-1}_{\mathscr{A}\mathscr{P}}(\alpha)$ is weak, that is the mapping $T\mathscr{P}_\alpha \longrightarrow T^*\mathscr{P}_\alpha : v \longrightarrow v \rfloor \omega_{\mathscr{P}_\alpha}$ is one–to–one but not onto. Consequently, not every function on $\mathscr{P}_\alpha$ admits the Hamiltonian vector field. If f and h are functions on $\mathscr{P}_\alpha$ admitting Hamiltonian vector fields ξ_f and ξ_h, respectively, then $[\xi_f, \xi_h]$ is the Hamiltonian vector field of the Poisson bracket $\{f,h\}$ of f and h, c.f. Eq. (6.3.3). Hence, smooth functions on $\mathscr{P}_\alpha$ which admit Hamiltonian vector fields form a Poisson algebra. We shall refer to this Poisson algebra as the *Poisson algebra* of $(\mathscr{P}_\alpha, \omega_{\mathscr{P}_\alpha})$. The *Poisson algebra* of $\mathscr{P}$ consists of smooth functions on $\mathscr{P}$ such that their restrictions to the fibres of $\pi_{\mathscr{A}\mathscr{P}}$ admit Hamiltonian vector fields. The Poisson bracket of such functions is defined by the Poisson brackets in the fibres of $\pi_{\mathscr{A}\mathscr{P}}$, c.f. Eq. (6.3.15). We denote the Poisson algebra of $\mathscr{P}$ by *Poiss*($\mathscr{P}$).

For each infinitesimal symmetry $\xi \in \mathscr{g}$, the corresponding momentum $J^\xi_{\mathscr{Z}}$ pushes forward to a function $J^\xi_{\mathscr{P}}$ on $\mathscr{P}$ such that

$$J^\xi_{\mathscr{P}} = \langle \Theta, \xi_{\mathscr{P}} \rangle . \tag{5.1}$$

As before, we shall refer to $J^\xi_{\mathscr{P}}$ as the *momentum* corresponding to ξ. The map $J^{\mathscr{g}}_{\mathscr{P}} : \mathscr{P} \longrightarrow \mathscr{g}^*$ such that, for each $\xi \in \mathscr{g}$,

$$J^\xi_{\mathscr{P}} = \langle J^{\mathscr{g}}_{\mathscr{P}}, \xi \rangle , \tag{5.2}$$

will be called the *momentum map* for the action of $\mathscr{G}$ in $(\mathscr{P}, d\Theta)$. Since symmetries preserve the Hamilton form Θ, it follows from the discussion in Section 6.3 that $J^{\mathscr{g}}_{\mathscr{P}}$ is equivariant. Theorem 7.4.9 ensures that, for each $\xi \in \mathscr{g}$ and each $\alpha \in \mathscr{A}$, the restriction to $\mathscr{P}_\alpha$ of *ver* $\xi_{\mathscr{P}}$ is the Hamiltonian vector field of the restriction of $J^\xi_{\mathscr{P}}$ to $\mathscr{P}_\alpha$. Hence, the momenta corresponding to the infinitesimal symmetries belong to the Poisson algebra of $\mathscr{P}$.

Localizable infinitesimal symmetries form a Lie subalgebra $\mathscr{h}$ of $\mathscr{g}$.

7.5.1 Proposition. $\mathscr{h}$ is an ideal in $\mathscr{g}$.

Proof. Given $\xi \in \mathscr{h}$ and $\zeta \in \mathscr{g}$, let U and U′ be any two open sets in X with disjoint closures. Since $\xi \in \mathscr{h}$, there exists $\xi' \in \mathscr{g}$ satisfying Eqs. (2.13) and (2.14). Since U is open it follows that in $\pi_{XZ}^{-1}(U)$

$$[\xi',\zeta]_Z = [\xi'_Z,\zeta_Z] = [\xi_Z,\zeta_Z] = [\xi,\zeta]_Z.$$

Similarly, in $\pi_{XZ}^{-1}(U')$

$$[\xi',\zeta]_Z = [\xi'_Z,\zeta_Z] = 0.$$

The vanishing of $[\xi',\zeta]_Z$ on ∂Z is a consequence of vanishing of ξ'_Z on ∂Z and the tangency of ζ_Z to ∂Z. Hence, $[\xi,\zeta]$ is localizable, which implies that $\mathscr{h}$ is an ideal in $\mathscr{g}$.■

The restriction to $\mathscr{h}$ of the momentum map $J^{\mathscr{g}}_{\mathscr{P}} : \mathscr{P} \longrightarrow \mathscr{g}^*$ gives rise to a map $J^{\mathscr{h}}_{\mathscr{P}}$: $\mathscr{P} \longrightarrow \mathscr{h}^*$ which is a *momentum map* for the action of $\mathscr{H}$ in $(\mathscr{P},d\Theta)$ according to the terminology introduced in Section 6.3. For each $\xi \in \mathscr{h}$,

$$\langle J^{\mathscr{h}}_{\mathscr{P}},\xi\rangle = J^{\xi}_{\mathscr{P}} \tag{5.3}$$

is the momentum associated to ξ. The Second Noether Theorem (Prop. 7.3.5) implies that the constraint set $\mathscr{C}$ is contained in the zero level of $J^{\mathscr{h}}_{\mathscr{P}}$,

$$\mathscr{C} \subset (J^{\mathscr{h}}_{\mathscr{P}})^{-1}(0) \tag{5.4}$$

In the following we assume that the constraint set $\mathscr{C}$ coincides with the zero level of $J^{\mathscr{h}}_{\mathscr{P}}$,

$$\mathscr{C} = (J^{\mathscr{h}}_{\mathscr{P}})^{-1}(0) \tag{5.5}$$

and interpret the space

$$\mathscr{S} = \mathscr{C} / \mathscr{H} \tag{5.6}$$

of the $\mathscr{H}$ orbits in $\mathscr{C}$ as the space of physical states of the system. The assumption that $\mathscr{C} = (J^{\mathscr{h}}_{\mathscr{P}})^{-1}(0)$ enables us to discuss the structure of $\mathscr{S}$ in terms of the action of $\mathscr{H}$ as in Section 6.7. We assume further that $\mathscr{H}$ acts in $\mathscr{T}$ by the identity transformation.

This implies that the projection map $\pi_{\mathcal{T}\mathcal{P}}: \mathcal{P} \longrightarrow \mathcal{T}$ induces a projection $\pi_{\mathcal{T}\mathcal{S}}$: $\mathcal{S} \longrightarrow \mathcal{T}$. We assume that $\pi_{\mathcal{T}\mathcal{S}}$ is a fibration. Atlas Hypothesis ensures that $\mathcal{H}$ acts transitively on the fibres of $\pi_{\mathcal{T}\mathcal{A}}: \mathcal{A} \longrightarrow \mathcal{T}$. For each $\alpha \in \mathcal{A}$, we denote by $\mathcal{C}_\alpha$ the intersection of $\mathcal{C}$ with $\mathcal{P}_\alpha$,

$$\mathcal{C}_\alpha = \mathcal{C} \cap \mathcal{P}_\alpha, \tag{5.7}$$

and by $\mathcal{H}_\alpha{}^{\circ}$ the isotropy group of α,

$$\mathcal{H}_\alpha{}^{\circ} = \{g \in \mathcal{H} \mid g_{\mathcal{A}}(\alpha) = \alpha\}. \tag{5.8}$$

Let $\mathcal{H}_\alpha{}'$ denote the normal subgroup of $\mathcal{H}_\alpha{}^{\circ}$ consisting of localizable symmetries which act in $\mathcal{P}_\alpha$ by the identity transformation,

$$\mathcal{H}_\alpha{}' = \{g \in \mathcal{H}_\alpha \mid g_{\mathcal{P}}(\pi) = \pi \quad \forall \quad \pi \in \mathcal{P}_\alpha\}, \tag{5.9}$$

and $\mathcal{H}_\alpha{}''$ the quotient group

$$\mathcal{H}_\alpha{}'' = \mathcal{H}_\alpha{}^{\circ} / \mathcal{H}_\alpha{}'. \tag{5.10}$$

The action of $\mathcal{H}_\alpha{}^{\circ}$ in $\mathcal{P}_\alpha$ induces an action of $\mathcal{H}_\alpha{}''$. For each $g \in \mathcal{H}_\alpha{}^{\circ}$, the action of the corresponding element $[g] \in \mathcal{H}_\alpha{}''$ is given by the diffeomorphism $g_{\mathcal{P}_\alpha}$ obtained by restricting $g_{\mathcal{P}}$ to $\mathcal{P}_\alpha$. Similarly, for each $[\xi]$ in the Lie algebra $h_\alpha{}''$ of $\mathcal{H}_\alpha{}''$ corresponding to $\xi \in h_\alpha{}^{\circ}$, the action of $[\xi]$ on $\mathcal{P}_\alpha$ is given by the vector field $\xi_{\mathcal{P}_\alpha}$ obtained by restricting $\xi_{\mathcal{P}}$ to $\mathcal{P}_\alpha$. Restricting $\xi_{\mathcal{P}} \lrcorner \, d\Theta + d\langle\Theta,\xi_{\mathcal{P}}\rangle = 0$ to $\mathcal{P}_\alpha$ we obtain

$$\xi_{\mathcal{P}_\alpha} \lrcorner \, \omega_{\mathcal{P}_\alpha} = d\langle\Theta,\xi_{\mathcal{P}}\rangle \mid \mathcal{P}_\alpha. \tag{5.11}$$

Hence, the action of $\mathcal{H}_\alpha{}''$ in $(\mathcal{P}_\alpha,\omega_\alpha)$ is Hamiltonian with a momentum map J_α : $\mathcal{P}_\alpha \longrightarrow h_\alpha{}''^{*}$ such that

$$\langle J_\alpha(\pi),[\xi]\rangle = \langle J^{h}_{\mathcal{P}}(\pi),\xi\rangle \tag{5.12}$$

for every $\pi \in \mathcal{P}_\alpha$ and every $\xi \in h_\alpha{}^{\circ}$. Clearly, the zero level of J_α contains the intersection $\mathcal{C}_\alpha$ of $(J^{h}_{\mathcal{P}})^{-1}(0)$ with $\mathcal{P}_\alpha$. We assume further that

$$\mathcal{C}_\alpha = J_\alpha^{-1}(0). \tag{5.13}$$

Since $\mathcal{H}$ acts transitively on the fibres of $\pi_{\mathcal{T}\mathcal{A}}$, for each $\tau \in \mathcal{T}$, the space $\mathcal{S}_\tau$ of the $\mathcal{H}$

orbits in $\mathscr{C}_T$ can be identified with the space of $\mathscr{H}_\alpha^{\,o}$ orbits in $\mathscr{C}_\alpha$ for any $\alpha \in \mathscr{A}$. On the other hand, the orbits of $\mathscr{H}_\alpha^{\,o}$ and $\mathscr{H}_\alpha''$ coincide in $\mathscr{C}_\alpha$. Hence, $\mathscr{S}_T$ can be identified with the space of orbits of a Hamiltonian action of $\mathscr{H}_\alpha''$ in a symplectic manifold $(\mathscr{P}_\alpha, \omega_\alpha)$ which are contained in the zero level of the momentum map J_α,

$$\mathscr{S}_T = J_\alpha^{-1}(0)/\mathscr{H}_\alpha'' .$$

In a sufficiently regular case we can expect that the structure of $\mathscr{S}$ is analogous to that obtained by the Marsden–Weinstein reduction in the case of a finite number of degrees of freedom (c.f. Section 6.7). That is $\mathscr{S}$ is a smooth manifold and $\pi_{\mathscr{T}\mathscr{S}}$: $\mathscr{S} \longrightarrow \mathscr{T}$ is a symplectic fibration. In general, zero levels of momentum maps for Hamiltonian group actions in symplectic manifolds can have singularities. It can be shown in the category of Hilbert manifolds that, under appropriate assumptions on the group action, the singularities of the momentum map are at most quadratic.[1] In relativistic field theories these assumptions are usually satisfied. Moreover, the constraint set $\mathscr{C}$ has an open dense regular component $\mathscr{C}^{\,o}$ stable under the action of the symmetry group, and the space $\mathscr{S}^o = \mathscr{C}^{\,o}/\mathscr{H}$ of the $\mathscr{H}$ orbits in $\mathscr{C}^{\,o}$ is open and dense in $\mathscr{S}$. The projection map $\pi_{\mathscr{T}\mathscr{S}}$ induces a symplectic fibration of $\mathscr{S}^o$ over $\mathscr{T}$.

Let $\mathscr{J}$ be the ideal in the Poisson algebra *Poiss*($\mathscr{P}$) of $\mathscr{P}$ generated by the momenta $J^\xi_{\mathscr{P}}$ for $\xi \in \mathscr{h}$,

$$\mathscr{J} = \{ \sum_{\xi \in \mathscr{h}} f_\xi J^\xi_{\mathscr{P}} \mid f_\xi \in Poiss(\mathscr{P})\}, \qquad (5.14)$$

where $f_\xi \neq 0$ for only a finite number of $\xi \in \mathscr{h}$. Every function $f \in \mathscr{J}$ vanishes on $\mathscr{C} = (J^{\mathscr{h}}_{\mathscr{P}})^{-1}(0)$. Conversely, we assume that every function in Poiss($\mathscr{P}$) vanishing on $(J^{\mathscr{h}}_{\mathscr{P}})^{-1}(0)$ is contained in $\mathscr{J}$. This assumption allows us to identify $\mathscr{J}$ with the space of functions in *Poiss*($\mathscr{P}$) which vanish on $\mathscr{C}$, and *Poiss*($\mathscr{P}$)/$\mathscr{J}$ with the space of restrictions to $\mathscr{C}$ of functions in *Poiss*($\mathscr{P}$). Let ρ : *Poiss*($\mathscr{P}$) $\longrightarrow$ *Poiss*($\mathscr{P}$)/$\mathscr{J}$ be the restriction map associating to each $f \in$ *Poiss*($\mathscr{P}$) its restriction to $\mathscr{C}$. The action of $\mathscr{H}$

---

[1]Arms, Marsden and Moncrief [1981].

in Poiss($\mathscr{P}$) preserves $\mathscr{J}$, and it induces an action of $\mathscr{H}$ in *Poiss*($\mathscr{P}$)/$\mathscr{J}$. We denote by $\mathcal{O}$ the space of the $\mathscr{H}$ invariant elements of Poiss($\mathscr{P}$)/$\mathscr{J}$,

$$\mathcal{O} = \{\rho(f) \mid g^*_{\mathscr{P}} f - f \in \mathscr{J} \ \forall \ g \in \mathscr{H}\} \tag{5.15}$$

Since elements of *Poiss*($\mathscr{P}$)/$\mathscr{J}$ are identified with function on $\mathscr{C}$, elements of $\mathcal{O}$ can be identified with functions on the space $\mathscr{S}$ of the $\mathscr{H}$ orbits in $\mathscr{C}$.

7.5.2. Proposition $\mathcal{O}$ has a Poisson algebra structure such that for every $f,f' \in \rho^{-1}(\mathcal{O})$,

$$\{\rho(f),\rho(f')\} = \rho(\{f,f'\}). \tag{5.16}$$

Proof. The Poisson bracket *in Poiss*($\mathscr{P}$) is defined in terms of the Poisson brackets in fibres of $\pi_{\mathscr{T}\mathscr{P}}$ which are symplectic manifolds. For any $\tau \in \mathscr{T}$, the fibre $\mathscr{S}_\tau$ can be identified with $J_\alpha^{-1}(0)/\mathscr{H}_\alpha''$, where $\mathscr{H}_\alpha''$ is given by Eq. (5.10), and J_α is given by Eq. (5.12). Consider the ideal $\mathscr{J}_\alpha$ in the Poisson algebra *Poiss*($\mathscr{P}_\alpha$) of $(\mathscr{P}_\alpha, \omega_{\mathscr{P}_\alpha})$ generated by the momenta $J_\alpha^{[\xi]}$ for $[\xi] \in \mathscr{h}_\alpha''$. For every $f \in \rho^{-1}(\mathcal{O})$, the restriction $\rho(f) \mid \mathscr{S}_\tau$ of $\rho(f)$ to $\mathscr{S}_\tau$ coincides with the element $\rho_\alpha(f \mid \mathscr{P}_\alpha) \in$ *Poiss*($\mathscr{P}_\alpha$)/$\mathscr{J}_\alpha$, where ρ_α is the canonical projection from *Poiss*($\mathscr{P}_\alpha$) to *Poiss*($\mathscr{P}_\alpha$)/$\mathscr{J}_\alpha$. Since $\rho(f)$ is $\mathscr{H}$ invariant, $\rho_\alpha(f \mid \mathscr{P}_\alpha)$ is $\mathscr{H}_\alpha''$ invariant. By Proposition 6.3.1 the space $\mathcal{O}_\alpha$ of $\mathscr{H}_\alpha''$ invariant elements of *Poiss*($\mathscr{P}_\alpha$)/$\mathscr{J}_\alpha$ is a Poisson algebra, and

$$\{\rho_\alpha(f \mid \mathscr{P}_\alpha),\rho_\alpha(f' \mid \mathscr{P}_\alpha)\} = \rho_\alpha(\{f \mid \mathscr{P}_\alpha, f' \mid \mathscr{P}_\alpha\}) \tag{5.17}$$

for every $f,f' \in \rho^{-1}(\mathcal{O})$. Thus, we may define the Poisson bracket in $\mathcal{O}$ by setting

$$\{\rho(f),\rho(f')\} \mid \mathscr{S}_\tau = \{\rho_\alpha(f \mid \mathscr{P}_\alpha),\rho_\alpha(f' \mid \mathscr{P}_\alpha)\}. \tag{5.18}$$

By Atlas Hypothesis the group $\mathscr{H}$ acts transitively on $\mathscr{A}_\tau$, and it induces isomorphisms of the corresponding Poisson algebras. Since $\rho(f)$ and $\rho(f')$ are $\mathscr{H}$ invariant, the right hand side of Eq. (5.18) is independent of $\alpha \in \mathscr{A}_\tau$. Eq. (5.17) implies Eq. (5.16). ∎

In a sufficiently regular case, when $\pi_{\mathcal{T}\mathcal{S}} : \mathcal{S} \longrightarrow \mathcal{T}$ is a smooth symplectic fibration and, therefore, $\mathcal{S}$ is a Poisson manifold, functions in $\mathcal{O}$ are smooth and the Poisson algebra $\mathcal{O}$ coincides with the Poisson algebra of $\mathcal{S}$,

$$\mathcal{O} = Poiss(\mathcal{S}) \tag{5.19}$$

In a singular case we have no independent definition of the Poisson algebra of $\mathcal{S}$, while the Poisson algebra $\mathcal{O}$ is well defined.

Functions in $Poiss(\mathcal{P})$ have the physical interpretation of dynamical variables. Since the physical states are described by points in $\mathcal{C}$, measurements of dynamical variables in physical states cannot distinguish between functions which agree on $\mathcal{C}$. Moreover, all points in $\mathcal{C}$ on the same $\mathcal{H}$ orbit describe the same physical state. Thus, under the assumptions made above, observable dynamical variables are given by elements of the Poisson algebra $\mathcal{O}$.

7.5.3. Proposition. The map $\mathcal{g} \longrightarrow \mathcal{O}\colon \xi \longrightarrow \rho(J^{\xi}_{\mathcal{P}})$ is a Lie algebra homomorphism.

Proof. For every $\xi,\zeta \in \mathcal{g}$,

$$\begin{aligned}\{J^{\xi}_{\mathcal{P}},J^{\zeta}_{\mathcal{P}}\} &= <dJ^{\zeta}_{\mathcal{P}},\, ver\,\xi_{\mathcal{P}}> \\ &= <dJ^{\zeta}_{\mathcal{P}},\, \xi_{\mathcal{P}} - hor\,\xi_{\mathcal{P}}> \\ &= J^{[\xi,\zeta]}_{\mathcal{P}} - <dJ^{\zeta}_{\mathcal{P}},\, hor\,\xi_{\mathcal{P}}>.\end{aligned}$$

The First Noether theorem and the definition of the constraint set imply that $<dJ^{\zeta}_{\mathcal{P}},\, hor\,\xi_{\mathcal{P}}> | \,\mathcal{C} = 0$. By hypothesis, every function in $Poiss(\mathcal{P})$ which vanishes on $\mathcal{C}$ is contained in $\mathcal{J}$. Applying the projection $\rho : \mathrm{Poiss}(\mathcal{P}) \longrightarrow \mathrm{Poiss}(\mathcal{P})/\mathcal{J}$ to the above equation, and taking into account Eq. (5.16) we obtain

$$\{\rho(J^{\xi}_{\mathcal{P}}),\rho(J^{\zeta}_{\mathcal{P}})\} = \rho(J^{[\xi,\zeta]}_{\mathcal{P}}) \;.$$

which copletes the proof. ∎

REFERENCES

ARMS, J. [1981], "The Structure of the Solution Set for the Yang–Mills Equations", *Math. Proc. Camb. Phil. Soc.*, 90, 361–372.

ARMS, J., MARSDEN, J. and MONCRIEF. V. [1981], "Symmetry and Bifurcations of Momentum Mappings", *Comm. Math. Phys.*, 78, 455–478.

BINZ, E. and ŚNIATYCKI, J. [1986], "Conservation Laws in Spacetimes with Boundary", *Class. Quantum Grav.*, 3, 1191–1197

BERGMANN, P.G. and GOLDBERG. I. [1954], "Dirac Bracket Transformations in Phase Space", *Phys. Rev.*, 98, 531–538

CARTAN, E. [1933], "*Les espaces metriques fondés sur la notion d'aire*", Actualités Sc. et Industr., No. 72, Hermann, Paris.

DEDONDER,. Th. [1929], "Théorie invariantive du clacul des variations", *Bull. Acad. de Belg.*, Chap. 1.

FISCHER, A., MARSDEN, J and MONCRIEF, V. [1980], "The Structure of the Space of Solutions of Einstein's Equations. I. One Killing Field", *Ann. Inst. H. Poincaré*, 33, 147–194.

GARCIA, P.L. and PEREZ–RENDON, A. [1971], "Symplectic Approach to the Theory of Quantized Fields. II", *Archive Rat. Mech. Anal.*, 43 ,101–124.

GAWĘDZKI, K. [1972], "On the Geometrization of the Canonical Formalism in the Classical Field Theory", *Rep. Math. Phys.*, 3, 307–325

GIMMSY [1988], see GOTAY, ISENBERG, MARSDEN, MONTGOMERY, ŚNIATYCKI and YASSKIN [1988].

GOLDSCHMIDT, H. and STERNBERG, S. [1973], "The Hamilton–Jacobi Formalism in the Calculus of Variations", *Ann. Inst. Fourier (Grenoble)*, 23, 203–267.

GOTAY, M., ISENBERG, J., MARSDEN, J., MONTGOMERY, R., ŚNIATYCKI, J. and YASSKIN Ph. [1988], "Momentum Maps and the Hamiltonian Treatment of Classical Field Theories with Constraints", in preparation.

KIJOWSKI, J. and SZCZYRBA, W. [1975], "Multisymplectic Manifolds and a Geometric Construction of the Poisson Bracket in the Classical Field Theory", in *Géométrie symplectique et physique mathématique*, J.–M. Souriau (ed.), Publications CNRS No. 237, Paris, pp. 347–379.

KIJOWSKI, J. and TULCZYJEW, W.M. [1979], *A Symplectic Framework for Field Theories*, Lecture Notes in Physics, vol. 107, Springer, Berlin.

LEPAGE, Th.H.J. [1936], *Acad. Roy. Belg. Bull.*, Cl. Sci. V. Ser. 22, 716.

NOETHER, E. [1918], "Invariante Variationsprobleme", *Nachr. Gess. Wiss. Göttingen*, 235–257.

ŚNIATYCKI, J. [1970a], "On the Geometric Structure of Classical Field Theory in Lagrangian Formulation", *Proc. Camb. Phil. Soc.*, 68, 475–483.

ŚNIATYCKI, J. [1970b], "On the Canonical Formulation of General Relativity", Proceedings of *Journées relativistes 1970*, pp. 127–135, Faculté des Sciences, Caen.

TULCZYJEW, W.M. [1968], Warsaw seminars on "*Geometry of Phase Space*", unpublished.

Chapter 8

YANG–MILLS THEORY

Yang–Mills fields are connections in a principal fibre bundle over a space–time manifold. Matter fields are sections of an associated bundle. Dynamics of interacting Yang–Mills and matter fields is given by a Lagrangian invariant under the group of automorphisms of the principal bundle which preserve the bundary conditions and cover isometries of the space time (conformal isometries in the absence of matter fields). Localizable symmetries are gauge transformations, that is automorphisms of the principal bundle which reduce to the identity on the boundary and cover the identity transformation of space–time. The presence of gauge transformations gives rise to constraints which can be studied in terms of the formalism developed in the preceding chapter.

In this chapter we show that Yang–Mills theory admits a covariant Hamiltonian description of dynamics developed in the preceding section. The conformal invariance of the Yang–Mills Lagrangian in absence of matter enables us to relate a Yang–Mills theory on a manifold with boundary to a Yang–Mills theory on Minkowski space.

Electrodynamics can be interpreted as a Yang–Mills theory for the structure group U(1). In fact, Yang–Mills theory is a generalization of this formulation of electrodynamics to non–abelian structure groups. Since a lot of our understanding of Yang–Mills theory is based on the analogy with electrodynamics we give a detailed discussion of some aspects of electrodynamics. In particular we study the notions of electric and magnetic charges in electrodynamics and their generalization to Yang–Mills theory. We relate the charges to symmetries of boundary conditions for Yang–Mills fields; matter fields are assumed to vanish on the boundary.

8.1 Field equations.

Yang–Mills fields fields are connections in a principal G bundle P over the space time manifold X.[1] We assume here that X is an oriented and time oriented Lorentzian manifold with a Lorenzian metric g_X of signature $(-,+,+,+)$, and that the Lie algebra $\mathfrak{g}$ of G is endowed with a metric $g_{\mathfrak{g}}$ invariant under the adjoint action of G. Usually connections are described in terms of a connection form ω on P or the corresponding horizontal distribution *hor* $TP = \{u \in TP \mid \omega(u) = 0\}$. The curvature of a connection ω in P is the horizontal part of the exterior differential of ω, $\Omega = d\omega + [\omega,\omega]$, where $[,]$ denotes the Lie bracket in $\mathfrak{g}$. If σ_{PX} is a local section od P, the pull back of the connection form ω is a $\mathfrak{g}$–valued 1–form

$$A = \sigma^{*}_{PX}\omega \tag{1.1}$$

called a Yang–Mills potential, and the pull back of the curvature form Ω is a $\mathfrak{g}$–valued 2–form

$$F = \sigma^{*}_{PX}\Omega \tag{1.2}$$

called the field strength. They are related by,

$$F = dA + [A,A]. \tag{1.3}$$

In field theory one describes connections in terms of Yang–Mills potentials (A) corres–ponding to some (fixed) family of trivializations of P, and we follow this approach. Abusing somewhat the terminology we sometimes refer to a Yang–Mills potential A as a connection in P, and to the field strength F as the curvature of A.

The metrics in X and $\mathfrak{g}$ give rise to a scalar product $<,>$ in the space of $\mathfrak{g}$–valued forms on X. The dynamics of Yang–Mills field is determined by a Lagrangian $L[\omega]$

[1]The case when $G = U(1)$ corresponds to electromagnetic field, c.f. Sec. 8.5. The generalization of the local gauge invariance to the structure group SU(2) was given by Yang and Mills, [1954], and further generalized by Utiyama, [1956], and Sakurai, [1960]. The reformulation of the Utiyama theory in terms of connections in principal fibre bundles was given in Kerbrat–Lunc [1964], and Trautman [1967].

given by

$$L[\omega] = (1/4)\langle F,F\rangle, \tag{1.4}$$

It should be noted that $L[\omega]$ is a function on X depending only on ω, and not on the choice of a trivialization σ_{PX} of P used in Eq. (1.2). The Euler–Lagrange equations corresponding to the Yang–Mills Lagrangian $L[\omega]$ are given by

$$D^*F = 0, \tag{1.5}$$

where D denotes the covariant derivative relative to the connection determined by A, and * is the Hodge operator defined by the orientation of X and the metric g_X. Eqs. (1.3) and (1.5) are called Yang–Mills equations.

Connections in a principal bundle P over X can be thought of as sections of the connection bundle of P, that is a bundle Y over X with the total space given by

$$Y = J^1P/G \tag{1.6}$$

and the projection $\pi_{XY}:X \longrightarrow Y$ induced by the projection $\pi_{XP}:P \longrightarrow X$. In the following we shall identify sections σ_{YX} of the connection bundle with the corresponding connections and their connection forms. There exists a unique Lagrangian form Λ on the first jet bundle Z of Y such that, for each section σ_{YX} of the connection bundle

$$j^1\sigma_{YX}{}^*\Lambda = L[\omega]\,vol[g_X], \tag{1.7}$$

where $vol[g_X]$ denotes the volume form on X defined by the orientation of X and the metric g_X.

Let (x^μ) be coordinate system in X with domain U, $\sigma_{PX}:U \longrightarrow P$ a local trivialization of P, and (e_a) a basis in $\mathfrak{g}$, a = 1,...,n. Given a connection form ω, its Yang–Mills potential A relative to σ_{PX} can be expressed in terms of functions $A^a_\mu(x^\alpha)$ such that

$$A = A^a_\mu(x^\alpha)e_a dx^\mu. \tag{1.8}$$

Hence, we can use local coordinates (x^μ, y^a_μ) in Y, defined in $\pi_{XY}{}^{-1}(U)$, such that a local representation of the section σ_{YX} of Y corresponding to the Yang–Mills potential form A is given by

$$y^a_\mu = A^a_\mu(x^\nu). \tag{1.9}$$

Similarly,

$$F = \tfrac{1}{2} F^a_{\mu\nu} e_a dx^\mu \wedge dx^\nu, \tag{1.10}$$

where

$$F^a_{\mu\nu} = A^a_{\nu,\mu} - A^a_{\mu,\nu} + c^a{}_{bc} A^b_\mu A^c_\nu , \tag{1.11}$$

$A^a_{\nu,\mu}$ denote the derivatives of A^a_ν with respect to x^μ, and c^a_{bc} denote the structure constants of $\mathfrak{g}$ relative to the basis (e_a). Let $(g_{\mu\nu})$ be the components of g_X relative to the coordinates (x^μ),

$$g_X = g_{\mu\nu} dx^\mu \otimes dx^\nu, \tag{1.12}$$

$(g^{\mu\nu})$ the inverse of $(g_{\mu\nu})$, (g_{ab}) the components of $g_{\mathfrak{g}}$ in the basis (e_a),

$$g_{\mathfrak{g}}(e_a, e_b) = g_{ab}, \tag{1.13}$$

and (g^{ab}) the inverse of (g_{ab}). Following the standard convention of field theory, we shall use the matrices $(g_{\mu\nu})$ and $(g^{\mu\nu})$ to lower and to raise the indices related to the chart (x^μ), respectively. Similarly, the indices related to the basis (e_a) in $\mathfrak{g}$ will be raised by multiplication by the matrix (g^{ab}), and lowered with the help of (g_{ab}). The local expression for $j\sigma_{YX}^* \Lambda$ is given by

$$j^1\sigma_{YX}^* \Lambda = (1/4) F^a_{\mu\nu} F^{\mu\nu}_a (-\mathit{det}\, g_X)^{1/2} d_4x , \tag{1.14}$$

where $\mathit{det}\, g_X$ is the determinant of $(g_{\mu\nu})$.

Let $(x^\mu, y^a_\mu, z^a_{\mu\nu})$ be the coordinates in the first jet bundle Z of Y induced by the coordinates (x^μ, y^a_ν) in Y. We have the following local expression for the Yang–Mills Lagrangian form Λ on Z:

$$\Lambda = (1/4)(-\mathit{det} g_X)^{1/2}(z^a_{\nu\mu} - z^a_{\mu\nu} + c^a_{bc} y^b_\mu y^c_\nu)(z^{\nu\mu}_a - z^{\mu\nu}_a + c^{ef}_a y^\mu_e y^\nu_f) d_4x \tag{1.15}$$

The DeDonder form Ω corresponding to the Yang–Mills Lagrangian has the following local representation

$$\Omega = (-det g_X)^{1/2}\, v_a^{\mu\nu}\, dy_\mu^a \wedge d_3x_\nu + \\ -\tfrac{1}{4}(-det g_X)^{1/2}\{v_a^{\mu\nu}\, v_{\mu\nu}^a + 2c_{bc}^a\, v_a^{\mu\nu}\, y_\mu^b\, y_\nu^c\} d_4x \tag{1.16}$$

where $v_a^{\mu\nu}$ is the derivative, with respect to $z_{\mu\nu}^a$ of the coefficient of $(-det g_X)^{1/2}\, d_4x$ on the right hand side of Eq. (1.15).

$$v_a^{\mu\nu} = (z_a^{\mu\nu} - z_a^{\nu\mu} + c_a^{bc}\, y_b^\nu\, y_c^\mu) \tag{1.17}$$

Let σ_{ZX} be a local section of the source map $\pi_{ZX} : X \longrightarrow Z$ given by

$$y_\mu^a = A_\mu^a(x) \quad \text{and} \quad z_{\mu\nu}^a = \phi_{\mu\nu}^a(x). \tag{1.18}$$

The DeDonder equation (7.1.14) implies that

$$\phi_{\mu\nu}^a - \phi_{\nu\mu}^a = A_{\mu,\nu}^a - A_{\nu,\mu}^a\,, \tag{1.19}$$

so that

$$v_a^{\mu\nu} = F_a^{\nu\mu}(x) \tag{1.20}$$

on the image of σ_{ZX}, and

$$F_{a;\nu}^{\mu\nu} + c_{ac}^b\, F_b^{\mu\nu}\, A_\nu^c = 0 \tag{1.21}$$

where the semicolon denotes the covariant differentiation of tensors with respect to the Levi–Civita connection of g_X. Eq. (1.21) is equivalent to Eq. (1.5). Hence, the DeDonder equation for σ_{ZX} implies that its projection σ_{YX} satisfies the Yang–Mills equations. Thus, Yang–Mills theory satisfies Hypothesis 7.1.3.

Matter fields ψ in Yang–Mills theory are usually described in terms of sections of vector bundles associated to P which vanish on the boundary ∂X of X. The total Lagrangian of matter fields ψ interacting with a Yang–Mills field ω is the sum of the Yang–Mills Lagrangian $L[\omega]$, given by Eq. (1.4), and the matter Lagrangian $L[\omega,\psi]$ which is usually homogeneous in ψ.[2] The Lagrange–Euler equations for the interacting

---

[2]For example see the Dirac Lagrangian in Sec. 8.5.

system are the matter field equations and the Yang–Mills equations (1.5) modified by the presence of matter,

$$D^*F + j = 0 , \qquad (1.22)$$

where j is a $\mathfrak{g}$–valued 3–form on X given by the derivative of the matter Lagrangian eith respect to the Yang–Mills field. In terms of a local coordinates we have

$$j = j_a^\mu e^a d_3 x^\mu , \qquad (1.23)$$

where

$$j_a^\mu = \frac{\partial L[\omega,\psi]}{\partial A_\mu^a} . \qquad (1.24)$$

8.2 Gauge transformations and conservation laws

If σ_{PX} and σ'_{PX} are two sections of P over the same domain U in X, then there exists a mapping $h:U \longrightarrow G$ such that, for each $x \in U$,

$$\sigma_{PX}(x) = h(x)\sigma'_{PX}(x) \; . \tag{2.1}$$

The pull backs of a connection form ω by σ_{PX} and σ'_{PX} are related as follows,

$$\sigma_{PX}^{\ *}\omega = \mathrm{ad}(h^{-1})\sigma'_{PX}{}^{*}\omega + h^{*}\beta \, , \tag{2.2}$$

where β is the canonical $\mathfrak{g}$–valued left invariant form on G such that

$$\beta(\xi) = \xi \tag{2.3}$$

for every ξ in $\mathfrak{g}$. Similarly, the pull backs of the curvature form are related by

$$\sigma_{PX}^{\ *}\Omega = \mathrm{ad}(h^{-1})\sigma'_{PX}{}^{*}\Omega \, . \tag{2.4}$$

In field theory, transitions between different local descriptions of the connection and curvature forms given by Eqs. (2.2) and (2.4) are called *gauge transformations.*

The independence of $L[\omega] = (1/4)<\sigma_{PX}^{\ *}\Omega, \sigma_{PX}^{\ *}\Omega>$ from the choice of a trivializing section σ_{PX}, referred to as the *gauge invariance* of the Yang–Mills Lagrangian, implies the L[A] is intrinsically defined in terms of the structure of the theory, and it is invariant under the automorphisms of this structure. Since the theory is formulated in terms of a principal G bundle P over an oriented Lorentzian manifold X, automorphisms of its structure are the automorphisms h_P of P, which cover orientation preserving isometries h_X of X. Let us recall that an automorphism of a principal fibre bundle P over X is a diffeomorphism h_P of P commuting with the right action of G in P,

$$h_P(pg) = h_P(p)g \tag{2.5}$$

for every $g \in G$ and $p \in P$. Thus, h_P preserves the fibration of P, so that there exists a unique diffeomorphism h_X of X such that

$$\pi_{XP}h_P = h_X\pi_{XP} \; . \tag{2.6}$$

Automorphisms of a principal bundle P act naturally in the connection bundle Y of P, and this action extends to the jet bundle Z of Y. For each automorphisms h_P of P, we denote by h_Y and h_Z the corresponding diffeomorphisms of Y and Z, respectively. Since the Lagrangian form Λ defined is defined in terms of L and the volume form $vol[g_X]$ on X, it is preserved by the action of all automorphisms h_P of P which induce orientation preserving isometries in X,

$$h_Z^* \Lambda = \Lambda . \tag{2.7}$$

Moreover, because of the way the metric on X appears in the construction of L and the volume form $vol[g_X]$, the Lagrangian form Λ is also preserved by the automorphisms of P which induce conformal transformations in X, that is diffeomorphisms h_X such that

$$h_X^* g_X = k^2 g_X \tag{2.8}$$

for a nonvanishing function k on X.

Automorphisms of P which induce the identity transformation in X form a normal subgroup $Aut_X P$ of the group of all automorphisms of P. The group $Aut_X P$ is isomorphic to the group of sections of the bundle P[G] associated to P with typical fibre G. We find it convenient to identity these groups. For each $h_P \in Aut_X P$ and each section σ_{PX} of P with domain U, $h_P \circ \sigma_{PX}$ is also a section of P with domain U so that there exists a mapping $h : U \longrightarrow G$ such that

$$h_P(\sigma_{PX}(x)) = \sigma_{PX}(x)h(x) \tag{2.9}$$

for every $x \in U$. Hence the mapping h appearing in Eq. (2.1) can be considered as a local description of an automorphism h_P, and the gauge transformations (2.2) and (2.4) are the local descriptions of the action of h_P on the connection and curvature forms, respectively. The interpretation of gauge transformations in terms of the action of $Aut_X P$ is more convenient for our purposes.[1]

The boundary conditions B considered here are given by specifying the pull back $\omega_{\partial P}$ of the connection form to ∂P. Hence, the symmetry group $\mathscr{G}$ of the theory

[1]The relationship of these two interpretations of gauge transformations is the same as the relationship between the passive and the active interpretations of Lorentz transformations discussed in relativistic dynamics and quantum theory.

contains the group $Aut(P;\omega_{\partial P},[g_X])$ of automorphisms of P which preserve the boundary condition $\omega_{\partial P}$ and the conformal class $[g_X]$ of the metric g_X in X. The Lie algebra $aut(P;\omega_{\partial P},[g_X])$ of $Aut(P;\omega_{\partial P},[g_X])$ consists of vector fields ξ_P on P which push forward to vector fields $\xi_{\partial P}$ on ∂P preserving the connection form $\omega_{\partial P}$,

$$\pounds_{\xi_{\partial P}} \omega_{\partial P} = 0, \tag{2.10}$$

and project to conformal Killing vector fields ξ_X on X,

$$\pounds_{\xi_X} g_X = k\, g_X, \tag{2.11}$$

for some function k on X.

The group $\mathscr{H}$ of localizable symmetries contains automorphisms in $aut_X P$ which induce the identity transformation in $T^* \partial P$. We have interpreted the gauge transformations in terms of the action of $Aut_X P$ on connection forms. Hence, the group of localizable symmetries contains all the gauge transformations which preserve the Dirichlet data. The question arises whether all localizable symmetries of the Yang–Mills theory are of this form.

The Lie algebra $\mathscr{h}$ of $\mathscr{H}$ contains the Lie algebra of vector fields ξ_P on P which vanish on ∂P and have vanishing projections to X. For a space–time conformally equivalent to Minkowski space, the vanishing of the momentum map on the elements of $\mathscr{h}$ gives all the constraints of the theory.[2]

Let ξ_P be an infinitesimal automorphism of P, ω a connection form on P, and A $= A^a_\mu e_a dx^\mu$ its Yang–Mills potential. If

$$\xi_P = \xi^\mu \frac{\partial}{\partial x^\mu} + \xi^a e_a \tag{2.12}$$

then the horizontal and the vertical components of ξ_P relative the connection ω are given by

[2]See Eardley and Moncrief [1982a,b] and the references quoted there.

$$hor\, \xi_P = \xi^\mu \frac{\partial}{\partial x^\mu} + A^a_\mu\, \xi^\mu\, e_a \tag{2.13}$$

$$ver\, \xi_P = (\xi^a - A^a_\mu\, \xi^\mu) e_a \tag{2.14}$$

The coordinate expression for the vector field ξ_Y on Y corresponding to ξ_P is

$$\xi_Y = \xi^\mu \frac{\partial}{\partial x^\mu} + (\xi^a_{\ ,\mu} + c^a_{bc}\, y^b_\mu\, \xi^c - y^a_\nu\, \xi^\nu_{\ ,\mu}) \frac{\partial}{\partial y^a_\mu}. \tag{2.15}$$

The left interior product of ξ_Z with the DeDonder form Ω, given by Eq. (1.16), is

$$\begin{aligned}\xi_Z \rfloor \Omega = (-det g_X)^{1/2}\{&(\xi^a_{\ ,\mu} + c^a_{bc}\, y^b_\mu\, \xi^c - y^a_\lambda \xi^\lambda_{\ ,\mu}) v^{\mu\nu}_a\, d_3x_\nu + \\ &-\tfrac{1}{4}(v^a_{\mu\nu}\, v_a^{\ \mu\nu} + 2c^a_{bc}\, v^{\mu\nu}_a\, y^b_\mu\, y^c_\nu)\xi^\lambda d_3x_\lambda - v^{\mu\nu}_a\, \xi^\lambda\, dy^a_\mu \rfloor d_2x_{\nu\lambda}\}\end{aligned} \tag{2.16}$$

We want to evaluate the pull back of $\xi_Z \rfloor \Omega$ by $j^1\sigma_{YX}$ such that $v^a_{\mu\nu} = F^a_{\nu\mu}(x^\alpha)$ and $y^a_\mu = A^a_\mu(x^\alpha)$. Taking into account the identity

$$dx^\lambda \wedge d_2x_{\mu\nu} = \delta^\lambda_\nu d_3x_\mu - \delta^\lambda_\mu d_3x_\nu \tag{2.17}$$

we obtain

$$\begin{aligned} j^1\sigma_{YX}{}^*(\xi_Z \rfloor \Omega) = \{&(-det g_X)^{1/2} F^{\mu\nu}_a (\xi^a - A^a_\lambda \xi^\lambda)\}_{,\nu} d_3x_\mu + \\ &+ (-det g_X)^{1/2}\{D_\mu F^{\mu\nu}_a (\xi^a - A^a_\lambda \xi^\lambda) + (-F^{\mu\nu}_a F^a_{\mu\lambda} + \tfrac{1}{4}\,\delta^\nu_\lambda\, F^{\alpha\beta}_a\, F^a_{\alpha\beta})\xi^\lambda\} d_3x_\nu \end{aligned} \tag{2.18}$$

where D_μ denotes the derivative with respect to x^μ which is covariant with respect to the connection ω in P and the Levi–Civita connection of g_X in X.

Let M be a Cauchy surface in X with boundary $\partial M \subset \partial X$ and the future oriented unit normal vector η_M tangent to ∂X along ∂M. We assume that the coordinates (x^μ) in X, used in Eq. (2.18), are chosen in such a way that $x^0 | M = 0$ and $\eta_M = \frac{\partial}{\partial x^0}$. The integral of $(j^1\sigma_{YX})^*(\xi_Z \rfloor \Omega)$ over M gives the value of the conserved momentum J^ξ on the Cauchy data of σ_{YX} on M, cf. Eq. (7.2.12),

$$J^{\xi}(j^1\sigma_{YX}|M) = \int_{\partial M} (-\mathit{detg}_X)^{1/2} F^{k0}_a (\xi^a - A^a_\lambda\,\xi^\lambda) d_2x_{0k} +$$

$$+ \int_M (-\mathit{detg}_X)^{1/2}\, D_m F^{m0}_a (\xi^a - A^a_\lambda\,\xi^\lambda) d_3x_0 + \int_{\partial M} (-\mathit{detg}_X)^{1/2}\, F^{m0}_a\, F^a_{mn}\, \xi^n\, d_3x_0 +$$

$$+ \tfrac{1}{2} \int_M (-\mathit{detg}_X)^{1/2} (- F^{0m}_a F^a_{0m} + \tfrac{1}{2} F^{mn}_a\, F^a_{mn}) \xi^0\, d_3x_0 \tag{2.19}$$

In order to express the physical and geometric interpretation of the right hand side of Eq. (2.19), we denote by A_M the pull back of A to M, by F_M the pull back of F to M, and by E_M the pull back of $\eta_M \lrcorner F$ to M, so that

$$E_M(u) = (\eta_M \lrcorner F)(u) \tag{2.20}$$

for every $u \in TM$. The $\mathfrak{g}$–valued 1–form A_M on M describes a connection in the pull back of P to M. The 2–form F_M corresponds to the curvature of this connection, it is called the "magnetic" part of F relative to M. The 1–form E_M on M describes the "electric" part of F relative to M. This terminology stems from the standard decomposition of electromagnetic field, that is F for the structure group U(1), into the electric and magnetic field relative to an observer c.f. Sec. 8.5. The restriction of ξ_Y to $\sigma_{YX}(M)$ can be decomposed into its vertical and horizontal components relative to the connection given by A. The vertical component of ξ_Y corresponds to an infinitesimal automorphism of P tangential to the fibres of π_{XP} which can be identified with a section of the adjoint bundle $P(\mathfrak{g})$, c.f. Sec. 3.3. We denote by *ver* ξ the $\mathfrak{g}$–valued function on M associated to this section by the trivializing section σ_{XP} used here. In terms of local coordinates

$$\mathit{ver}\,\xi = (\xi^a - A^a_\lambda \xi^\lambda) e_a . \tag{2.21}$$

The metric g_M on M, given by the pull back of g_X to M, gives rise to the Hodge * operator in the space of $\mathfrak{g}$–valued forms on M. Denoting by ξ_M the component of ξ_X tangential to M,

$$\xi_X | M = - g_X(\xi_X, \eta_M)\eta_M + \xi_M \tag{2.22}$$

we can rewrite Eq. (2.19) in the form

$$J^{\xi}(j^1\sigma_{YX}|M) = \int_{\partial M} g_g(*E_M, ver\,\xi) + \tag{2.23}$$

$$-\int_M g_g(D^*E_M, ver\,\xi) - \int_M \langle E_M, \xi_M \lrcorner F_M\rangle vol[g_M] +$$

$$+\frac{1}{2}\int_M \{\langle E_M,E_M\rangle + \frac{1}{2}\langle F_M,F_M\rangle\}\xi_X \lrcorner vol[g_X]$$

If ξ_Y is vertical along M, then the last two integrals vanish and the corresponding momentum

$$J^{\xi} = \int_{\partial M} g_g\,(*E_M, ver\,\xi) - \int_M g_g\,(D^*E_M, ver\,\xi) \tag{2.24}$$

has an interpretation of the conserved charge corresponding to ξ. If ξ is an infinitesimal gauge symmetry, $\xi \in \mathscr{h}$, then $ver\,\xi$ vanishes on ∂X and the first integral vanishes. The Second Noether Theorem ensures that, for $\xi \in \mathscr{h}$, J^{ξ} vanishes on the Cauchy data of solutions of the field equations. Hence, we obtain the constraint equation

$$D^*E_M = 0. \tag{2.25}$$

It coincides with the pull back to M of the Yang–Mills equation (1.5). For the Cauchy data satisfying the constraint equation the second integral in Eq. (2.24) vanishes and the conserved charge is given by

$$J^{\xi} = \int_{\partial M} g_g(*E_M, ver\,\xi) \tag{2.26}$$

If ξ_Y is horizontal along M then the first two integrals in Eq. (2.23) vanish. In the case when ξ_X is normal to M the third integral also vanishes. If

$$\xi_X|M = \eta_M \tag{2.27}$$

then

$$J^{\xi} = \frac{1}{2}\int_M \{\langle E_M,E_M\rangle + \frac{1}{2}\langle F_M,F_M\rangle\} vol[g_M] \tag{2.28}$$

is the energy of the classical state $j^1\sigma_{YX}|M$. For horizontal ξ_Y, and ξ_X tangent to M, the only surviving term in Eq. (2.23) is

$$J^{\xi} = - \int_M \langle E_M, \xi_M \,\lrcorner\, F_M \rangle \, vol[g_M] \tag{2.29}$$

which describes an analogue of a linear or an angular momentum.

8.3 Hamiltonian formulation

A typical Cauchy surface in X is a 3–dimensional compact submanifold of X, with boundary ∂M contained in ∂X, and such that the pull–back of g_X to M is a Riemannian metric g_M on M. Following the notation of Sec. 7.3, we denote by $\mathcal{T}$ the space of parametrized Cauchy surfaces given by embeddings of M into X, by $\mathcal{D}$ the space of the Dirichlet data, and by $\mathcal{Z}$ the space of the Cauchy data on Cauchy surfaces in $\mathcal{T}$. For each $\tau \in \mathcal{T}$, Dirichlet data $\delta \in \mathcal{D}_\tau$ correspond to a connection ω in P defined along $\tau(M)$, which can be represented as a pair (α_τ, A_τ), where α_τ is the restriction of the connection A to the directions normal to $\tau(M)$, and A_τ is the induced connection in the pull–back of P to $\tau(M)$. In the following we shall identify δ with (α_τ, A_τ).

$$\delta = (\alpha_\tau, A_\tau) . \tag{3.1}$$

Similarly, Cauchy data γ in $\mathcal{Z}_\tau$ are 1–jets of connections in P defined along $\tau(M)$. They describe the curvature Ω of the connection ω defined along $\tau(M)$ and the 1–jet $j^1\alpha_\tau$ of α_τ. We denote by F the pull back of Ω by σ_{PX}. It is a $\mathfrak{g}$–valued form on X defined along $\tau(M)$. Let η_τ denote the future oriented unit normal to $\tau(M)$ and E_τ the left interior product of η_τ and F pulled back to $\tau(M)$,

$$E_\tau(u) = (\eta_\tau \lrcorner F)(u) \tag{3.2}$$

for every $u \in T\tau(M)$. We shall identify the element γ of $\mathcal{Z}_\tau$ with the triplets $(j^1\alpha_\tau, A_\tau, E_\tau)$,

$$\gamma = (j^1\alpha_\tau, A_\tau, E_\tau) . \tag{3.3}$$

Vectors $\zeta \in T_\gamma \mathcal{Z}_\tau$ can be decomposed into their components in the directions $j\alpha_\tau$, A_τ, and E_τ, denoted by $j^1\dot{\alpha}_\tau$, $\dot{A}_\tau$ and $\dot{E}_\tau$ respectively,

$$\zeta = (j^1\dot{\alpha}_\tau, \dot{A}_\tau, \dot{E}_\tau) . \tag{3.4}$$

Let us note that the component $\dot{A}_\tau$ in the direction A_τ is a $\mathfrak{g}$–valued form on $\tau(M)$. Eqs. (3.4), (1.16), and the definition of the Cartan form Ξ, Eq. (7.3.2), yield

$$\Xi(\zeta) = \int_{\tau(M)} \pi_{XP}*\langle E_\tau, \dot{A}_\tau\rangle vol[g_\tau] , \tag{3.5}$$

where $vol[g_\tau]$ is the induced volume form of $\tau(M)$. Let ξ_P be an infinitesimal automorphism of P such that the induced vector field ξ_Z preserves the boundary condition B, so that ξ_P induces a vector field $\xi_{\mathscr{Z}}$ on $\mathscr{Z}$. Given $\gamma \in \mathscr{Z}_\tau$, we decompose the restriction of ξ_P to $\pi_{XP}^{-1}(\tau(M))$ into its vertical and horizontal components with respect to the connection ω along $\tau(M)$ defined by γ,

$$\xi_P \,|\, \pi_{XP}^{-1}(\tau(M)) = ver\, \xi_P + hor\, \xi_P . \tag{3.6}$$

Also, we decompose the projection ξ_X of ξ_P to X into its components normal and tangential to $\tau(M)$,

$$\xi_X \,|\, \tau(M) = \xi_\tau - g_X(\eta_\tau, \xi_X)\eta_\tau . \tag{3.7}$$

Eq. (2.19) gives the following expression for the value of Ξ on $\xi_{\mathscr{Z}}(\gamma)$,

$$\begin{aligned} \Xi(\xi_{\mathscr{Z}}(\gamma)) = & \int_{\partial M} g_g({}^*E_\tau, ver\, \xi_P) + \\ & - \int_M g_g(D^*E_\tau, ver\, \xi_P) - \int_M \langle E_\tau, \xi_\tau \lrcorner F_\tau\rangle\, vol[g_\tau] + \\ & - \tfrac{1}{2}\int_M \{\langle E_\tau, E_\tau\rangle + \tfrac{1}{2}\langle F_\tau, F_\tau\rangle\}, \xi_X \lrcorner\, vol[g_X] , \end{aligned} \tag{3.8}$$

where F_τ is the field strength of A_τ. Since every vector in $T_\gamma \mathscr{Z}$ can be expressed as the sum of a vector in $T_\gamma \mathscr{Z}_\tau$ and the value at γ of $\xi_{\mathscr{Z}}$, for a suitably chosen infinitesimal automorphism ξ_P of P, if follows that Eqs. (3.5) and (3.8) determine the Cartan form completely.

Eq. (3.5) shows that the reduced phase bundle $\mathscr{R}$, defined in Hypothesis 6.3.1, is parametrized by the variables (E_τ, A_τ). Similarly, we obtain from Eq. (3.8) that the primary constraint bundle $\mathscr{P}$, defined in Hypothesis 6.3.3, is parametrized by the variables $(\alpha_\tau, E_\tau, A_\tau)$. The fibres of $\pi_{\mathscr{R}\mathscr{P}}: \mathscr{P} \to \mathscr{R}$ are parametrized by the variable α_τ which describes a partial connection in $P\,|\,\tau(M)$ in the direction normal to $\tau(M)$. For each $\tau \in \mathcal{T}$, such partial connections in $P|\tau(M)$ form an affine space. The partial conections in $P|\tau(M)$ compatible with the boundary conditions form a convex subset $\mathcal{A}_\tau$ of this affine space. The union of $\mathcal{A}_\tau$, over $\tau \in \mathcal{T}$, has the structure of a manifold fibred over $\mathcal{T}$, and $\mathcal{P}$ is the fibre product of $\mathcal{A}$ and $\mathcal{R}$,

$$\mathcal{P} = \mathcal{A} \times_{\mathcal{T}} \mathcal{R} . \tag{3.9}$$

This product structure is stable under the action of the symmetry group $\mathcal{G}$ of the Yang–Mills theory.

8.3.1 Proposition. For each $\tau \in \mathcal{T}$ and each $\alpha,\alpha' \in \mathcal{A}_\tau$ there exists a symmetry $g \in \mathcal{H}$ such that $g_{\mathcal{P}}(\alpha,\rho) = (\alpha',\rho)$ for all $\rho \in \mathcal{R}_\tau$.

Proof. Let η be a vector field in X such that, for each $m \in M$, $\eta(\tau(m))$ is the unit normal vector of $\tau(M)$ at $\tau(m)$, and φ_t the local one parameter group of local diffeomorphisms of X generated by η. Since M is compact, there exists $\epsilon > 0$ such that the mapping φ: $(-\epsilon,\epsilon) \times \tau(M) \longrightarrow X$ given by $\varphi(t,\tau(m)) = \varphi_t(\tau(M))$ is a diffeomorphism onto its image.

Let ω be a connection form in P which agrees with α on lifts of $\eta \mid \tau(M)$ to P, $\eta^\#$ the lift of η to P horizontal with respect to ω, and $\varphi_t^\#$ the one parameter local group of local diffeomorphisms of P generated by $\eta^\#$. We denote by P_τ the pull–back of P to $\tau(M)$ and by $\varphi^\#$: $(-\epsilon,\epsilon) \times P_\tau \longrightarrow P$ the mapping given by $\varphi^\#(t,p) = \varphi_t^\#(p)$. Since $\varphi_\tau^\#$ covers φ_τ, it follows that $\varphi^\#$ is a diffeomorphism onto its image. Moreover, $\varphi^\#$ intertwines the action of G, $\varphi^\#(t,pg) = \varphi^\#(t,p)g$ for every $g \in G$. Hence, we can use $\varphi^\#$ to identify $(-\epsilon,\epsilon) \times P_\tau$ with its image in P. Under this identification the vector field $\eta^\#$ in P corresponds to the differentiation ∂_t with respect to the parameter $t \in (-\epsilon,\epsilon)$. This implies that $\alpha(\partial_t) = 0$.

Vector fields tangent to the fibres of a principal fibre bundle can be identified with $\mathfrak{g}$–valued functions on the bundle. Under this identification $\alpha'(\eta^{\#})$ can be considered as a vector field on P_τ generating a one parameter group Ψ_t of automorphisms of P_τ which covers the identity in $\tau(M)$. Let $u : (-\epsilon,\epsilon) \longrightarrow \mathbb{R}$ be a smooth function with compact support such that $u(t) = t$ for t in a neighbourhood of $0 \in (-\epsilon,\epsilon)$. The mapping $\Psi: (-\epsilon,\epsilon) \times P_\tau \longrightarrow (-\epsilon,\epsilon) \times P_\tau$ given by $\Psi(t,p) = (t,\Psi_{u(t)}(p))$ is an automorphism of $(-\epsilon,\epsilon) \times P_\tau$ covering the identity transformation in the complement of *support* $u \times P_\tau$. Hence it extends to a global automorphism g_P of P which is the identity in the complement in P of *support* $u \times P_\tau$. Clearly, g_P acts in P_τ as the identity transformation. Since every $\rho \in \mathcal{R}_\tau$ is given by a pair (E_τ, A_τ) of $\mathfrak{g}$–valued equivariant forms on P_τ it follows that the induced action of g_P in $\mathcal{R}_\tau$ is given by the identity transformation, $g_{\mathcal{R}}(\rho) = \rho$ for each $\rho \in \mathcal{R}_\tau$. Moreover,

$$\begin{aligned} g_{\mathcal{A}}(\alpha)(\partial_t) &= \alpha(Tg_P(\partial_t)) = \alpha(\partial_t + \alpha'(\eta^{\#})) \\ &= \alpha(\partial_t) + \alpha'(\eta^{\#}) = \alpha'(\partial_t) \end{aligned}$$

since $\alpha(\partial_t) = 0$, $\eta^{\#} = \partial_t$, and the value of a connection form on a vertical vector identified with an element of the Lie algebra $\mathfrak{g}$ is equal to this element of $\mathfrak{g}$. Since α, α' are $\mathfrak{g}$–valued form vanishing on lifts of vector tangent to $\tau(M)$ and g_P covers the identity in $\tau(M)$, it follows that $g_{\mathcal{A}}(\alpha) = \alpha'$. Hence $g_P(\alpha,\rho) = (\alpha',\rho)$ for every $\rho \in \mathcal{R}_\tau$, which completes the proof.

■

Thus, the Yang–Mills theory is quasi–regular and it satisfies parts (i) and (ii) of Atlas Hypothesis 7.4.8. Part (iii) of Atlas Hypothesis can be verified by direct computations. Hence, the Yang–Mills theory admits a covariant Hamiltonian formulation of dynamics described in Corollary 7.4.7.

Let $\xi \in \mathcal{G}$ be such that ξ_X is a hypersurface orthogonal Killing vector field in X, and $c_{\mathcal{G}}$ an integral curve of $\xi_{\mathcal{G}}$ associating to each $t \in$ *domain* $c_{\mathcal{G}}$ a Cauchy surface $c_{\mathcal{G}}(t)(M)$ orthogonal to ξ_X. A lift $c_{\mathcal{A}}$ of $c_{\mathcal{G}}$ to an integral curve of $\xi_{\mathcal{A}}$ defines a partial connection in P in the direction of ξ_X. Usually in Yang–Mills theory one chooses this lift in such a way that ξ_P is a horizontal vector field in P, such a choice is called a *temporal gauge condition.* In this case, for every $\alpha \in$ *range* $c_{\mathscr{A}}$, the evolutions in $(\mathscr{P}_\alpha, \omega_\alpha)$ is given by the Hamiltonian

$$H = \frac{1}{2}\int_{\tau(M)} \{<E_\tau, E_\tau> + \frac{1}{2}<F_\tau, F_\tau>\}\xi_X \lrcorner\, vol[g_X]. \tag{3.10}$$

The symplectic form ω_α can be obtained from Eq. (3.5). The Hamiltonian vector field of H has components $\dot{E}_\tau$, and $\dot{A}_\tau$ given by

$$\dot{E}_\tau = -D * F_\tau, \quad \dot{A}_\tau = E_\tau. \tag{3.11}$$

The constraint equations, given by the vanishing of J_ξ for all $\xi \in \mathscr{h}$, are

$$D * E_\tau = 0. \tag{3.12}$$

Eqs. (3.11) and (3.12) are equivalent to the Yang–Mills equations (1.5).

For Yang–Mills fields on Minkowski space–time, c.f. Section (8.4), and boundary conditions corresponding to a product connection, one can show in the category of Hilbert manifolds that the constraint set $\mathscr{C}$ is given by the vanishing of the momentum map $J^{\mathscr{h}}_{\mathscr{P}}: \mathscr{P} \to \mathscr{h}^*$,

$$\mathscr{C} = (J^{\mathscr{h}}_{\mathscr{P}})^{-1}(0). \tag{3.13}$$

That is, for each Cauchy data γ satisfying the constraint equations (3.12), there exists a curve satisfying the Hamiltonian equations (3.11) which passes through γ.[1] If the boundary conditions are non–trivial we do not have a similar existence theorem for a pure Yang–Mills field. However, there are existence theorems for coupled Yang–Mills and Higgs fields.[2]

---

[1]Finite time evolution of Cauchy data in Yang–Mills theory was studied in Segal [1979], Arms [1981], and Eardley and Moncrief [1982a,b].

[2]Jaffe and Taubes, [1980].

8.4 <u>Minkowski space</u>

Up to now we have discussed the Yang–Mills theory on manifolds with boundary such that the Cauchy surfaces were compact manifolds with boundary. However, in physics one is interested in Yang–Mills fields on Minkowski space. In this case Cauchy surfaces are non–compact and the boundary conditions are replaced by asymptotic conditions at infinity. For pure Yang–Mills fields one can use the conformal invariance of the theory to relate Minkowski space and asymptotic conditions to a manifold with boundary and the corresponding boundary conditions.

Minkowski space is an affine space $\mathring{X}$ endowed with a Lorentzian metric $g_{\mathring{X}}$ invariant under translations in $\mathring{X}$. The structure of infinity in $(\mathring{X},g_{\mathring{X}})$ is conveniently discussed in terms of conformal compactifications of $(\mathring{X}'g_{\mathring{X}})$. Let f be an embedding of $(\mathring{X},g_{\mathring{X}})$ into a time orientable Lorentzian manifold $(\mathring{X},g_{\mathring{X}})$ such that

$$f^* g_{\mathring{X}} = K^2\, g^0_X \tag{4.1}$$

for some positive function K on $\mathring{X}$ such that, for each geodesics $c : \mathbb{R} \longrightarrow \mathring{X}$ of $g_{\mathring{X}}$, there exist limits in $\mathring{X}$ of $f(c(t))$ as $t \longrightarrow +\infty$ and $t \longrightarrow -\infty$. The set B of limit points of $f(c(t))$ as $t \longrightarrow +\infty$ and $t \longrightarrow -\infty$ for all geodesics c of $g_{\mathring{X}}$ is the conformal boundary of Minkowski space determined by the embedding f.

8.4.1. <u>Proposition</u>. Let f_1 and f_2 be two conformal embeddings of Minkowski space–time into time orientable Lorentzian manifolds (X_1,g_1) and (X_2,g_2), respectively, such that, for every geodesics c of g_X, $f_1(c(t))$ and $f_2(c(t))$ have limits as $t \longrightarrow \infty$ and as $t \longrightarrow -\infty$. Then, there exists a neighbourhood U_i of the closure of $f_i(X)$ in X_i and a diffeomorphism $h : U_1 \longrightarrow U_2$ such that

$$h^* g_2 | U_2 = \mathring{K}^2\, g_1 | U_1$$

for some positive function $\mathring{K}$ on U_1, and

$$f_2 = h \circ f_1.$$

Proof can be found in Schmidt [1979].

As a consequence of Proposition. 8.4.1 it follows that conformal extensions of Minkowski space into time orientable Lorentian manifolds yield diffeomorphic boundaries. Hence, in order to analyze the nature of the conformal boundary of Minkowski space, it suffices to exhibit the boundary B corresponding to any embedding f.

We shall describe the conformal boundary of $\mathring{X}$ in terms of a conformal embedding $f : \mathring{X} \longrightarrow \tilde{X}$, where $\tilde{X} = \mathbb{R}^1 \times S^3$ and $g_{\tilde{X}}$ is a Lorentian metric in $\tilde{X}$ given by the pull back g_S of the Riemannian metric in S^3, induced by the Euclidean metric in $\mathbb{R}^4$, minus the pull back $(d\tau)^2$ of the Euclidean metric on $\mathbb{R}^1$,

$$g_{\tilde{X}} = -(d\tau)^2 + g_S. \tag{4.2}$$

Using local coordinates $(\tau,\alpha,\beta,\gamma) \in \mathbb{R}^1 \times S^3$, such that $(\cos\alpha, \sin\alpha\cos\beta, \sin\alpha\sin\beta\cos\gamma, \sin\alpha\sin\beta\sin\gamma)$ is a point in $S^3 \subset \mathbb{R}^4$ associated to (α,β,γ), we can write $g_{\tilde{X}}$ in the form

$$g_{\tilde{X}} = -(d\tau)^2 + (d\alpha)^2 + (\sin\alpha)^2[(d\beta)^2 + (\sin\beta)^2(d\gamma)^2] . \tag{4.3}$$

Let (t,r,θ,φ) be coordinates in $\mathring{X}$, identified with $\mathbb{R}^1 \times \mathbb{R}^3$, such that t is an affine coordinate in the first factor, and (r,θ,φ) are spherical polar coordinates in $\mathbb{R}^3$. Using these coordinates we can write

$$g_{\mathring{X}} = -dt^2 + dr^2 + r^2[(d\theta)^2 + (\sin\theta)^2(d\varphi)^2]. \tag{4.4}$$

The embedding $f : \mathring{X} \longrightarrow \tilde{X}$ considered here is given in terms of local coordinates by

$$\left.\begin{array}{l} \tau \circ f = \arctan(t+r) + \arctan(t-r). \\ \alpha \circ f = \arctan(t+r) - \arctan(t-r) \\ \beta \circ f = \theta, \text{ and } \gamma \circ f = \varphi . \end{array}\right\} \tag{4.5}$$

One can verify by direct calculations that f sataisfies all required conditions. In particular Eq. (4.1) is satisfied with

$$K^2 = 4[(t+r)^2 + 1]^{-1}[(t-r)^2 + 1]^{-1} . \tag{4.6}$$

Observe that $K \longrightarrow 0$ when t, r, (t−r) or (t+r) tend to $\pm\infty$.

The range of $(\tau\mathrm{of}, \alpha\mathrm{of})$ is given by $-\pi < \tau\mathrm{of} + \alpha\mathrm{of} < \pi$, and $-\pi < \tau\mathrm{of} - \alpha\mathrm{of} < \pi$, $\alpha\mathrm{of} > 0$. Hence, the corresponding conformal boundary B consists of five strata

$$i_- = \{x \in \breve{X} \mid \tau(x) = -\pi,\ \alpha = 0\}$$
$$i_+ = \{x \in \breve{X} \mid \tau(x) = \pi,\ \alpha = 0\}$$
$$i_0 = \{x \in \breve{X} \mid \tau(x) = 0,\ \alpha = \pi\}$$
$$\mathcal{J}^+ = \{x \in \breve{X} \mid \tau(x) = \alpha(x) = \pi,\ \tau(x) > 0,\ \alpha(x) < \pi\}$$
$$\mathcal{J}^- = \{x \in \breve{X} \mid \tau(x) - \alpha(x) = \pi,\ \tau(x) > 0,\ \alpha(x) < \pi\}.$$

Since the coordinates (α,β,γ) on S^3 are singular when $\alpha = 0$ or $\alpha = \pi$, it follows that the first 3 strata consist of single points. The remaining two strata $\mathcal{J}^+$ and $\mathcal{J}^-$ are co–dimension one submanifolds of $\breve{X}$.

Let c be a geodesics of $g_{\breve{X}}$. If it is time–like, then f∘c extends to a curve joining i_- to i_+. If it is null, then f∘c extends to a curve joining $\mathcal{J}^-$ to $\mathcal{J}^+$. If c is space–like, then f∘c extends to a loop at i_0. Hence, the point i_+ represents future time–like infinity, i_- represents the past time–like infinity, and i_- represents the past time–like infinity, and i_0 represents the space–like infinity. The future null infinity is described by $\mathcal{J}^+$ and the past null infinity corresponds to $\mathcal{J}^-$.

Cauchy surfaces for relativistic field equations in Minkowski space are space–like hyperplanes. Since every space–like hyperplane $\breve{M}$ in $\breve{X}$ is filled by space–like geodesics, the closure operation in $\breve{X}$ attaches to $f(\breve{M})$ a single point i_0. Thus, a conformal boundary of Minkowski space is not of the type we need when we want to set up a Hamiltonian formulation of dynamics in spaces of Cauchy data on space–like Cauchy surfaces. In order to obtain the required boundary we attach to $\breve{X}$ the hyperboloid H of space–like directions in $T_{i_0}\breve{X}$, which can be identified with the hyperboloid of unit space–like vectors in $T_{i_0}\breve{X}$,

$$H = \{v \in T_{i_0}\breve{X} \mid g_{\breve{X}}(v,v) = 1\}. \tag{4.7}$$

Let $h : (0,\epsilon) \times H \longrightarrow \mathring{X}$ be given by $f(h(t,v)) = exp(tv)$, where *exp* denotes the exponential map from $T_{i_0}\mathring{X}$ to $\mathring{X}$ defined in terms of the metric $g_{\mathring{X}}$, and $\epsilon > 0$ is such that exp maps $(0,\epsilon)\times H$ diffeomorphically onto an open subset of $\mathring{X}$. Let X be the union of $[0,\epsilon)\times H$ with $\mathring{X}$, in which the points of $(0,\epsilon)\times H$ are identified with their images in $\mathring{X}$ under the mapping h.

8.5.2. Proposition. X constructed above has the structure of a manifold with boundary. The boundary ∂X of X has a conformal structure of a hyperbolic manifold of signature 1.

Proof follows from the construction of X, and the fact that ∂X is diffeomorfic to H. The conformal structure of H is independent of the metric $g_{\mathring{X}}$ in its conformal class, and it induces a conformal structure in ∂X

■

We have constructed a manifold X with boundary ∂X such that its interior is conformally isomorphic to Minkowski space $\mathring{X}$, and the conformal factor K, given by Eq. (4.6), extends to a function on X which vanishes on the boundary. A coordinate system in X, convenient for a discussion of the boundary ∂X, is obtained from the Cartesian coordinates (x^μ) in $\mathring{X}$ by the transformation

$$\tilde{x}^\mu = x^\mu/\rho^2 \tag{4.8}$$

where $\rho(x)$ is the distance of the point x from the origin of the coordinate system:

$$\rho(x) = \sqrt{\mathring{g}_{\mu\nu}x^\mu x^\nu} \tag{4.9}$$

where $\mathring{g}_{\mu\nu}$ are the components of $g_{\mathring{X}}$ with respect to the coordinates (x^μ). The coordinate transformation (4.8) corresponds to a diffeomorphism f of the domain in $\mathring{X}$ consisting of the points which are spatially separated from the origin,

$$U = \{x \in \mathring{X} \mid \rho(x)^2 > 0\}. \tag{4.9}$$

It is a conformal diffeomorphism f of U with the conformal factor $K = \rho^2$,

$$f^* g_{\mathring{X}} = \rho^4 g_{\mathring{X}} \tag{4.10}$$

If we identify $\mathring{X}$ with its image $f(\mathring{X})$ in $\tilde{X}$ under the conformal embedding f discussed above, we see that the new coordinates $(\tilde{x}^\mu)$ extend to i_0 in such a way that $\tilde{x}^\mu(i_0) = 0$. Space–like directions at i_0 are given by $\tilde{x}^\mu/\tilde{\rho}$ where $\tilde{\rho}(x) = (\mathring{g}_{\mu\nu}\tilde{x}^\mu\tilde{x}^\nu)^{1/2} = 1/\rho(x)$. Hence $\tilde{x}^\mu/\tilde{\rho} = x^\mu/\rho$, so that one can describe the boundary ∂X of X in terms of space–like directions x^μ/ρ as $\rho \longrightarrow \infty$.

As an application of the coordinate transformation (4.8), let us consider a family of hyperplanes in $\mathring{X}$ given by $x^0 = t$. In terms of the coordinates $\tilde{x}^\mu$ they are given by $\tilde{x}^0\,\tilde{\rho}^{-2} = t$ or

$$\tilde{x}^0 - t\mathring{g}_{\mu\nu}\tilde{x}^\mu\,\tilde{x}^\nu = 0. \tag{4.11}$$

The differential of the left hand side of Eq. (4.11) at i_0, (i.e. $\tilde{x}^\mu = 0$), is independent of t, which means that all the hyperplanes $x_0 = t$ in $\mathring{X}$ extended to X have the same intersection with ∂X. Hence, the vector field $\dfrac{\partial}{\partial x^0}$ in $\mathring{X}$ extends to a vector field in X which vanishes on ∂X. Similarly, infinitesimal space–like translations in $\mathring{X}$ extend to vector fields on X which vanish on ∂X. On the other hand, infinitesimal rotations and boosts in Minkowski space extend to vector fields on X which do not vanish on ∂X.

As we have seen, the Lagrangian form for pure Yang–Mills fields is invariant under automorphisms of P which induce conformal transformations in X, c.f. discussion preceding Eq. (2.8). In particular, the Yang–Mills equations Eq. (1.3) and (1.5) are invariant under conformal diffeomorphisms.

In local coordinates (x^μ) we have

$$*F = \tfrac{1}{2}\,\epsilon_{\mu\nu\lambda\rho}\,g^{\lambda\sigma}\,g^{\rho\tau}\,F_{\sigma\tau}\,\sqrt{-\det(g_{\alpha\beta})}\;dx^\mu \wedge dx^\nu \tag{4.12}$$

Under a conformal diffeomorphism of X with a conformal factor K^2, the contravariant components of the metric are multiplied by K^{-2}, while the determinant of the metric is multiplied by K^{2n}, where $n = \dim X$. Hence, for $n = 4$, which is the dimension of physical space–time, the dual $*F$ of F is independent of the metric g_X in its conformal class. Therefore, the Yang–Mills theory in Minkowski space with boundary conditions given at the hyperboloid H of space–like directions at spatial infinity is equivalent to the Yang–Mills theory on the manifold X with boundary ∂X obtained from Minkowski space in the manner described above.

In the presence of matter fields the total Lagrangian form need not be invariant under conformal transformations. However, if one assumes stronger boundary conditions on matter fields, requiring that they should have compact intersections with Cauchy surfaces, for every matter field configuration, one can find a conformal transformation which reduces to the identity in the support of the matter field, and still enables one to attach the boundary at spatial infinity. Under these conditions one can study asymptotic behaviour of Yang–Mills fields in terms of the formulation developed for manifolds with boundary.

8.5. Maxwell–Dirac theory

Maxwell fields corresponding to a decomposition of Minkowski space $(\mathfrak{X}, g_{\mathfrak{X}})$ into a product of time T and space M,

$$\mathfrak{X} = T \times M, \tag{5.1}$$

are a pair of time dependent vector fields on M: an electric field $\vec{E}$ and a magnetic field $\vec{H}$. In the system of units in which the velocity of light and the electric and magnetic permeability of vacuum are equal 1, Maxwell's equations read

$$\dot{\vec{H}} + curl\, \vec{E} = 0, \quad div\, \vec{H} = 0 \tag{5.2}$$

and

$$-\dot{\vec{E}} + curl\, \vec{H} = \vec{j}, \quad div\, \vec{E} = \rho \tag{5.3}$$

where dot denotes the differentiation with respect to time, *curl* and *div* operators are defined with respect to the Euclidean metric g_M in M, and ρ and $\vec{j}$ are sources of the electromagnetic fields given by the distribution of charge and current, respectively. The electric and the magnetic fields combine into a 2–form F on $\mathfrak{X}$ given by

$$F = dt \wedge E_i dx^i - \tfrac{1}{2}\, \epsilon_{ijk}\, H^i\, dx^j \wedge dx^k \tag{5.4}$$

where (x^i) are Cartesian coordinates in M and (E^i) and (H^i) the components of $\vec{E}$ and $\vec{H}$, respectively. F is called the electromagnetic field strength. The covariant form of Maxwell's equations is

$$dF = 0 \tag{5.5}$$

and

$$\delta F = {*j} \tag{5.6}$$

where the codifferential δ is defined in terms of the Hodge * operator corresponding to the Minkowski metric $g_{\mathfrak{X}}$ in $\mathfrak{X}$, and j, given by

$$j = \rho d_3 x + \tfrac{1}{2}\, j^k\, \epsilon_{ijk}\, dt \wedge dx^i \wedge dx^j, \tag{5.7}$$

describes the distribution of charges and currents in space–time. Recall that the

* operator acting on p–forms on an m–dimensional manifold with metric with k negative eigenvalues satisfies the identity

$$*^2 = (-1)^{p(m-p)+k}, \tag{5.8}$$

and the operators d and δ are related by

$$\delta = *d*(-1)^{m(p+1)+k}, \quad d = *\delta*(-1)^{pm+1+k} \tag{5.9}$$

Moreover, in the case under consideration with m = 4 and p = 2, the form $*F$ is independent of the metric in the conformal class of $g_{\mathring{X}}$ used in the definition of the * operator. Hence, the vacuum Maxwell equations,

$$dF = 0 \quad \text{and} \quad d*F = 0, \tag{5.10}$$

are conformally invariant.

In Minkowski space every closed form is exact. Hence, Eq. (5.5) implies that there exists a 1–form A on $\mathring{X}$, called an electromagnetic potential, such that

$$F = dA \tag{5.11}$$

The electromagnetic potential A in Eq. (5.11) is determined by F up to an exact form $d\varphi$, where φ is a function on $\mathring{X}$. The transformation

$$A \longrightarrow A' = A - d\varphi \tag{5.12}$$

is called in electrodynamics a gauge transformation of the second kind. Using the gauge freedom given by Eq. (5.12) we can choose φ so that

$$A(\partial_t) = 0 \tag{5.13}$$

where ∂_t is the vector field along the time axis in the decomposition (5.1). In this case, referred to as a *temporal gauge*, we can interpret A as a time dependent form A_M on M.

Vacuum Maxwell equations (5.10) give rise to second order equations for electromagnetic potentials. In the temporal gauge they take the form

$$\ddot{A}_M - \delta_M d_M A_M = 0 \tag{5.14}$$

and

$$\delta_M \dot{A}_M = 0 \tag{5.15}$$

where δ_M and d_M are operators on forms on M related by the Hodge * operator defined in terms of the metric g_M on M. Because of the conformal invariance of the vacuum Maxwell equations we can consider M to be the interior of a manifold with boundary ∂M obtained by attaching to M the sphere of directions at infinity.

The Hodge decomposition on manifolds with boundary yields $A_M = \beta + d\gamma$, where $\delta\beta = 0$ and $i^*_{M\partial M}(*\gamma) = 0$.[1] Eqs. (5.14) and (5.15) are satisfied if $\ddot{\beta} = \delta d\beta$ and $d\dot{\gamma}$ is independent of time and co–exact. Hence $d\gamma$ is uniquely determined by the Cauchy data

$$A_M(0) = \beta(0) + d\gamma(0), \quad \dot{A}_M(0) = \dot{\beta}(0) + d\dot{\gamma}(0),$$

and β satisfies the wave equation, $\ddot{\beta} = \Delta\beta$. Let μ and ν be the unique harmonic forms such that $\mu|\partial M = \beta(0)|\partial M$ and $\nu|\partial M = \dot{\beta}(0)|\partial M$. The form $\theta = \beta - \mu - t\nu$ satisfies the wave equation, $\ddot{\theta} = \Delta_M\theta$, and it vanishes on ∂M. Therefore, $\theta = \Sigma(a_\lambda \cos \lambda t + b_\lambda \sin \lambda t)\omega_\lambda$, where $\Delta_M\omega_\lambda = -\lambda^2\omega_\lambda$ and $\omega_\lambda|\partial M = 0$. The coefficients a_λ and b_λ in the expansion of θ are uniquely determined by the Cauchy data for θ. Hence, for every Cauchy data $(A_M(0), \dot{A}_M(0))$, there exists a unique solution A_M of the Maxwell equations.

It should be noted that the above discussion implies that the admissible boundary data $A_M|\partial M$ are linear functions of time. Since we are concerned only with evolutions in the direction of symmetries of the theory, one might expect that the boundary data should be independent of time. The linear dependence of the boundary data on time is the consequence of the temporal gauge condition. If $A|\partial X$ is independent of time, but it does not satisfy the temporal gauge condition, then the gauge transformed potential, which satisfies the temporal gauge condition, will have boundary data depending linearly on time.

In the Maxwell–Dirac theory matter is described by a spinor field Ψ on $(\mathring{X}, g_{\mathring{X}})$ corresponding to a wave function of an electron. It satisfies the Dirac equation

---

[1] C.f. Morrey [1966].

$$[\gamma^\mu(\partial_\mu - \mathrm{i}eA_\mu) + \mathrm{m}]\Psi = 0 \tag{5.16}$$

where e and m are the charge and the mass of the electron, respectively, and γ^μ are the Dirac matrices satisfying the anticommutation relations

$$\gamma^\mu\gamma^\nu + \gamma^\nu\gamma^\mu = 2\,\mathrm{g}^{\mu\nu}\ .$$

Though the physical significance of the Dirac equation is in the realm of quantum mechanics, we can treat it as a field equation for a classical field Ψ interacting with the electromagnetic field A. This point of view is usually adopted in the process of second quantization. The Dirac equation can be obtained from the Dirac Lagrangian

$$\mathrm{L}[\bar{\Psi},\Psi] = -\tfrac{1}{2}\bar{\Psi}[\gamma^\mu(\partial_\mu - \mathrm{i}eA_\mu) + \mathrm{m}]\Psi + \mathrm{c.c.}\ , \tag{5.17}$$

where c.c. denotes the complex conjugate terms, $\bar{\Psi} = \mathrm{i}\Psi^\dagger\gamma^0$, and $\Psi^\dagger$ is the Hermitian conjugate of Ψ. The Lagrangian for interacting Maxwell and Dirac fields is the sum of the Dirac Lagrangian $\mathrm{L}[\bar{\Psi},\Psi]$ and the Maxwell Lagrangian,

$$\mathrm{L}[A] = \tfrac{1}{4}\, F_{\mu\nu}\, F^{\mu\nu} \tag{5.18}$$

The Euler–Lagrange equations corresponding to $\mathrm{L} = \mathrm{L}[A] + \mathrm{L}[\bar{\Psi},\Psi]$ are the Dirac equation (5.16) and

$$\partial_\nu\, F^{\mu\nu} = j^\mu\ , \tag{5.19}$$

where

$$j^\mu = \mathrm{e}\,\bar{\Psi}\,\gamma^\mu\,\Psi \tag{5.20}$$

has the physical interpretation of the current 4–vector. Eq. (5.19) is a contravariant coordinate version of Eq. (5.6) with

$$\mathrm{j} = j^\mu\partial_\mu \,\lrcorner\, \mathrm{vol}[\mathrm{g}_{\mathrm{X}}]\ . \tag{5.21}$$

The 3–form j describes the distribution of charges and currents in space–time. In particular, for each Cauchy surface M, the pull–back j_{M} of j to M gives the distribution of the electric charge on M. If U is a relatively compact domain in M, the integral of j_{M} over U measures the amount of charge contained in U. Equation (5.6) and Stokes' Theorem yield

$$\int_{\mathrm{U}} \mathrm{j}_{\mathrm{M}} = \int_{\partial\mathrm{U}} {}^*F = \int_{\partial\mathrm{U}} \mathrm{E} \,\lrcorner\, \mathrm{vol}[\mathrm{g}_{\mathrm{M}}] \tag{5.22}$$

where ∂U is the boundary of U. Thus we have obtained the classical result that the total charge contained in U is given by the flux of the electric field through the boundary of U. If the integral of j_M over M converges, then it describes the total charge of the system. The conservation of the total charge is a consequence of the invariance of the theory under the one parameter group of gauge transformations of the first kind acting on the Maxwell–Dirac fields as follows:

$$A \longrightarrow A, \quad \Psi \longrightarrow exp(-\mathrm{ies})\Psi, \quad \overline{\Psi} \longrightarrow exp(\mathrm{ies})\overline{\Psi}, \tag{5.23}$$

when s is a real parameter.

The Maxwell–Dirac theory can be reformulated in terms of connections in a principal bundle and sections of an associated bundle, giving rise to a prototype of the Yang–Mills theory. In this formulation the structure group G is the group U(1) of complex numbers of modulus 1,

$$\mathrm{U}(1) = \{\mathrm{z} \in \mathbb{C} \mid |\mathrm{z}| = 1\}. \tag{5.24}$$

Since Minkowski space $\mathfrak{X}$ is contractible, every principal fibre bundle P over $\mathfrak{X}$ admits a global section $\sigma_{\mathrm{P}\mathfrak{X}} : \mathfrak{X} \longrightarrow \mathrm{P}$, which enables one to identify P with the product of Minkowski space and the structure group. In the case under consideration we have G = U(1) so that

$$\mathrm{P} = \mathrm{X} \times \mathrm{U}(1) \tag{5.25}$$

The Lie algebra u(1) of U(1) is one dimensional. It can be identified with purely complex numbers $\mathrm{i}\mathbb{R}$ by associating to each ia the one parameter group $s \longrightarrow exp(\mathrm{isa})$. Under this identification the imaginary unit i plays the role of a basis in u(1)

In order to define the Yang–Mills Lagrangian one needs an invariant metric on the Lie algebra of the structure group. We choose a metric $g_{u(1)}$ such that the vector $\mathrm{ei} \in \mathrm{u}(1)$ is normalized to unity, here e is the charge parameter introduced in the Dirac equation,

$$g_{u(1)}(\mathrm{ei},\mathrm{ei}) = 1 \tag{5.26}$$

Given a connection form ω in P, we denote by A its pull back by $\sigma_{P\mathfrak{X}}$,

$$\mathrm{A} = \sigma^*_{P\mathfrak{X}}\,\omega \tag{5.27}$$

and by A the component of A with respect to the normalized basis ei in u(1),

$$\mathrm{A} = \mathrm{ei}A\,. \tag{5.28}$$

A is a 1–form on $\mathfrak{X}$ which will be identified with the electromagnetic potential introduced in Eq. (5.11). Since the structure group U(1) is abelian, the field strength F corresponding to the Yang–Mills potential A is the exterior derivative of A. Eqs. (5.11) and (1.3) yield

$$\mathrm{F} = \mathrm{ei}F\,. \tag{5.29}$$

Therefore, the identification of the component of $\sigma^*_{P\mathfrak{X}}\,\omega$ with respect to the basis vector ei with the electromagnetic potential A leads to the equality of the Yang–Mills Lagrangian L[ω] given by Eq. (1.4) and the Maxwell Lagrangian L[A] given by Eq. (5.18).

Irreducible unitary representations of U(1) are one dimensional. They are of the form $\rho_n\colon z \longrightarrow z^n$, where n is an integer. Let Y_{matter} be an associated bundle of P with typical fibre $\mathbb{C}^4$ on which U(1) acts by the fundamental representation ρ_1. We can identify the Dirac field Ψ with sections of Y_{matter}. Under this identification the expression $(\partial_\mu - \mathrm{ie}A_\mu)\Psi$ in the Dirac Lagrangian, Eq. (5.17), is the covariant derivative $D_\mu\Psi$ of the section corresponding to Ψ,

$$D_\mu\Psi = \partial_\mu\Psi + \mathrm{ie}A_\mu\Psi\,. \tag{5.30}$$

The Dirac Lagrangian can be generalized to arbitrary structure group.[2]

The abelian nature of the structure group U(1) sets the Maxwell–Dirac theory apart from the Yang–Mills theories with non–abelian structure groups. In particular, it enables one to reinterpret the notion of gauge transformations of the first kind in the framework of the general Yang–Mills theory. The theory of interacting Maxwell–Dirac

[2]C.f. Bleecker [1981].

fields is invariant under the gauge transformations of the second kind

$$A \longrightarrow A - d\varphi,\ \ \Psi \longrightarrow exp(-ie\varphi)\Psi,\ \ \bar{\Psi} \longrightarrow exp(ie\varphi)\bar{\Psi} \tag{5.31}$$

where φ is a smooth function on space–time. The identification of eA with a u(1)–valued form $A = \sigma^*_{P\mathfrak{X}}\omega$ corresponding to a connection form ω and a trivializing section $\sigma_{P\mathfrak{X}}$ enables one to interpret the transformation (5.31) as the action of an element of the group $Aut_{\mathfrak{X}}P$ of automorphisms of P covering the identity transformation in $\mathfrak{X}$. This group is isomorphic to the group of sections of the group bundle P[U(1)] associated to P with typical fibre U(1). Since P is trivial, the trivializing section $\sigma_{P\mathfrak{X}}$ used here gives rise to an isomorphism between $Aut_{\mathfrak{X}}P$ and the group of U(1)–valued functions on $\mathfrak{X}$. Similarly, the Lie algebra $aut_{\mathfrak{X}}P$ is isomorphic to the Lie algebra of sections of the adjoint bundle P[u(1)] of P. The trivializing section $\sigma_{P\mathfrak{X}}$ induces an isomorphism between P[u(1)] and the Lie algebra of u(1)–valued functions on $\mathfrak{X}$. These isomorphisms lead to the interpretation of $exp(i\varphi e)$ appearing in (5.31) as the value at s = 1 of the one parameter subgroup $s \longrightarrow exp(is\varphi e)$ in $Aut_{\mathfrak{X}}P$ generated by the element of $aut_{\mathfrak{X}}P$ corresponding to the section of P[u(1)] given by the map φe from $\mathfrak{X}$ to u(1). Since u(1) is abelian, the covariant differential of the section of P[u(1)] corresponding to φe is given by the exterior differential of φe,

$$D\varphi e = d\varphi e. \tag{5.32}$$

Hence $A \longrightarrow A - d\varphi$ corresponds to $A \longrightarrow A - D\varphi e$, which describes the result of the action of $exp(i\varphi e) \in Aut_{\mathfrak{X}}P$ on connections in P, c.f. discussion is Section 8.2. Under this interpretation, gauge transformations of the first kind correspond to automorphisms generated by sections $\xi : \mathfrak{X} \longrightarrow P[u(1)]$ which are covariantly constant on $\mathfrak{X}$,

$$D\xi = 0\ . \tag{5.33}$$

In particular, the one parameter group of gauge transformations of the first kind given by (5.23) is generated by the section ξ associating to each $x \in \mathfrak{X}$ the basic vector

$e \in u(1)$. Following the general formalism we denote by J^{ξ} the conserved momentum associated to ξ. The discussion in Sec. 4 shows that J^{ξ} is given by Eq. (2.26), where ∂M is the sphere of directions at infinity. In order to rewrite Eq. (2.26) in terms of the variables used here, note that Eqs. (2.20), (5.28) and (5.11) imply that $^*E_M$ coincides with the pull back of $^*Fe$ to M. Taking into account the normalization of e given by Eq. (5.26), and the relation between F and $\vec{E}$ given by Eq. (5.4) we can rewrite Eq. (2.26) in the form

$$J^{\xi} = \int_{\partial M} {}^*F = \int_{\partial M} \vec{E} \lrcorner \operatorname{vol}[g_M] , \tag{5.34}$$

obtaining the usual expression for the total electric charge.

We see that the total electric charge in Maxwell–Dirac theory is the conserved momentum correesponding to an infinitesimal symmetry ξ given by a covariantly constant section of P[u(1)] normalized to 1 by the metric $g_{u(1)}$. The following alternative characterization of ξ is useful for non–abelian gauge theories. Since ξ is an infinitesimal symmetry of the theory it has to preserve the boundary condition given by a connection $\omega | \partial X$ along the boundary ∂X attached to space–time. Hence ξ has to be covariantly constant along ∂X,

$$D_{\partial X} \xi = 0 \tag{5.35}$$

where $D_{\partial X}$ denotes the covariant differential in ∂X with respect to the pull back $\omega_{\partial X}$ of $\omega | \partial X$ to ∂X. Eq. (5.32) implies that ξ has to be constant along ∂X. For each Cauchy surface M with boundary $\partial M \subset \partial X$, Eq. (5.33) implies that the pull back ξ_M of ξ to M is harmonic, that is it satisfies the Laplace–Beltrami equation

$$\Delta_M \, \xi_M = 0, \tag{5.36}$$

where Δ_M is the Laplace–Beltrami operator acting on the space of sections of the pull–back of P[u(1)] to M. Since the Dirichlet problem for the Laplace–Beltrami operator considered here has a unique solution, the solution ξ of Eq. (5.35) generating the group of gauge transformations of the first kind is uniquely characterized by its boundary values and by the fact that it is harmonic on every Cauchy surface.

The relativistic form of Maxwell equations in presence of matter is asymmetric with respect to interchange of d and δ. The codifferential of F corresponds to the distribution of electric charges and currents. One can introduce a 3–form k describing the distribution of magnetic charges and currents in space–time, and consider a symmetric generalization of Maxwell equations.

$$\delta F = {}^*\mathrm{j} \quad \text{and} \quad \mathrm{d}F = \mathrm{k}, \tag{5.37}$$

Integrating the second of above equations over M we obtain an expression for the total magnetic charge K contained in M,

$$\mathrm{K} = \int_{\partial \mathrm{M}} \mathrm{F} \tag{5.38}$$

If the form k vanishes in a neighbourhood of ∂X, then the total magnetic K is independent of time. Point particles with non–zero magnetic changes are called magnetic monopoles. They have not been observed in experiments. However, they receive a great deal of attention in theoretical considerations since their analogues appear in the Yang–Mills theory for non–abelian structure groups.

The symmetric form of Maxwell equations, given by Eq. (5.37), does not allow for an interpretation of the electromagnetic field in terms of connections in a principal fibre bundle over Minkowski space $\mathfrak{X}$. However, such an interpretation is possible in the complement $\check{\mathfrak{X}}$ of the support of k,

$$\check{\mathfrak{X}} = \mathfrak{X} - \mathit{support}\,\mathrm{k}. \tag{5.39}$$

Let $\check{F}$ denote the pull–back of F to $\check{\mathfrak{X}}$. Clearly, it is a closed form, and it is exact if and only if its de Rham cohomology class $[F]$ in $\mathrm{H}^2(\check{\mathfrak{X}},\mathbb{R})$ vanishes. It follows from the discussion of Chern classes in Chapter 5 that there exists a principal U(1) bundle $\check{\mathfrak{P}}$ over $\check{\mathfrak{X}}$ with connection form $\tilde{\omega}$ such that $\mathrm{d}\tilde{\omega} = \mathrm{ie}\,\pi_{\check{\mathfrak{X}}\check{\mathfrak{P}}}^{\;*}F$, if and only if

$$2\pi\mathrm{e}[\check{F}] \in \mathrm{H}^2(\check{\mathfrak{X}},\mathbb{Z}), \tag{5.40}$$

where $\mathrm{H}^2(\check{\mathfrak{X}},\mathbb{Z})$ is the second cohomology group of $\check{\mathfrak{X}}$ with integer coefficients, considered as a subgroup of $\mathrm{H}^2(\mathrm{X},\mathbb{R})$. Evaluating $[\check{F}]$ on the class of the boundary ∂M of M we obtain a relation between the electric charge e and the magnetic charge K,

$$2\pi eK = n \in \mathbb{Z} \tag{5.41}$$

known as the Dirac quantization condition.[3]

---

[3]Dirac [1931].

8.6. Yang–Mills charges

Let us return to a Yang–Mills theory with a non–abelian structure group G. We have seen that charges are the conserved quantities corresponding to the vertical components of infinitesimal automorphisms of P preserving the boundary conditions, c.f. Sec. 8.2. Vertical components of infinitesimal automorphisms of P correspond to elements of the Lie algebra $aut_X P$ of the group $Aut_X P$. This Lie algebra is isomorphic to the Lie algebra of sections of the adjoint bundle $P[\mathfrak{g}]$, that is the bundle associated to P with typical fibre $\mathfrak{g}$ and the adjoint action of G on $\mathfrak{g}$. It is convenient to identify these two Lie algebras. Let ξ be an infinitesimal symmetry corresponding to a section of $P[\mathfrak{g}]$. The action of ξ on the space of connections is given by $\omega \longrightarrow \omega' = \omega - \text{hor}\, d\xi^{\#}$, where $\xi^{\#} : P \longrightarrow \mathfrak{g}$ is the equivariant map corresponding to the section of $P[\mathfrak{g}]$ defined by ξ. Thus, if A is the Lie algebra valued form on X associated to ω by a trivialization of P, then the form A′ associated to ω' is given by $A' = A - D\xi$, where $D\xi$ is the covariant differential of ξ, treated as a section of $P[\mathfrak{g}]$, with respect to the connection ω. Hence, ξ preserves the boundary conditions $\omega_{\partial P}$ if and only if the pull back of $D\xi$ to ∂X vanishes. This implies that the restriction $\xi_{\partial P}$ of ξ to ∂P is an infinitesimal automorphisms of ∂P preserving $\omega_{\partial P}$. We denote by $Aut_{\partial X}(\partial P;\omega_{\partial P})$ the group of automorphisms of ∂P which induce the identity transformation in ∂X and preserve the connection $\omega_{\partial P}$. Let $Aut_X(P;\omega_{\partial P})$ denote the group of automorphisms of P which cover the identity transformation in X and preserve the connection $\omega_{\partial P}$ in ∂P, and $Aut_{X\,\partial P} P$ the group of automorphisms of P which induce the identity transformations in X and ∂P. Clearly, $Aut_{X\,\partial P} P$ is a normal subgroup of $Aut_X(P;\omega_{\partial P})$ and $Aut_{\partial X}(\partial P;\omega_{\partial P}) = Aut_X(P;\omega_{\partial P})/Aut_{X\,\partial P} P$. Similarly, the Lie algebra $aut_{\partial X}(\partial P;\omega_{\partial P})$ of $Aut_{\partial X}(\partial P;\omega_{\partial P})$ is the quotient of the corresponding Lie algebras $aut_{\partial X}(\partial P;\omega_{\partial P}) = aut_X(P;\omega_{\partial P})/aut_{X\,\partial P}\, P$. The group $Aut_{\partial X}(\partial P;\omega_{\partial P})$ is a Lie group isomorphic to a subgroup of the structure group G.

Let M be a Cauchy surface in X with boundary $\partial M \subset \partial X$, $\partial P_{\partial M}$ the pull back of ∂P to ∂M, and $\omega_{\partial P_{\partial M}}$ the pull back to $\partial P_{\partial M}$ of the connection form $\omega_{\partial P}$ in ∂P. As before, we denote by $aut_{\partial M}(\partial P_{\partial M};\omega_{\partial M})$ the Lie algebra of infinitesimal automorphisms of $\partial P_{\partial M}$ covering the identity in ∂M and preserving the connection $\omega_{\partial P_{\partial M}}$. Clearly, the restriction of an element of $aut_{\partial X}(\partial P;\omega_{\partial P})$ to $\partial P_{\partial M}$ is contained in $aut_{\partial M}(\partial P_{\partial M};\omega_{\partial P_{\partial M}})$. Since we are concerned only with evolutions in directions of one parameter groups of symmetries, we assume in this section that every element of $aut_{\partial M}(\partial P_{\partial M};\omega_{\partial P_{\partial M}})$ is given by the pull back to $\partial P_{\partial M}$ of an element of $aut_{\partial X}(\partial P,\omega_{\partial P})$.

For each $\xi \in aut_X(P;\omega_{\partial P})$, Eq. (2.26) shows that the value of the conserved charge J^ξ on the Cauchy data satisfying the constraint equation depends on ξ only through its restriction to $\partial P_{\partial M}$. Hence, the assumption about $aut_{\partial M}(\partial P_{\partial M},\omega_{\partial M})$ made here ensures that the values of the conserved charges on the Cauchy data satisfying the constraint equations are labelled by elements of $aut_{\partial M}(\partial P_M,\omega_{\partial M})$. For each $\beta \in aut_{\partial M}(\partial P_M,\omega_{\partial M})$, considered as a section of $\partial P_{\partial M}[\mathfrak{g}]$, we denote by J_β the function on the set of the Cauchy data on M, which satisfy the constraint equations, given by

$$J_\beta = \int_{\partial M} g_{\mathfrak{g}}(^*E_M,\beta) \tag{6.1}$$

If β is the restriction to ∂M of $\xi \in aut_X(P,\omega_{\partial P})$, considered as a section of $P[\mathfrak{g}]$, then J_β is the same as the restriction of the conserved charge J^ξ to the set of the Cauchy data on M which satisfy the constraint equations,

$$J_\beta = J^\xi = \int_{\partial M} g_{\mathfrak{g}}(^*E_M,\xi). \tag{6.2}$$

Applying Stokes' Theorem we can rewrite Eq. (6.2) in the form

$$J_\beta = \int_M d(g_g(*E_M, \xi_M)), \tag{6.3}$$

where ξ_M is the pull back of ξ to a section of $P_M[g]$. One would like to interpret the integrand in Eq. (6.3) as the density of charge corresponding to β in the state described by the Cauchy data on M for the Yang–Mills field and the matter fields under consideration. However, for this interpretation to be well defined, one needs an intrinsic way of association ξ_M to β and the Cauchy data on M.

In electrodynamics one requires that ξ_M should be the pull–back to M of an infinitesimal symmetry ξ given by a covariantly constant section of P[u(1)] normalized to 1 by the metric $g_{u(1)}$. This is equivalent to the requirement that ξ_M should be harmonic and the pull back of ξ_M ato ∂M should be covariantly constant and normalized to 1. If the structure group is non–abelian, for a large class of connections in P_M, only the zero section of $P_M[g]$ is covariantly constant. On the other hand, under appropriate regularity conditions, the Dirichlet problem for the Laplace–Beltrami operator Δ_M in $P_M[g]$,

$$\Delta_M \xi_M = 0, \quad \xi_M | \partial M = \beta \tag{6.4}$$

has a unique solution. Substituting this solution into the integrand in Eq. (6.3) we obtain a 3–form $d(g_g(*E_M, \xi_M))$ on M, intrinsically defined in terms of β and the Cauchy data for the Yang–Mills field, which describes the local distribution of the conserved charge corresponding to β. In the presence of matter we obtain

$$d(g_g(*E_M, \xi_M)) = g_g(j_M, \xi_M) + g_{ab} D_M \xi_M^a \wedge *E_M^b, \tag{6.5}$$

where j_M is the pull–back to M of the matter current j given by Eq. (1.23), and D_M is the covariant differential with respect to the connection in P_M given by the Cauchy data for the Yang–Mills field. The first term on the right hand side of Eq. (6.5) gives the charge density of the matter fields, while the second term describes the distribution of charge carried by the Yang–Mills field.

The charges discussed above are Yang–Mills analogues of the electric charge in electrodynamics. In analogy with Eq. (5.38), for each $\beta \in aut_{\partial X}(\partial P, \omega_{\partial P})$, one can introduce the magnetic charge corresponding to β given by

$$K_\beta = \int_{\partial M} g_g(F,\beta) \tag{6.6}$$

Since $\partial M \subset \partial X$ we can rewrite Eq. (6.6) in the form

$$K_\beta = \int_{\partial M} g_g(F_{\partial X},\beta) \tag{6.7}$$

where $F_{\partial X}$ is the pull–back of F to ∂X. Since $F_{\partial X}$ and β are covariantly constant, it follows that K_β is independent of ∂M, and it depends only on the boundary condition $\omega_{\partial P}$ and $\beta \in aut_{\partial X}(\partial P, \omega_{\partial P})$. Thus, the magnetic charges provide a characterization of the boundary conditions.

The charges discussed here depend on the boundary conditions assumed for the Yang–Mills field. At present we have no general existence theorem for mixed problems for Yang–Mills equations. The theorems available in the literature prove the existence of solutions of the evolution equations for the Cauchy data which belong to appropriate Sobolev spaces on a hyperplane in Minkowski space[1]. The assumption that the Yang–Mills potentials are square integrable implies that the connection approaches a product connection at infinity. This corresponds to a trivial boundary condition, that is a product connection on the boundary. There are some examples of solutions of Yang–Mills equations satisfying non–trivial boundary conditions[2]. However, the most commonly studied Yang–Mills fields with non–trivial asymptotic behaviour are solutions of Yang–Mills–Higgs equations in which the auxiliary Higgs field, introduced

---

[1]c.f. Segal [1979], and Eardley and Moncrief [1982].

[2]See for example Actor [1979].

in order to obtain massive vector bosons, does not vanish at infinity. The physical significance of the Higgs fields is not quite clear, and there are several attempts to to generate massive vector bosons without an introduction of additional dynamical fields[3].

[3]See for example Jaffe and Taubes [1980], Quigg [1983], Cheng and Li [1984], and references quoted there. For an alternative approach see Kerbrat, Kerbrat–Lunc and Śniatycki [1988 a] and [1988 b].

References

ACTOR, A. [1979], "Classical Solutions of SU(2) Yang–Mills Theories", *Rev. Mod. Phys.*, 51, 461–525.

ARMS, J. [1981], "The Structure of the Solution Set for the Yang–Mills Equations", *Math. Proc. Camb. Phil. Soc.*, 90, 361–372.

BLEECKER, D. [1981], *Gauge Theory and Variational Principles*, Addison–Wesley, Reading, Mass.

CHENG, T.–P., and LI, L.–F. [1984], *Gauge Theory of Elementary Particle Physics*, Clarendon Press, Oxford.

DIRAC, P.A.M. [1931], "Quantised Singularities in the Electromagnetic Field", *Proc. Roy. Soc.*, A133, 60–70.

EARDLEY, D.M. and MONCRIEF, V. [1982], "The Global Existence of Yang–Mills Higgs Fields in 4–dimensional Minkowski Space", *Comm. Math. Phys.*, 83, 171–179.

JAFFE, A. and TAUBES, C. [1980], *Vortices and Monopoles*, Birkhauser, Boston.

KERBRAT, Y. and KERBRAT–LUNC, H. [1986], "Spontaneous Symmetry Breaking and Principal Fibre Bundles", *J. G. P.*, 3, 221–230.

KERBRAT, Y., KERBRAT–LUNC, H. and ŚNIATYCKI, J. [1988 a], "How to Get Masses from Kaluza–Klein Theory", Department of Mathematics and Statistics, University of Calgary, preprint No 665.

KERBRAT, Y., KERBRAT–LUNC, H. and ŚNIATYCKI, J. [1988 b], "Geometry of Symmetry Breaking in Electroweak Interactions", in preparation.

KERBRAT–LUNC, H. [1964], "Introduction mathématique à l'étude du champ Yang–Mills sur un espace temps courbé", *C.R. Acad. Sc. Paris*, 259, 3449–3450.

MEYER, M.E. [1981], "The Geometry of Symmetry Breaking in Gauge Theories", *Acta Phys. Austr.* (Suppl.) 23, 477–479.

MORREY, C.B. Jr. [1966], *Multiple Integrals in the Calculus of Variations*, Springer, Heidelberg.

QUIGG, C. [1983], *Gauge Theories of Strong, Weak and Electromagnetic Interactions*, Benjamin.

SAKURAI, J.J. [1960], "Theory of Strong Interactions", *Ann. Phys.* 11, 1–48.

SCHMIDT, B. [1979], "Asymptotic Structure of Isolated Systems", in *Isolated Gravitating Systems in General Relativity*, Proceedings of the International School of Physics "Enrico Fermi", J. Ehlers (ed.), pp. 11–49, North Holland, Amsterdam.

SEGAL, I. [1979], "The Cauchy Problem for the Yang–Mills Equations", *J. Funct. Anal.* 33, 175.

ŚNIATYCKI, J. [1988], "Gauge Invariance, Boundary Conditions and Charges", *Rep. Math. Phys.*, to appear.

TRAUTMAN, A. [1967], "The Application of Fibre Bundles in Physics", mimeographed lecture notes, Kings College, London.

TRAUTMAN, A. [1970], "Fibre Bundles Associated with Space–Time", *Rep. Math. Phys.*, 1, 29–62.

UTIYAMA, R. [1956], "Invariant Theoretical Interpretation of Interactions", *Phys. Rev.*, 101, 1957–1607.

YANG, C.N. and MILLS, R.L. [1954], "Conservation of Isotopic Spin and Isotopic Gauge Invariance, *Phys. Rev.* 96, 191–195.

Chapter 9

GENERAL RELATIVITY

In general relativity the dynamical variables are Lorentzian metrics on the space–time manifold. They satisfy Einstein equations which are derivable from a variational principle with a Lagrangian depending on the second derivatives of the metric. Following Palatini, we consider metrics and torsion free connections as independent variables, and obtain a variational principle with a Lagrangian depending on first derivatives. The corresponding DeDonder form pushes forward to the fibre product of the bundles of Lorentzian metrics and of torsion free connections, which enables us to apply the general theory developed for the first order Lagrangians.

The boundary conditions for Einstein equations considered here consist of specifying a Lorentzian metric along the boundary, such that the induced metric on the boundary is Lorentzian. With these boundary conditions general relativity admits a covariant Hamiltonian formulation of dynamics, which is a modification of the ADM formalism.

The symmetry group of general relativity contains diffeomorphisms of the space–time manifold which preserve the boundary conditions. Localizable infinitesimal symmetries are given by vector fields which vanish on the boundary to the first order. The corresponding constraints are given by the vanishing of the superhamiltonian and the supermomentum. The conserved momenta on the set of Cauchy data satisfying the constraint conditions correspond to the Killing vectors of the boundary condition.

A conformal compactification of the space–time manifold and attaching the hyperboloid of space–like directions at spatial infinity give an embedding of asymptotically flat space–times into a manifold with boundary. This enables us to extend the general formalism to asymptotically flat space times, and to discuss conservation laws for isolated gravitating systems.

9.1 Field equations

The dynamical variable of general relativity is the gravitational field described by a Lorentzian metric g on the space–time manifold X. It satisfies Einstein equations

$$Ric[\nabla[g]] - \frac{1}{2}\, gR[g] = 8\pi T[g,\psi], \tag{1.1}$$

where $\nabla[g]$ is the Levi–Civita connection of g, $Ric[\nabla[g]]$ is the Ricci tensor of $\nabla[g]$,

$$R[g] = tr_g\, Ric[\nabla[g]] \tag{1.2}$$

is the scalar curvature of $\nabla[g]$, and $T[g,\psi]$ is the energy momentum tensor corresponding to a distribution of matter described by a matter field ψ.[1]

In absence of matter the metric g satisfies the vacuum Einstein equations

$$Ric[\nabla[g]] = 0 \tag{1.3}$$

These equations are derivable from a variational principle with Hilbert's action integral

$$A[g] = \frac{1}{16\pi}\int_X R[g]\, vol[g], \tag{1.4}$$

where $vol[g]$ is the volume form on X defined by the orientation of X and the metric g.[2] Since R[g] depends on second derivatives of g or cannot apply directly the general theory developed in Chapter 6.[3] A variational formulation of vacuum Einstein equations in terms of a first order Lagrangian is due to Palatini.[4] In Palatini's approach the action integral is

$$A[g,\Gamma] = \frac{1}{16\pi}\int_X tr_g\, Ric[\Gamma]\, vol[g], \tag{1.5}$$

where a metric g and a torsion free connection Γ are treated as independent variables.

---

[1]Einstein [1915]

[2]Hilbert [1915]

[3]The common way of getting around the problem of second derivatives in the Hilbert Lagrangian is to integrate by parts the terms involving second derivatives. If one ignores the boundary terms, one obtains an equivalent action integral with the integrand independent of the second derivatives of the metric, and one can apply the usual theory for first order Lagrangians. However, the new Lagrangian obtained in this way is no longer an intrinsic geometric object, it depends on the coordinate system used in the process of integration.

[4]A. Palatini, [1919].

The variation of $A[g,\Gamma]$ with respect to Γ vanishes if and only if Γ is the Levi–Civita connection of g,

$$\Gamma = \nabla[g], \tag{1.6}$$

and the variation of $A[g,\Gamma]$ with respect to g vanishes if and only if the Ricci tensor of Γ vanishes,

$$Ric[\Gamma] = 0 \tag{1.7}$$

Equations (1.6) and (1.7) and vacuum Palatini equations. They are equivalent to the vacuum Einstein equations.

Lorentzian metrics on X are sections of a subbundle Y of $\otimes^2 T^* X$ consisting of symmetric forms of signature (−,+,+,+). Similarly, torsion free connections are sections of a subbundle V of the connection bundle of the frame bundle of X. We denote by Z the fibre product of Y and V,

$$Z = Y \times_X V \tag{1.8}$$

and by π_{XZ} the projection map from Z to X. A section σ_{ZX} of π_{XZ} is a pair (g,Γ), where g is a Lorentzian metric and Γ is a torsion free connection on X. The Palatini action integral corresponds to a unique Lagrangian form Λ on J^1Z such that, for every section $\sigma_{ZX} = (g,\Gamma)$ of π_{XZ},

$$\overset{*}{\sigma}_{ZX}\Lambda = \frac{1}{16\pi}\, tr_g\, Ric[\Gamma]\, vol[g] \tag{1.9}$$

Since $Ric[\Gamma]$ is linear in the derivatives of Γ, and the derivatives of g do not appear on the right hand side of Eq. (1.9), it follows that the Palatini Lagrangian form Λ is linear along the fibres of the target map $J^1Z \longrightarrow Z$.

9.1.1. Proposition. The DeDonder form on J^1Z corresponding to the Palatini Lagrangian pushes forward to a form Ω on Z.

Proof follows directly from the linearity of Λ along the fibres of the target map $J^1Z \longrightarrow Z$ and Eq. (6.1.10). ∎

In the following we shall refer to the form Ω on Z, obtained by pushing forward the DeDonder form on J^1Z, as the *DeDonder form of General Relativity.*[5]

Most computations in general relativity can be conveniently performed in terms of local coordinates. Let $(x^\mu, y_{\mu\nu}, v^\lambda_{\mu\nu})$, $\mu,\nu,\lambda = 0,1,2,3$, be local coordinates in Z adapted to the structure of Z given in Eq. (1.8). That is (x^μ) are local coordinates in X, and $(y_{\mu\nu})$ and $(v^\lambda_{\mu\nu})$ are the induced coordinates in typical fibres of Y and V, respectively. A section $\sigma_{ZX} = (g,\Gamma)$ of π_{XZ} is given locally by specifying the coordinates $y_{\mu\nu}$ and $v^\lambda_{\mu\nu}$ as functions of the coordinates x^μ,

$$y_{\mu\nu} = g_{\mu\nu}(x), \quad v^\lambda_{\mu\nu} = \Gamma^\lambda_{\mu\nu}(x) \tag{1.10}$$

where $g_{\mu\nu}$ are the components of g and $\Gamma^\lambda_{\mu\nu}$ are the components of Γ with respect to the coordinates (x^μ). A connection Γ is the Levi Civita connection of g if and only if

$$\Gamma^\lambda_{\mu\nu} = \frac{1}{2} g^{\lambda\rho}(g_{\mu\rho,\nu} + g_{\nu\rho,\mu} - g_{\mu\nu,\rho}). \tag{1.11}$$

The Ricci tensor $Ric[\Gamma]$ of a connection Γ has component $R_{\mu\nu}$ given by

$$R_{\mu\nu} = \Gamma^\rho_{\mu\nu,\rho} - \Gamma^\rho_{\mu\rho,\nu} + \Gamma^\rho_{\mu\nu}\Gamma^\sigma_{\rho\sigma} - \Gamma^\rho_{\mu\sigma}\Gamma^\sigma_{\sigma\rho} \tag{1.12}$$

The Palatini Lagrangian is given by

$$tr_g Ric[\Gamma]\, vol[g] = \sqrt{-det(g_{\alpha\beta})}\, R\, d_4x \tag{1.13}$$

where

$$R = g^{\mu\nu} R_{\mu\nu} \tag{1.14}$$

and $(g^{\mu\nu})$ is the inverse of the matrix $(g_{\mu\nu})$,

$$g^{\mu\nu} g_{\nu\lambda} = \delta^\mu_\lambda.$$

Introducing new coordinates $(u^{\mu\nu})$ in a typical fibre of Y, related to $(y_{\mu\nu})$ by

[5]The DeDonder form and the corresponding Cauchy data formulation of General Relativity were introduced in Śniatycki [1970]. See also Szczyrba [1976].

$$u^{\mu\nu}y_{\nu\lambda} = \sqrt{-det(y_{\alpha\beta})}\,\delta^{\mu}_{\lambda}$$

and following the construction of the DeDonder form given in Sec. 6.1 we obtain

$$\Omega = u^{\mu\nu}(dv^{\lambda}_{\mu\nu} - \delta^{\lambda}_{\nu}dv^{\sigma}_{\mu\sigma}) \wedge d_3x_{\lambda} + u^{\mu\nu}(v^{\lambda}_{\mu\nu}v^{\sigma}_{\lambda\sigma} - v^{\lambda}_{\mu\sigma}v^{\sigma}_{\lambda\nu})d_4x\,. \tag{1.15}$$

9.1.2. Proposition. A section $\sigma_{ZX} = (g,\Gamma)$ satisfies the Palatini equations (1.6) and (1.7) if and only if it satisfies the DeDonder equations:

$$\overset{*}{\sigma}_{ZX}(\xi_Z \lrcorner\, d\Omega) = 0 \tag{1.16}$$

for every vector field ξ_Z on Z.

Proof follows by direct computation using equations (1.15), (1.11) and (1.12). ∎

The mapping assigning to each metric g on X its Levi–Civita connection $\nabla[g]$ gives rise to a bundle isomorphism $\alpha : J^1Y \longrightarrow Z$ such that, for each Lorentzian metric $g : X \longrightarrow Y$,

$$\alpha \circ j^1g = (g,\nabla[g]) \tag{1.17}$$

The pull–back of the DeDonder form Ω on Z by α leads to a DeDonder formulation of general relativity in the space J^1Y of 1–jets of Lorentzian metrics. Actually, it is convenient to use α to identify J^1Y with Z.

Einstein equations in the presence of a matter field ψ can be obtained from a variational principle with the Hilbert action integral

$$A[g,\psi] = \frac{1}{16\pi}\int_X R[g]\,vol[g] + \int_X L[g,\psi]\,vol[g], \tag{1.18}$$

where $L[g,\psi]$ is a Lagrangian of the matter field. The energy momentum tensor $T[g,\psi]$ is obtained from the derivative of $\int L[g,\psi]\,vol[g]$ with respect to g. A relativistic matter Lagrangian $L[g,\psi]$ depends locally on the metric g, the matter field ψ, and the covariant derivatives $\nabla[g]\psi$ of ψ with respect to the Levic Civita connection $\nabla[g]$ of g,

$$L[g,\psi](x) = \lambda(g(x),\psi(x),\nabla[g]\psi(x)) \tag{1.19}$$

for every $x \in X$. The dependence of the Lagrangian on $\nabla[g]\psi$ implies that it may depend on the first derivatives of the metric g, since $\nabla[g]$ depends on the derivatives of g. Unless the dependence of L on the first derivatives of g is linear, the energy momentum tensor depends on second derivatives of g and it may affect the character of Einstein equations. In order to avoid this situation we restrict our considerations to matter fields with Lagrangians depending linearly on first derivatives of the metric, e.g. scalar fields and Yang–Mills fields which have Lagrangians independent of the derivatives of the metric, or spinor fields with Lagrangians linear in derivatives.

Let Y_{matter} be a bundle over X such that the matter field ψ corresponds to a section of Y_{matter}. The DeDonder form for the gravitational field interacting with the matter field ψ is the sum of pull–backs to the fibre product over X of Z and J^1Y_{matter} of the form Ω given above and the DeDonder form for the matter field.

9.2 Conservation laws and constraints.

The Hilbert Lagrangian and the Palatini Lagrangian are invariant under the action of the group of diffeomorphisms of the space time manifold X. Diffeomorphisms of X have natural actions in the bundles Y and Z. Moreover, the DeDonder form Ω on Z corresponding to the Palatini Lagrangian is defined intrinsically in terms of the given structure. Hence, Ω is invariant under the action in Z of the group of diffeomorphisms of X.

Infinitesimal diffeomorphisms of X are given by vector fields ξ_X on X. The actions of ξ_X in Y and Z give rise to vector fields ξ_Y and ξ_Z on Y and Z, respectively, such that $\xi_X = \pi_{XY*}\xi_Y$ and $\xi_Y = \pi_{YZ*}\xi_Z$. The diffeomorphisms of X acting in Z preserve Ω. Hence, it follows that, for each infinitesimal diffeomorphisms ξ_X,

$$\pounds_{\xi_Z} \Omega = 0. \tag{2.1}$$

A local coordinate description of ξ_Z in terms of the coordinates $(x^\mu, u^{\mu\nu}, v^\lambda_{\mu\nu})$ in Z, introduced in Sec. 9.1, is as follows. If ξ^μ are the components of ξ_X with respect to the coordinates (x^μ) in X, then

$$\xi_Z = \xi^\mu \frac{\partial}{\partial x^\mu} + \xi^{\mu\nu} \frac{\partial}{\partial u^{\mu\nu}} + \xi^\lambda_{\mu\nu} \frac{\partial}{\partial v^\lambda_{\mu\nu}} \tag{2.2}$$

where

$$\xi^{\mu\nu} = u^{\mu\alpha}\xi^\nu{}_{,\alpha} + u^{\alpha\nu}\xi^\mu{}_{,\alpha} - u^{\mu\nu}\xi^\alpha{}_{,\alpha} \tag{2.3}$$

and

$$\xi^\lambda_{\mu\nu} = v^\alpha_{\mu\nu}\xi^\lambda{}_{,\alpha} - v^\lambda_{\alpha\nu}\xi^\alpha{}_{,\mu} - v^\lambda_{\mu\alpha}\xi^\alpha{}_{,\nu} - \xi^\lambda{}_{,\mu\nu} \tag{2.4}$$

Taking into account Eq. (1.21) giving the coordinate description of Ω one can verify Eq. (2.1) by direct computations.

We consider here boundary conditions B given by specifying along ∂X a metric in X, that is by specifying a section $\sigma_{\partial Y \partial X}$ of $\pi_{\partial X \partial Y}$: $\partial Y \longrightarrow \partial Z$,

$$B = \pi_{\partial Y \partial Z}^{-1}(\sigma_{\partial Y \partial X}(\partial X)). \tag{2.5}$$

We denote by $g_{\partial X}$ the induced metric in ∂X and $\eta_{\partial X}$ the unit normal vector field of ∂X oriented towards the interior of X. We assume $g_{\partial X}$ has signature $(-,+,+)$.

Each connection Γ in X gives rise to the corresponding Weingarten map $W_{\partial X}$ and the second fundamental form $h_{\partial X}$ of ∂X. Recall that the Weingarten map of ∂X is the linear map $W_{\partial X} : T\partial X \longrightarrow T\partial X$ such that, for each vector $v \in T\partial X$, the tangential to ∂X component of the covariant derivative of $\eta_{\partial X}$ in the direction v is given by $W_{\partial X}(v)$. Similarly, the second fundamental form of ∂X is a symmetric bilinear form $h_{\partial X}$ on ∂X such that, for each pair (ξ,ζ) of vector fields on X tangent to ∂X, the normal to ∂X component of the covariant derivative of ζ with respect to ξ is given by $h_{\partial X}(\xi_{\partial X},\zeta_{\partial X})\eta_{\partial X}$, where $\xi_{\partial X}$ and $\zeta_{\partial X}$ are the restrictions of ξ and ζ to ∂X, respectively. It should be noted that the notions of a Weingarten map and a second fundamental form do not require that the connection Γ should be the Levi Civita connection of a metric. If Γ is such that the covariant derivatives of the metric of along ∂X in the directions tangent to ∂X vanish, then the corresponding Weingarten map and the second fundamental forms are related as follows. For every pair of vectors $u,v \in T\partial X$,

$$h_{\partial X}(u,v) = -\,g_{\partial X}(u,W_{\partial X}(v)) \tag{2.6}$$

We denote by i_{ZB}: $B \longrightarrow Z$ the inclusion map, and by $\pi_{\partial XB}$: $B \longrightarrow \partial X$ the projection induced by $\pi_{\partial X \partial Z}$: $\partial Z \longrightarrow \partial X$. Let $P : B \longrightarrow \mathbb{R}$ be a function such that, for every section $\sigma_{ZX} = (g,\Gamma)$ satisfying the boundary condition B,

$$P\circ(\sigma_{ZX}|\,\partial X) = tr_{g_{\partial X}} h_{\partial X} - tr\, W_{\partial X} \tag{2.7}$$

where $W_{\partial X}$ and $h_{\partial X}$ are the corresponding Weingarten map and second fundamental form, and $tr_{g_{\partial X}}$ denotes the trace with respect to the metric $g_{\partial X}$. The function P on B and the volume form $vol[g_{\partial X}]$ on ∂X give rise to a 3–form Π on B defined by

$$\Pi = P \cdot \pi_{\partial XB}^{*} vol[g_{\partial X}] \tag{2.8}$$

9.2.1. Proposition. For each vector field ξ_X on X preserving the boundary condition B,

$$£_{\xi_B} \Pi = 0, \tag{2.9}$$

where ξ_B is the restriction of ξ_Z to B. Moreover,

$$i_{ZB}^{*} \Omega = d\Pi. \tag{2.10}$$

Proof. Eq. (2.9) follows from the fact that Π is intrinsically defined in terms of the data defining the boundary conditons and ξ_X preserves these data. Choosing a coordinate system in X such that ∂X is given by $x^1 = 0$, $\eta_{\partial X} = \frac{\partial}{\partial x^1}$, and g along ∂X has components $g_{11} = 1$, $g_{a1} = 0$, a = 0,2,3, we obtain from Eq. (1.15)

$$i_{ZB}^{*} \Omega = \sqrt{-det(g_{cd})}\,(g^{ab} dv^1_{ab} - dv^a_{1a}) \wedge d_3x_1 =$$

$$= d\{(g^{ab} v^1_{ab} - v^a_{1a}) \sqrt{-det(g_{cd})}\, d_3x_1\} .$$

However, in this coordinate system, the components W^b_a of the Weingarten map $W_{\partial X}$ corresponding to a connection Γ are given by

$$W^b_a = \Gamma^b_{1a} \tag{2.11}$$

Similarly, the components h_{ab} of the second fundamental form $h_{\partial X}$ corresponding to Γ are

$$h_{ab} = \Gamma^1_{ab}. \tag{2.12}$$

Hence, the coordinate expression for P is

$$P = g^{ab} v^1_{ab} - v^a_{1a} \tag{2.13}$$

and, therefore, $i_{ZB}^{*} \Omega = d\Pi$, which completes the proof. ∎

It follows from Prop. 9.2.1 that the symmetry group $\mathcal{G}$ of the theory contains diffeomorphisms of X which preserve the boundary conditions. The Lie algebra $\mathfrak{g}$ of infinitesimal symmetries is given by vector fields ξ_X on X for which

$$\pounds_{\xi_X} g_{\partial X} = 0 \qquad (2.14)$$

and

$$\pounds_{\xi_X} \eta_{\partial X} = 0 \qquad (2.15)$$

According to Prop. 6.2.3, the constants of motion corresponding to infinitesimal symmetries $\xi \in \mathcal{g}$ are given by

$$J^{\xi}(\sigma_{ZX}|M) = \int_M \sigma_{ZX}^{*}(\xi_Z \lrcorner \Omega) + \int_{\partial M} \sigma_{ZX}^{*}(\xi_B \lrcorner \Pi) \quad (2.16)$$

9.2.2. Proposition. Let M be a Cauchy surface in X with boundary $\partial M \subset \partial X$, and $\sigma_{ZX} = (g,\Gamma)$ a section of π_{XZ} satisfying the boundary condition B, and such that the induced metric g_M on M is positive definite, M is normal to ∂X along ∂M and the metric g is invariant under parallel transport along curves in M. For every infinitesimal symmetry ξ,

$$\begin{aligned} J^{\xi}(\sigma_{ZX}|M) = &\int_M \{H\, \xi_X \lrcorner\, vol[g] + \langle J,\xi_M\rangle vol[g_M]\} + \\ &+ \int_{\partial M} \{tr\, W_{\partial M} - tr_{g_{\partial M}} h_{\partial M}\} \xi_X \lrcorner\, vol[g_{\partial X}] + \qquad (2.17) \\ &\int_{\partial M} \{h_{\partial X}(\eta_M,\xi_{\partial M}) - g_{\partial X}(\eta_M, W_{\partial X}(\xi_M))\} vol[g_{\partial M}]. \end{aligned}$$

Here H is a function on M, called a *superhamiltonian*, given by

$$H = R[g_M] - \{tr(W_M^2) - (tr\, W_M)^2\}, \qquad (2.18)$$

where W_M is the Weingarten map of M defined by the connection Γ along M and a unit normal vector field η_M. J is a 1–form on M, called a *supermomentum*, given

$$J = 2\{div\, W_M - d(tr\, W_M)\}. \qquad (2.19)$$

ξ_M is the component of ξ_X tangent to M:

$$\xi_M(x) = \xi_X(x) + g(\xi_X(x),\eta_M(x))\eta_M(x) \tag{2.20}$$

for every $x \in M$. $W_{\partial M}$ and $h_{\partial M}$ are the Weingarten map and the second fundamental form of ∂M considered as a submanifold of M and oriented with the normal $\eta_{\partial X} | \partial M$.

Proof. Let (x^μ) be local coordinates in X, and $g_{\mu\nu}$ and $\Gamma^\lambda_{\mu\nu}$ the corresponding components of g and Γ, respectively. The pull back of $\xi_Z \lrcorner \Omega$ by σ_{ZX} can be evaluated in terms of the coordinate expressions (1.15), (2.2) and (2.3). We obtain

$$\overset{*}{\sigma}_{ZX}(\xi_Z \lrcorner \Omega) = \sqrt{-det(g_{\alpha\beta})}\,\{R\xi^\lambda - g^{\mu\nu}\pounds_{\xi_X}\Gamma^\lambda_{\mu\nu} + g^{\nu\lambda}\pounds_{\xi_X}\Gamma^\mu_{\nu\mu}\}d_3x_\lambda \tag{2.21}$$

Since

$$\pounds_{\xi_X}\Gamma^\lambda_{\mu\nu} = \xi^\lambda_{\ ;\mu\nu} - R^\lambda_{\mu\nu\rho}\xi^\rho \tag{2.22}$$

where ; denotes the covariant differentiation with respect to the connection Γ, we can rewrite Eq. (2.21) in the form

$$\overset{*}{\sigma}_{ZX}(\xi_Z \lrcorner \Omega) = \tag{2.23}$$

$$= \sqrt{-det(g_{\alpha\beta})}\{R\xi^\lambda + g^{\mu\nu}R^\lambda_{\mu\nu\rho}\xi^\rho - g^{\nu\lambda}R^\mu_{\nu\mu\rho}\xi^\rho - g^{\mu\nu}\xi^\lambda_{\ ;\mu\nu} + g^{\nu\lambda}\xi^\mu_{\ ;\nu\mu}]d_3x_\lambda\,.$$

If the coordinates (x^μ) are chosen in such a way that M is given by the equation $x^0 = 0$ and $\eta_M = \frac{\partial}{\partial x^0}$, then the components of g and Γ satisfy

$$g^{00} = -1,\quad g^{0k} = 0,\quad \Gamma^0_{0k} = 0,\quad \Gamma^n_{0m} = W^n_{m},\quad \Gamma^0_{mn} = h_{mn},$$

where $k,m,n = 1,2,3$, and W^n_m and h_{mn} are the components of W_M and h_M, respectively. The pull back of $\overset{*}{\sigma}_{ZX}(\xi_Z \lrcorner \Omega)$ to M is given by

$$i_{XM}^{\ *}(\sigma_{ZX}^{\ *}(\xi_Z \lrcorner \Omega)) = \tag{2.24}$$

$$\sqrt{-\det(g_{\alpha\beta})}\{g^{mn}R^k_{mkn}\xi^0 + g^{mn}R^0_{mnk}\xi^k + R^m_{0mk}\xi^k - \xi^k_{\ ;0k} - g^{mn}\xi^0_{imn}\}d_3x_0\,.$$

By direct computation we can verify that

$$g^{mn}R^k_{mkn} = H$$

$$R^{0}_{mnk} = h_{mk|n} - h_{mn|k}$$
$$R^{m}_{0mk} = W^{m}_{k|m} - W^{m}_{m,k}$$

where $|n$ denotes the covariant derivative in M with respect to x^n relative to the induced connection in M. Hence

$$i_{XM}^{*}(\sigma_{ZX}^{*}(\xi_Z \lrcorner \Omega)) =$$

$$= \sqrt{-det(g_{\alpha\beta})}\,\{H\,\xi^0 + <J,\xi_M> - \xi^k_{;0k} - g^{mn}\xi^0_{;mn}\}d_3x_0\,,$$

which can be rewritten in the form

$$i_{XM}^{*}(\sigma_{ZX}^{*}(\xi_Z \lrcorner \Omega)) = \sqrt{-\det(g_{\alpha\beta})}\,\{H\,\xi^0 + <J,\xi_M>\}d_3x^0$$
$$- d\{\sqrt{-\det(g_{\alpha\beta})}\,(\xi^m_{;0} + g^{mn}\xi^0_{;n})d_2x_{0m}\}\,, \qquad (2.25)$$

because the covariant derivatives of $g_{\alpha\beta}$ is directions tangent to M vanish by hypothesis. On the other hand, Eqs. (2.8) and (2.13) yield

$$i_{\partial X \partial M}^{*}(\sigma_{ZX}^{*}(\xi_B \lrcorner \Pi)) = \sqrt{-det(g_{\alpha\beta})}\,(g^{ab}\Gamma^1_{ab} - \Gamma^a_{1a})\xi^0 d_2x_{10}$$

where we have assumed that our coordinate system is chosen in such a way that ∂X is given by $x^1 = 0$, $\eta_{\partial X} = \dfrac{\partial}{\partial x^1}$ and, therefore, $\xi^1 = 0$. We have

$$J^{\xi}(\sigma_{ZX}|M) = \int_M \{H\,\xi \lrcorner\, vol[g] + <J,\xi_M> vol[g_M]\} +$$
$$+ \int_{\partial M} \sqrt{-det(g_{\alpha\beta})}\,\{-\xi^1_{;0} - g^{11}\xi^0_{;1} - (g^{ab}\Gamma^1_{ab} - \Gamma^a_{1a})\xi^0\}d_2x_{01}$$
$$= \int_M \{H\,\xi \lrcorner\, vol[g] + <J,\xi_M> \mathrm{vol}[g_M]\} +$$
$$+ \int_{\partial M} \{-\Gamma^1_{0A}\,\xi^A - \Gamma^0_{1A}\,\xi^A - \xi^0(g^{AB}\Gamma^1_{AB} - \Gamma^A_{1A})\}vol[g_{\partial M}]$$

where the indices A,B = 2,3, and we have used the assumption that $\pounds_{\xi_X} g$ vanishes along ∂X so that, for each $x \in \partial M$,

$$0 = £_{\xi_X} g_{01}(x) = g_{01,a}\xi^a + g_{00}\xi^0_{,1} + g_{11}\xi^1_{,0},$$

which implies that $\xi^0_{,1} = 0$. The last equality implies Eq. (2.7), which completes the proof. ■

The group of localizable symmetries of general relativity contains the identity component $\mathscr{H}$ of the group diffeomorphisms of X which induce the identity transformation in ∂Y. According to the Second Noether Theorem, for each infinitesimal localizable symmetry ξ, each section $\sigma_{ZX} = (g,\Gamma)$ satisfying the field equations, and each Cauchy surface M, the momentum $J_\xi(\sigma_{ZX}|M)$ vanishes. It follows from Prop. 9.2.2 that if $\sigma_{ZX} = (g,\Gamma)$ satisfies the field equations then, for each Cauchy surface M, the corresponding superhamiltonian and supermomentum vanish,

$$H = 0 \tag{2.26}$$

$$J = 0 \tag{2.27}$$

Eqs. (2.26) and (2.27) are the constraint equations of general relativity. Substituting them to Eq. (2.17) and noting that if $\sigma_{ZX} = (g,\Gamma)$ satisfies the field equations then Γ is the Levi Civita connection of g, we obtain an expression for the value of a conserved momentum J^ξ at the Cauchy data of a solutions σ_{ZX} on a Cauchy surface M,

$$\begin{aligned} J^\xi(\sigma_{ZX}|M) = 2\int_{\partial M\cdot}(tr\, W_{\partial M})\xi_X \lrcorner\, vol[g_{\partial X}] \\ - 2\int_{\partial M} g_{\partial X}(\eta_M, W_{\partial X}(\xi_M))vol[g_{\partial M}]. \end{aligned} \tag{2.28}$$

If the restriction $\xi_{\partial\xi}$ of ξ_X to ∂X is normal to ∂M, $g_{\partial X}(\xi_{\partial X},\xi_{\partial X}) = -1$, and $i_{\partial X\partial M}{}^*(\xi_{\partial X} \lrcorner\, vol[g_{\partial X}]) = -\, vol[g_{\partial M}]$, then the second integral in Eq. (1.18) vanishes and

$$J^\xi(\sigma_{ZX}|M) = -2\int_{\partial M} tr\, W_{\partial M}\, vol[g_{\partial M}] \tag{2.29}.$$

In this case, $J^{\xi}(\sigma_{ZX}|M)$ is interpreted as the energy E of the gravitational field in the state $\sigma_{ZX}|M$. It can be rewritten in the usual coordinate form

$$E = \int_{\partial M} \sqrt{\det(g_{rs})}\, g^{ij} g^{kn} (g_{ik,j} - g_{ij,k}) d_n^2 x \tag{2.30}$$

where (x^i), i = 1,2,3, are coordinates in M such that ∂M is given by $x^1 = 0$, and $d_n^2 x = \frac{1}{2} \epsilon_{nij} dx^i \wedge dx^j$.

If $\xi_{\partial X}$ is tangential to ∂M, then the first integral in Eq. (2.28) vanishes. In this case $J^{\xi}(\sigma_{ZX}|M)$ can be interpreted as an analogue of a linear momentum (or an angular momentum of the gravitational field in the state $\sigma_{ZX}|M$. Since

$$g_{\partial X}(\eta_M, W_{\partial X}(\xi_M)) = g(\eta_M, \nabla_{\xi_M} \eta_{\partial X}) = -g(\nabla_{\xi_M} \eta_M, \eta_{\partial X}) = -g(W_M(\xi_M), \eta_{\partial X})$$

and $\mathrm{vol}[g_{\partial M}] = \eta_{\partial X} \lrcorner \mathrm{vol}[g_M]$, it follows that

$$J^{\xi}(\sigma_{ZX}|M) = 2\int_{\partial M} W_M(\xi_M) \lrcorner \mathrm{vol}[g_M] \tag{2.31}$$

Its local coordinate expression is given by

$$J^{\xi}(\sigma_{ZX}|M) = 2\int_{\partial M} \sqrt{\det(g_{ij})}\, \Gamma^n_{0m}\, \xi^m d_n^2 x, \tag{2.32}$$

in agreement with the results given in the literature.[1]

In the presence of matter fields ψ the constraint equations (2.26) and (2.27) are replaced by

$$H = 16\pi \langle T, \eta_M \otimes \eta_M \rangle \tag{2.33}$$

and

$$J = 16\pi\, i_{XM}^{\ *} (\eta_M \lrcorner T) \tag{2.34}$$

where H and J are the superhamiltonian and the supermomentum given by Eqs. (2.18)

[1]Conservation laws in general relativity are usually studied for asymptotically flat space times which will be discussed in Section 9.4. Conservation laws for space–times with boundary were studied in Binz and Śniatycki [1987].

and (2.19), respectively, and T is the energy momentum tensor of the matter field.

If the matter field vanishes on ∂X then the conserved momenta for the interacting system are given by the same expression, Eq. (2.28), as in the case of the gravitational field in absence of matter.

9.3 Hamiltonian formulation

The boundary conditions $B \subset \partial Z$ discussed in the preceding section define a metric $g_{\partial X}$ in ∂X of signature $(-,+,+)$. We assume that ∂X is oriented and time oriented. That is, we have a choice of orientation in the cone in $T\partial X$ consisting of vector u such that $g_{\partial X}(u,u) < 0$. Positively oriented vectors are called future oriented. The group $\mathscr{G}$ of symmetries of the theory consists of diffeomorphisms of X which preserve the boundary condition B and the orientation and time orientation of ∂X. The subgroup $\mathscr{H}$ of localizable symmetrics contains diffeomorphisms of X which induce the identity transformation in ∂Y.

Following the notation of Sec. 7.3 we denote by M a typical Cauchy surface, and by $\mathscr{T}$ a manifold of embeddings $\tau : M \longrightarrow X$ such that $\tau(\partial M)$ is a space–like submanifold of ∂X separating ∂X into two disjoint components, which can be interpreted as the future and the past of ∂M, and the unit normal vector field of ∂X defined by the boundary conditions B is tangent to $\tau(M)$ along $\tau(\partial M)$.

Let $\gamma : M \longrightarrow Z$ be an embedding such that $\tau = \pi_{XZ} \circ \gamma \in \mathscr{T}$. It describes a set of Cauchy data on the Cauchy surface $\tau(M)$, which consists of a metric g and a connection Γ defined along $\tau(M)$. We assume that the metric g along $\tau(\partial M)$ agrees with the metric along ∂X given by the boundary conditions B. Moreover, we assume that g is invariant under the parallel transport along $\tau(M)$ given by the connection Γ, and that the metric g_τ in $\tau(M)$ induced by g is positive definite. A space of Cauchy data of general relativity is a manifold $\mathscr{Z}$ of such embeddings $\gamma : M \longrightarrow Z$, stable under the action of the group $\mathscr{G}$ of symmetries of the theory, and fibered over $\mathscr{T}$ by the projection $p_{\mathscr{T}\mathscr{Z}}: \mathscr{Z} \longrightarrow \mathscr{T}$, defined by $\pi_{\mathscr{T}\mathscr{Z}}(\gamma) = \pi_{XZ} \circ \gamma$ for every $\gamma \in \mathscr{Z}$. Given $\tau \in \mathscr{T}$, we denote by $\mathscr{Z}_\tau$ the fibre of $\pi_{\mathscr{T}\mathscr{Z}}$ over τ.

Every element $\gamma \in \mathscr{Z}_\tau$ corresponds to the quadruple $(\eta_\tau, \nabla_{\eta_\tau}, g_\tau, W_\tau)$, where η_τ is the unit normal to $\tau(M)$ relative to the metric g along $\tau(M)$ corresponding to γ, such that its restriction to $\tau(\partial M)$ is future oriented in ∂X, ∇_{η_τ} is the operator of covariant differentiation in the direction of η_τ defined by the connection Γ along $\tau(M)$ corresponding to γ, g_τ is the induced metric on $\tau(M)$ and W_τ is the Weingarten map of $\tau(M)$. In the following we use the identification

$$\gamma = (\eta_\tau, \nabla_{\eta_\tau}, g_\tau, W_\tau) \tag{3.1}$$

A vector $\zeta \in T_\gamma \mathcal{Z}_\tau$ can be decomposed into its components in the directions η_τ, ∇_{η_τ}, g_τ and W_τ, denoted by $\dot{\eta}_\tau$, $\dot{\nabla}_{\eta_\tau}$, $\dot{g}_\tau$ and $\dot{W}_\tau$, respectively,

$$\zeta = (\dot{\eta}_\tau, \dot{\nabla}_{\eta_\tau}, \dot{g}_\tau, \dot{W}_\tau) \tag{3.2}$$

Eqs. (1.21), (2.8), and the definition of the Cartan form Ξ corresponding to Ω, c.f. Eq. (6.3.2), yield

$$\Xi(\zeta) = \int_{\tau(M)} \{2 \operatorname{tr} \dot{W}_\tau + \operatorname{tr}_{g_\tau}(\dot{g}_\tau W_\tau)\} vol[g_\tau] \tag{3.3}$$

If ξ_X is a vector field on X such that its extension ξ_Z to Z preserves the boundary conditions B, then the value of Ξ on $\xi_Z \circ \gamma \in T_\gamma \mathcal{Z}$ can be obtained from Prop. 8.2.2. Eqs (2.16), (2.17), (2.29), and (2.31),

$$\Xi(\xi_Z \circ \gamma) = \int_{\tau(M)} \{H_\tau \xi_X \lrcorner\ vol[g] + \langle J_\tau, \xi_{\tau(M)} \rangle vol[g_\tau]\} +$$
$$+ 2\int_{\tau(\partial M)} \{tr\, W_{\tau(\partial M)} g_{\partial X}(\eta_\tau, \xi_X) vol[g_{\tau(\partial M)}] + W_\tau(\xi_{\tau(M)}) \lrcorner\ vol[g_\tau]\} \tag{3.4}$$

where H_τ and J_τ are the superhamiltonian and the supermomentum, respectively, defined by g_τ and W_τ, c.f. Eqs. (2.18) and (2.19), $\xi_{\tau(M)}$ is the component of ξ_X tangential to $\tau(M)$, Eq. (2.20), $W_{\tau(\partial M)}$ is the Weingarten map of $\tau(\partial M)$ embedded in $\tau(M)$. Since every vector in $T_\gamma \mathcal{Z}$ can be decomposed as the sum of a vector in $T_\gamma \mathcal{Z}_\tau$ and of $\xi_Z \circ \gamma$, for an appropriate ξ_X on X, Eqs. (3.3) and (3.4) determine Ξ uniquely.

Let $\mathcal{P}$ be the quotient manifold of $\mathcal{Z}$ given by the projection map $\pi_{\mathcal{P}\mathcal{Z}}: \mathcal{Z} \to \mathcal{P}$

which associates to each $\gamma = (\eta_\tau, \nabla_{\eta_\tau}, g_\tau, W_\tau) \in \mathscr{Z}$ the triplet $(\eta_\tau, g_\tau, W_\tau)$. It is fibered over $\mathscr{T}$ by a map $\pi_{\mathscr{T}\mathscr{P}}$ associating to each $(\eta_\tau, g_\tau, W_\tau)$ the parametrized Cauchy surface $\tau \in \mathscr{T}$. Eq. (3.3) and (3.4) show the Cartan form Ξ pushes forward to a form Θ on $\mathscr{P}$ so that $\mathscr{P}$ is the primary constraint bundle of general relativity. Further, let $\mathscr{R}$ be the quotient manifold of $\mathscr{P}$ given by the projection map $\pi_{\mathscr{R}\mathscr{P}}: \mathscr{P} \longrightarrow \mathscr{R}$ associating to each $(\eta_\tau, g_\tau, W_\tau) \in \mathscr{P}$ the pair $(g_\tau, W_\tau) \in \mathscr{R}$. It is the reduced phase bundle of the theory. The fibres of the projection map $\pi_{\mathscr{R}\mathscr{P}}: \mathscr{P} \longrightarrow \mathscr{R}$ are parametrized by vector fields η_τ transverse to $\tau(M)$ such that their restrictions to points in $\tau(\partial M)$ coincide with the future pointing unit normal of $\tau(\partial M)$ with respect to the metric $g_{\partial X}$. Let $\mathscr{A}$ be the space of embeddings $\alpha : M \longrightarrow TX$ covering the embeddings τ in $\mathscr{T}$ such that, for each $m \in M$, $\alpha(m)$ is a vector in $T_{\tau(m)}X$ transverse to $\tau(M)$, and $\alpha(m)$ is the future pointing unit normal of $\tau(\partial M)$, whenever $m \in \partial M$. Clearly,

$$\mathscr{P} = \mathscr{A} \times_{\mathscr{T}} \mathscr{R}, \tag{3.5}$$

and this product structure is stable under the action of the symmetry group $\mathscr{G}$ of general relativity. Hence, the first part of Hypothesis 6.4.5 holds.

9.3.1. <u>Proposition</u>. For every $\tau \in \mathscr{T}$ and every $\alpha, \alpha' \in \mathscr{A}_\tau$ such that

$$j^1\alpha | \tau(\partial M) = j^1\alpha' | \tau(\partial M), \tag{3.6}$$

there exists a localizable symmetry h in $\mathscr{H}$ such that $h_{\mathscr{A}}(\alpha) = \alpha'$ and $h_{\mathscr{R}} | \mathscr{R}_\tau$ is the identity transformation in $\mathscr{R}_\tau$.

<u>Proof</u>. Let $\alpha, \alpha' \in \mathscr{A}_\tau$ satisfy Eq. (3.5). Consider a vector field ξ on X such that $j^1\xi$ vanishes on ∂X, and a smooth function f with compact support such that $f | \tau(M) = 0$ and $\langle df, \alpha \rangle = 1$, where α is interpreted as a vector field in X defined along $\tau(M)$ and transverse to $\tau(M)$. The vector field $\zeta = f\xi$ has compact support and if generates a one parameter group h_X^t of diffeomorphisms of X. Since $j^1\zeta$ vanishes on ∂X it follows that h_X^t is a one parameter subgroup of the group of localizable symmetries of general

relativity. Since ζ vanishes on $\tau(M)$, the points of $\tau(M)$ are fixed by the action of h_X^t and, for each $x \in \tau(M)$, Th_X^t induces a one parameter group of linear transformations of T_xX. For every $u \in T_xX$, we have

$$\frac{d}{dt} Th_X^t(u) = \xi(x)<df, Th_X^t(u)>. \tag{3.6}$$

Let (x^μ) be coordinates in a neighbourhood U of x wuch that $x^0 = f|U$ and (x^k) are coordinates in $U \cap \tau(M)$. Denoting by $u^\mu(t)$ and ξ^μ the corresponding components of $Th_X^t(u)$ and ξ, respectively, we can rewrite Eq. (3.5) in the form

$$\frac{d}{dt} u^0(t) = \xi^0 u^0(t), \quad \frac{d}{dt} u^k(t) = \xi^k u^0(t).$$

Hence,

$$u^0(1) = exp(\xi^0)u^0(0), \quad u^k(1) = u^k(0) + exp(\xi^0)u^0(0)\xi^k/\xi^0 .$$

Setting $u^\mu(0)$ to be the components of α, that is $u^0(0) = 1$ and $u^k(0), = 0$, and $u^\mu(1)$ to be the components a^μ of α', and assuming that $a^0 \neq 1$, we can solve Eq. (3.7) for ξ^μ,

$$\xi^0 = \ell n\, a^0, \quad \xi^k = a^k a^0 \ell n(a^0) . \tag{3.8}$$

By assumption, α and α' induce the same orientation of $\tau(M)$ in X, since on ∂M they both agree with the future pointing unit normal vector of $\tau(\partial M)$ in ∂X. Hence, $a^0 > 0$, and $\ell n(a^0)$ exists. The assumption that $a^0 \neq 1$ is equivalent to the assumption that $\alpha - \alpha'$ is transverse to $\tau(M)$ at the points of the interior of $\tau(M)$. Moreover, since $j^1\alpha$ and $j^1\alpha'$ agree on ∂M, Eq. (3.8) implies that $j^1\xi$ vanishes on $\partial X \cap U$. Covering $\tau(M)$ by coordinate neighbourhoods of this type and using a partition of unity argument we obtain a vector field ξ on X, vanishing to first order on ∂X and such that the one parameter group h_X^t generated by $f\xi$ acts by the identity transformation on $\tau(M)$, and $Th_X^1 \circ \alpha = \alpha'$. This implies that $h^1_{\mathscr{A}}(\alpha) = \alpha'$ and $h^1_{\mathscr{R}}(\rho) = \rho$ for each $\rho \in \mathscr{R}_\tau$. If $\alpha - \alpha'$ is not transverse to $\tau(M)$ in the interior of $\tau(M)$, we need to choose α'' such that $\alpha'' - \alpha$ and $\alpha'' - \alpha'$ are transverse and find two elements of $\mathscr{H}$, one mapping α to α'' and the other mapping α'' to α'. ■

The statement of Proposition (9.3.1) is weaker than Part (ii) of Atlas Hypothesis, because of the condition (3.5). This condition is required because the boundary conditions we have assumed specify not only the induced metric on ∂X but also the unit normal of ∂X in X, otherwise the pull–back of Ω to the boundary would fail to be exact[1]. However, we shall see that this weaker version of Atlas Hypothesis suffices to ensure the existence of a covariant Hamiltonian formulation of dynamics of general relativity. Actually, we shall use only the following corollary to the proof of Proposition 9.3.1.

9.3.2. Corollary. For every $\tau \in \mathcal{T}$, every $\alpha, \alpha' \in \mathcal{A}_\tau$ and every open set $U \subset M$ such that $\bar{U} \cap \partial M = \phi$, there exists a localizable symmetry g such that $g_{\mathcal{A}}(\alpha)|U = \alpha'|U$, and $g_{\mathcal{R}}$ induces the identity transformation in $\mathcal{R}_\tau$.

Proof. If $\alpha - \alpha'$ is transverse to $\tau(U)$, choose a function k on X such that k vanishes in a neighbourhood of ∂X, and k agrees with f in a neighbourhood of $\tau(\bar{U})$ in X. Then, the value at 1 of the one parameter group of diffeomorphisms of X generated by $kf\xi$ is the required localizable symmetry. If $\alpha - \alpha'$ is not transverse to $\tau(U)$ choose $\alpha'' \in \mathcal{A}_\tau$ such that $\alpha - \alpha''$ and $\alpha'' - \alpha'$ are transverse to $\tau(U)$ and repeat the argument, obtaining g' and g'' in $\mathcal{H}$ such that $g'_{\mathcal{A}}(\alpha)|U = \alpha''|U$, $g''_{\mathcal{A}}(\alpha'')|U = \alpha|U$, and $g'_{\mathcal{R}}$ and $g''_{\mathcal{R}}$ induce the identity transformation in $\mathcal{R}_\tau$. Then $g = g'' \circ g'$ is the required localizable symmetry. ∎

In order to verify that *hor* $T\mathcal{P}$, defined by Eq. (7.4.9), is transverse to the fibres of $\pi_{\mathcal{A}\mathcal{P}}$ it suffices to show that, for every $\pi \in \mathcal{P}$ and every $\zeta \in T_\alpha \mathcal{A}$, where $\alpha = \pi_{\mathcal{A}\mathcal{P}}(\pi)$, there exists a lift of ζ to *hor* $T_\pi \mathcal{P}$. If ζ is tangent to the fibres of $\pi_{\mathcal{T}\mathcal{A}}$ then its horizontal lift is obtained from the product structure given by Eq. (7.4.7). If $\zeta = \xi_{\mathcal{A}}(\alpha)$, where $\xi_{\mathcal{A}}$ is the vector field on $\mathcal{A}$ defined by the action of a one parameter group of symmetries generated by a vector field ξ_X on X, then Theorem 7.4.9

[1]Recall that in Yang–Mills theory the boundary conditions specifying the pull–back of the connection to the boundary suffice to ensure the vanishing of the pull–back of Ω to the boundary.

yields $hor\ \xi_{\mathscr{P}}(\pi) = \xi_{\mathscr{P}}(\pi) - ver\ \xi_{\mathscr{P}}(\pi)$, where $ver\ \xi_{\mathscr{P}}(\pi)$ is the value at π of the Hamiltonian vector field of $\Theta(\xi_{\mathscr{P}})$ restricted to $\mathscr{P}_\alpha$. It can be shown by direct computation that, for every $\alpha \in \mathscr{A}$, $\Theta(\xi_{\mathscr{P}})|\mathscr{P}_\alpha$ admits a unique Hamiltonian vector field.

9.3.3. Proposition. Statements (i) and (ii) of Theorem 7.4.9 hold in general relativity.

Proof. (i) is a straightforward consequence of parts (i) and (iii) of Atlas Hypothesis. In order to prove (ii) it suffices to modify the proof of Theorem 7.4.9 as follows.

We have a curve $c_{\mathscr{P}}$, which is the integral curve of hor $\xi_{\mathscr{P}}$ passing through $\pi \in C$, and a curve $c'_{\mathscr{P}}$ corresponding to a solution of the field equations with initial data π, both projecting to the same curve $c_{\mathscr{R}}$ in $\mathscr{R}$. Corollary 9.3.2 ensures that, for each $t \in domain\ c_\tau$ and each open set $U \subset M$ with closure disjoint from ∂M, $\overline{U} \cap \partial M = \phi$, there exists a localizable symmetry $g \in \mathscr{H}$ such that $g_{\mathscr{P}}(c'_{\mathscr{P}}(t))|U = c_{\mathscr{P}}(t)|U$. We can continue the argument of the proof of Theorem 7.4.9 with ζ such that $support\ \zeta \subset U$. This will imply that σ_{ZX} satisfies the field equations in every open set with closure disjoint form domain $\sigma_{ZX} \cap \partial X$. Hence, σ_{ZX} satisfies the field equations at every point of the interior of its domain, as required. ∎

This proposition ensures that general relativity admits a covariant Hamiltonian formulation of dynamics described in Corollary 7.4.10. This formulation of dynamics is essentially the A.D.M. formalism, written in a covariant form, in which the asymptotic conditions are replaced by boundary conditions.[2]

The Second Noether Theorem ensures that the constraint set $\mathscr{C}$ is contained in the zero level of the momentum map $J^{\mathscr{h}}_{\mathscr{P}}: \mathscr{P} \to \mathscr{h}^*$, where $\mathscr{h}$ is the Lie algebra of

[2]Hamiltonian formulation of general relativity for asymptotically flat space times was first studied in Dirac [1958], and developed into the ADM formalism in Arnowitt, Deser and Misner [1962]. The Hamiltonian formulation of general relativity in terms of the manifolds of maps was given in Fischer and Marsden [1972].

localizable infinitesimal symmetries consisting of vector fields ξ_X on X with compact supports and vanishing to the first order on ∂X, $j^1\xi_X | \partial X = 0$. We do not have existence theorems for Einstein equations on manifolds with boundary which would ensure that $\mathscr{C} = (J^{\mathscr{h}}_{\mathscr{G}})^{-1}(0)$. For asymptotically that space times, which will be discussed in the following sections, it can be shown in the category of Hilbert manifolds that the constraint set is given by the vanishing of the superhamiltonian and the supermomentum[3], Eqs. (2.26) and (2.27), so that it coincides with the zero level of $J^{\mathscr{h}}_{\mathscr{G}}$.

---

[3]c.f. Fischer and Marsden [1979] and the references quoted there.

9.4 Asymptotically flat space times.[1]

In previous sections we studied space times with boundary. However, applications of general relativity to isolated gravitating systems require asymptotically flat space times. Roughly speaking, a space time is asymptotically flat if at infinity it looks like a Minkowski space.

Recall that a conformal compactification of a Minkowski space $(\overset{\circ}{X},\overset{\circ}{g})$, given by an embedding f of $\overset{\circ}{X}$ into a Lorentzian manifold (X',g') such that

$$f^{*}g' = K^{2}\overset{\circ}{g} \tag{4.1}$$

for some positive function K on $\overset{\circ}{X}$ vanishing at infinity, leads to a representation of the spatial infinity of $\overset{\circ}{X}$ by a single point $i \in X'$. By attaching to $\overset{\circ}{X}$ the hyperboloid H of unit space like vectors in T_iX',

$$H = \{u \in T_iX' \mid g'(u,u) = 1\}, \tag{4.2}$$

one obtains a manifold X with interior isomorphic to $\overset{\circ}{X}$ and boundary ∂X isomorhpic to H. X is the quotient of $\overset{\circ}{X} \cup [0,a) \times H$ by the equivalence relation identifying (t,u) with $f(exp(tu))$, where $exp : T_iX' \longrightarrow X'$ is the exponential map defined by the metric g', and a is a positive number such that the *exp* induces a diffeomorphism of the neighbourhood of $0 \in T_iX'$, consisting of vectors with length less than a, onto a neighbourhood i in X'.

For asymptotically flat space times we can restrict the above considerations to a neighbourhood of infinity. Thus, we consider a manifold X with boundary ∂X, and a flat Lorentzian metric $\overset{\circ}{g}$ in the interior of a neighbourhood U of ∂X, such that the geodesics of $\overset{\circ}{g}$ which extend to ∂X have infinite length. If g is an asymptotically flat Lorentzian metric on X then, for each geodesics x(r) in U which tends to a limit in ∂X as the distance parameter r tends to infinity,

$$g_{\mu\nu}(x(r)) - \overset{\circ}{g}_{\mu\nu}(x(r)) = O(\tfrac{1}{r}) \ , \tag{4.3}$$

[1]The presentation adopted in this section is based on Śniatycki [1988].

and

$$g_{\mu\nu,\lambda}(x(r)) = O\left[\frac{1}{r^2}\right] , \tag{4.4}$$

where $g_{\mu\nu}$ are the components of g in a coordinate system in which $\overset{o}{g}$ has components $\overset{o}{g}_{\mu\nu} = diag(-1,1,1,1)$, and $O(r^{-k})$ denotes terms whose supremum norms are bounded by r^{-k}, and whose first three derivatives decay accordingly. One can rewrite Eqs. (4.3) and (4.4) in a coordinate independent form as follows. For every triplet (ξ,η,ζ) of vector fields on U such that $\overset{o}{g}(\xi,\xi) = \overset{o}{g}(\eta,\eta) = \overset{o}{g}(\zeta,\zeta) = 1$,

$$[g(\xi,\eta)](x(r)) - [\overset{o}{g}(\xi,\eta)](x(r)) = O(\tfrac{1}{r}) , \tag{4.3a}$$

and

$$[(\overset{o}{\nabla}_\zeta g)(\xi,\eta)](x(r)) = O\left[\frac{1}{r^2}\right] , \tag{4.4a}$$

where $\overset{o}{\nabla}$ denotes the covariant derivative with respect to the Levi–Civita connection of $\overset{o}{g}$.

The metric $\overset{o}{g}$ on U gives rise to a connection on the restriction Z_U of Z, defined as follows. Every section (g_U,Γ_U) of Z_U can be uniquely expressed in the form $g_U = \overset{o}{g} + k_U$ and $\Gamma_U = \overset{o}{\Gamma} + S_U$, where $\overset{o}{\Gamma}$ is the Levi Civita connection of $\overset{o}{g}$, and k_U and S_U are tensor fields on U. A section (g_U,Γ_U) is horizontal if and only if k_U and S_U are covariantly constant with respect to $\overset{o}{\Gamma}$. Since $\overset{o}{\Gamma}$ is flat the connection on Z_U defined ;in this way is also flat. We denote by *hor* TZ_U the corresponding horizontal distribution on Z_U.

The pull back Ω_U of Ω to Z_U can be decomposed ;into its horizontal and vertical parts,

$$\Omega_U = hor\,\Omega_U + ver\,\Omega_U . \tag{4.4}$$

If (x^μ) are local coordinates in U such that the components of $\overset{o}{g}_N$ form a diagonal matrix with the entries (–1,1,1,1) on the diagonal, then the corresponding local representation of *hor* Ω_U and *ver* Ω_U is given by

$$hor\, \Omega_U = u^{\mu\nu}(\nu^\lambda_{\mu\nu}\, \nu^\sigma_{\lambda\sigma} - \nu^\lambda_{\mu\sigma}\, \nu^\sigma_{\lambda\nu}) d_4x\,, \tag{4.5}$$

$$ver\, \Omega_U = u^{\mu\nu}(d\nu^\lambda_{\mu\nu} - \delta^\lambda_\nu d\nu^\sigma_{\mu\sigma}) \wedge d_3x_\lambda\,, \tag{4.6}$$

c.f. Eq. (1.21). We can rewrite the expression for $ver\, \Omega_U$ in the form

$$ver\, \Omega_U = d\Pi_U - \Sigma \tag{4.7}$$

where

$$\Pi_U = u^{\mu\nu}(\nu^\lambda_{\mu\nu} - \delta^\lambda_\nu \nu^\sigma_{\mu\sigma}) d_3x_\lambda \tag{4.8}$$

and

$$\Sigma = (\nu^\lambda_{\mu\nu} - \delta^\lambda_\nu\, \nu^\sigma_{\mu\sigma}) du^{\mu\nu} \wedge d_3x_\lambda\,. \tag{4.9}$$

In order to elucidate a geometric interpretation of the forms Π_U and Σ consider a section $\sigma = (g,\Gamma)$ of π_{XZ} and a 3–dimensional submanifold M of U such that the metric g on X pulls back to a non–singular metric g_M on M. We denote by $W_M[g,\Gamma]$ and $h_M[g,\Gamma]$ the Weingarten map and the second fundamental form of M, respectively, defined in terms of the metric g and the connection Γ. Since a neighbourhood U of M is endowed with a flat metric $\overset{\circ}{g}$ and its Levi–Civita connection $\overset{\circ}{\Gamma}$, we can form three more Weingarten maps and second fundamental forms corresponding to the combinations $(g,\overset{\circ}{\Gamma})$, $(\overset{\circ}{g},\Gamma)$ and $(\overset{\circ}{g},\overset{\circ}{\Gamma})$.

9.4.1. Proposition. Let $i_{XM} : M \longrightarrow X$ denote the inclusion map. The pull back of Π_U by $\sigma_{ZX} \circ i_{XM}$ is given by

$$\begin{aligned} \overset{*}{i}_{XM}\, \overset{*}{\sigma}_{ZX}\, \Pi_U = \{&(tr_{g_M} h_M[g,\Gamma] - trW_M[g,\Gamma]) + \\ &- (tr_{g_M} h_M[g,\overset{\circ}{\Gamma}] - trW_M[g,\overset{\circ}{\Gamma}])\}\, vol\, [g_M]\,. \end{aligned} \tag{4.10}$$

Proof Eq. (4.8) holds in a coordinate system in which the components of $\overset{\circ}{\Gamma}$ vanish. Denoting by $g_{\mu\nu}$ and $\Gamma^\lambda_{\mu\nu}$ the components of g and Γ with respect to this coordinate system and by η^λ the components of the unit normal vector η_M to M defined in terms of the metric g we obtain

$$\overset{*}{i}_{XM}\ \overset{*}{\sigma}_{ZX}\ \Pi_U = g^{\mu\nu}(\Gamma^\lambda_{\mu\nu} - \delta^\lambda_\nu \Gamma^\sigma_{\mu\sigma}) g_{\lambda\rho}\ \eta^\rho\ vol[g_M]\ . \tag{4.11}$$

A covariant expression for $\overset{*}{i}_{XM}\ \overset{*}{\sigma}_{ZX}\ \Pi_U$, valid in any coordinate system is obtained from Eq. (4.11) by replacing $\Gamma^\lambda_{\mu\nu}$ by the difference $\Gamma^\lambda_{\mu\nu} - \hat{\Gamma}^\lambda_{\mu\nu}$, where $\Gamma^\lambda_{\mu\nu}$ now denote the components of Γ with respect to any arbitrary coordinate system and $\hat{\Gamma}^\lambda_{\mu\nu}$ are the components of $\hat{\Gamma}$ with respect to the same coordinates. We obtain a covariant equation

$$\begin{aligned}\overset{*}{i}_{XM}\ \overset{*}{\sigma}_{ZX}\ \Pi_U = \{&g^{\mu\nu}(\Gamma^\lambda_{\mu\nu} - \delta^\lambda_\nu \Gamma^\sigma_{\mu\sigma}) g_{\lambda\rho}\eta^\rho + \\ &- g^{\mu\nu}(\hat{\Gamma}^\lambda_{\mu\nu} - \delta^\lambda_\nu\ \hat{\Gamma}^\sigma_{\mu\sigma}) g_{\lambda\rho}\eta^\rho\} vol\,[g_M]\end{aligned} \tag{4.12}$$

Using a coordinate system adapted to M, and following the argument used in the proof of Prop. 8.2.1 we obtain Eq. (4.10). ■

If Γ is the Levi–Civita connection of g then $h_M[g,\Gamma]$ and $W_M[g,\Gamma]$ are not independent, c.f. Eq. (8.2.6). In particular $tr_{g_M}\ h_M[g,\Gamma] = tr\ W_M[g,\Gamma]$ is the mean curvature $H_M[g]$ of M embedded in (X,g). We can extend the notion of mean curvature to the case when Γ is not the Levi Civita connection of g, and define the mean curvature $H_M[g,\Gamma]$ of M embedded in (X,g,Γ) by

$$H_M[g,\Gamma] = \tfrac{1}{2}(tr_{g_M} h_M[g,\Gamma] - tr W_M[g,\Gamma])\ . \tag{4.13}$$

With this definition we obtain

$$\overset{*}{i}_{XM}\ \overset{*}{\sigma}_{ZX}\ \Pi_U = 2\{H_M[g,\Gamma] - H_M[g,\hat{\Gamma}]\}\ vol\,[g_M]\ . \tag{4.14}$$

Thus, the pull back of Π_U by $\sigma_{ZX} \circ i_{XM}$ is proportional to the difference between the mean curvatures of M defined by (g,Γ) and $(g,\hat{\Gamma})$. Comparing Eqs. (4.8) and (4.9) we see that the form Σ gives the partial differential of Π_U with respect to the variations of the metric.

Symmetries of the theory under considerations are given by diffeomorphisms of X which preserve the metric $\mathring{g}$ in U. Infinitesimal symmetries are given by vector fields ξ_X on X such that their restrictions ξ_U to U are Killing vector fields of $\mathring{g}$.

9.4.2. Proposition. Π_N is invariant under the action of the isometries of $\mathring{g}$. In particular, for every infinitesimal isometry ξ_U of $\mathring{g}$,

$$\pounds_{\xi_Z} \Pi_U = 0 \tag{4.15}$$

where ξ_Z is the natural extension of ξ_U to Z_U.

Proof. The form Π_U is defined intrinsically in terms of Ω_U and $\mathring{g}$. Since Ω_N is preserved by the diffeomorphisms of U, it follows that Π_U is preserved by the isometries of $\mathring{g}$. ■

Let M be a 3–dimensional submanifold of X with boundary $\partial M \subset \partial X$. We assume that $M \cap U$ is a flat space–like submanifold of $(U, \mathring{g})$, and that there exists a family $\{M_r\}$ of open submanifolds of M such that, for sufficiently large r, the boundary ∂M_r of $\overline{M}_r$ is a sphere in $(U \cap M\ ,\ \mathring{g}_{U \cap M})$ of radius r, $M_r \subset M_s$ for $r < s$, and $M = \cup M_r$. In analogy with Eq. (2.16), for each infinitesimal symmetry ξ, each Cauchy data $\sigma_{ZX}|M$, and each r, we set

$$J^{\xi}(\sigma_{ZX}|M_r) = \int_{M_r} \overset{*}{\sigma}_{ZX}(\xi_Z \lrcorner\, |\Omega) + \int_{\partial M_r} \overset{*}{\sigma}_{ZX}(\xi_Z \lrcorner\, |\Pi_U) \tag{4.16}$$

If the Cauchy data $\sigma_{ZX}|M = (g|M,\ \Gamma|M)$ satisfy the constraint equations (2.26) – (2.27), then the integral over M_r can be expressed as a surface integral over ∂M_r.

Taking into account Eq. (2.25) we obtain

$$J^{\xi}(\sigma_{ZX}|M) = \int_{\partial M_r} \sqrt{-\det(g_{\alpha\beta})}\{g^{\mu\lambda}\nabla_{\lambda}\xi^{\nu} + g^{\lambda\rho}\Gamma^{\mu}_{\lambda\rho}\xi^{\nu} - g^{\mu\lambda}\Gamma^{\rho}_{\lambda\rho}\xi^{\nu}\}d_2x_{\mu\nu}. \quad (4.17)$$

If the limit

$$J^{\xi}(\sigma_{ZX}|M) = \lim_{r\to\infty} J^{\xi}(\sigma_{ZX}|M_r) \quad (4.18)$$

exists, then it can be interpreted as the momentum associated to the infinitesimal symmetry ξ. If ξ_U coincides along $U \cap M$ with the future pointing unit normal vector of $U \cap M$ defined in terms of the flat metric $\overset{o}{g}$, then $J^{\xi}(\sigma_{ZX}|M)$ is the energy of the state $\sigma_{ZX}|M$. If ξ_U is an infinitesimal translation tangential to $M \cap U$, then J^{ξ} has the interpretation of a linear momentum. Similarly, if ξ_U is an infinitesimal rotation, then J^{ξ} is the corresponding angular momentum.

In order to relate the momenta introduced here to the ADM momenta which can be found in the literature we choose coordinates (x^{μ}) to the $U \cap M$ is given by $x^o = 0$, and introduce the ADM variables: the lapse function

$$N = (-g^{00})^{-1/2}, \quad (4.19)$$

the shift vector

$$N^i = g^{0i}N^2, \quad (4.20)$$

and the canonical momentum

$$\pi^i_j = \sqrt{-det(g_{\alpha\beta})}(g_{\mu}^{ik}\Gamma^0_{kj} - \delta^i_j g_M^{kn}\Gamma^0_{kn}), \quad (4.21)$$

where g_M^{ij} is the contravariant metric tensor of the metric $g_M = g_{ik}dx^i dx^k$ on M.

The energy of the state $\sigma_{ZX}|M$ is given by the limit as $r \longrightarrow \infty$ of

$$E(\sigma_{ZX}|M_r) = J^{\xi}(\sigma_{ZX}|M_r), \quad (4.22)$$

where ξ_U coincides along $U \cap M$ with the future pointing unit normal vector of $U \cap M$.

By direct computations we obtain

$$E(\sigma_{ZX}|M) = \int_{\partial M} N\sqrt{det(g_{ij})}(g^{km}\Gamma^n_{km} - g^{nm}\Gamma^k_{mk})d_2x_n +$$

$$+ \int_{\partial M_r} N^{-1}\sqrt{det(g_{ij})}(N^k\Gamma^n_{0k} - N^n\Gamma^k_{0k})d_2x_n. \tag{4.23}$$

If the normal directions to $U \cap M$ in U defined with respect to the metrics g and $\overset{o}{g}$ coincide, then the second integral on the right hand side vanishes, and the first integral gives the ADM expression for energy. The asymptotic conditions (4.3), (4.4) ensure that, in the limit as $r \to \infty$, the first integral converges and the second integral vanishes. Thus,

$$E(\sigma_{ZX}|M) = \lim_{r\to\infty} E(\sigma_{ZX}|M_r) \tag{4.24}$$

exists and it coincides with the ADM energy in the state $\sigma_{ZX}|M$.

Linear and angular momenta correspond to ξ_U such that, along $U \cap M$, $\xi^0 = 0$ and $\xi^k{}_{,0} = 0$. After rather lengthy computations involving integration by parts we obtain

$$J^\xi(\sigma_{ZX}|M_r) = -2\int_{\partial M_r} \pi^n_k\xi^k d_2x_n +$$

$$\int_{\partial M}(\pounds_{\xi_X}\sqrt{-det(g_{\alpha\beta})})N^{-2}N^n d_2x_n +$$

$$+ \int_{\partial M_r}(\sqrt{-det(g_{\alpha\beta})})_{,m}N^m N^{-2}\xi^n d_2x_n + 0(\tfrac{1}{r}).$$

The first integral corresponds to the ADM expression, and the remaining two integrals vanish if $N^m = 0$. If ξ_U corresponds to an infinitesimal translation, then its components

ξ^k are constant. In this case the asymptotic conditions ensure that, in the limit as $r \longrightarrow \infty$, the first integral converges and the remaining two integrals vanish. Hence $J^{\xi}(\sigma_{ZX}|M)$ exists and it coincides with the ADM linear momentum corresponding to ξ. In the case of angular momenta the situation is more complicated. If ξ_U corresponds to an infinitesimal rotation, then its components are of order $O(r)$, and the asymptotic conditions (4.3) – (4.4) fail to ensure the convergence of $J^{\xi}(\sigma_{ZX}|M_r)$ is the limit as $r \longrightarrow \infty$. We shall discuss this problem later.

If we restrict our considerations to the energy and the linear momenta of the gravitational field we obtain the following conservation law.

9.4.3. Proposition. If σ_{ZX} satisfies the field equations then the energy and the linear momenta are independent of the Cauchy surface M.

Proof. Let M and M′ be two Cauchy surfaces in X. We consider $(U,\overset{\circ}{g})$ as a neighbourhood of spatial infinity in a Minkowski space $(\overset{\circ}{X},\overset{\circ}{g})$. For sufficiently large r, the hyperboloid H_r of space–like vectors in $\overset{\circ}{X}$ which have length r, is contained in U and it intersects the Cauchy surfaces M and M′ along spheres $\partial M_r = M \cap H_r$ and $\partial M' = M' \cap H_r$. We denote by K_r the domain in H_r bound by ∂M_r and $\partial M'_r$ by M_r the domain in M bound by ∂M_r, and by M'_r the domain in M′ bound by $\partial M'_r$. Let D_r be the domain in X bound by M_r, M'_r and K,

$$\partial D_r = K_r + M'_r - M_r.$$

If σ_{ZX} satisfies the field equations then, for every vector field ξ_X on X which induces in U a Killing vector field ξ_U of $\overset{\circ}{g}$, we have

$$0 = \int_{D_r} \overset{*}{\sigma}_{ZX}(\xi_Z \lrcorner\, d\Omega) = -\int_{\partial D_r} \overset{*}{\sigma}_{ZX}(\xi_Z \lrcorner\, \Omega) =$$

$$-\int_{K_r} \overset{*}{\sigma}_{ZX}[\xi_Z \rfloor (hor\,\Omega_U + d\Pi_U - \Sigma)] - \int_{M'_r} \overset{*}{\sigma}_{ZX}(\xi_Z \rfloor \Omega)$$

$$+\int_{M_r} \overset{*}{\sigma}_{ZX}(\xi_Z \rfloor \Omega) =$$

$$= -\int_{K_r} \overset{*}{\sigma}_{ZX}[\xi_Z \rfloor (hor\,\Omega_U - \Sigma_U)] - J_\xi(\sigma_{ZX}|M'_r) + J_\xi(\sigma_{ZX}|M_r).$$

Taking into account the coordinate expressions for $hor\ \Omega_U$ and Σ we obtain after a simple calculation

$$J^\xi(\sigma_{ZX}|M'_r) - J^\xi(\sigma_{ZX}|M_r) =$$

$$-3\int_{K_r} \sqrt{-det(g_{\alpha\beta})}\, g^{\mu\nu}(\Gamma^\lambda_{\mu\nu}\Gamma^\sigma_{\lambda\sigma} - \Gamma^\lambda_{\mu\sigma}\Gamma^\sigma_{\lambda\nu})\xi^\alpha d_3x_\alpha$$

$$-\int_{K_r} (\Gamma^\lambda_{\mu\nu} - \delta^\lambda_\nu \Gamma^\sigma_{\mu\sigma})\ \xi_X\{\sqrt{-det(g_{\alpha\beta})}\, g^{\mu\nu}\}d_3x_\lambda$$

If ξ_U corresponds to an infinitesimal translation in $(U,\overset{\circ}{g})$, then $\xi^\alpha = O(1)$, and the integrands on the right hand side of Eq. (4.15) are of order $O(r^{-4})$. Since the volume of K_r is at most of order $O(r^3)$, the right hand side of Eq. (4.15) is of order $O(r^{-1})$, and it vanishes in the limit as $r \longrightarrow \infty$. This completes the proof. ∎

Let us return to the problem of angular momentum. We have seen that, for infinitesimal rotations ξ_U, the asymptotic conditions (4.3) and (4.4) fail to ensure the convergence of the right hand side of Eq. (4.25) in the limit as $r \longrightarrow \infty$. Moreover, for an infinitesimal rotation ξ_U, the right hand side of Eq. (4.26) need not tend to zero as $r \longrightarrow \infty$. Hence, the asymptotic conditions (4.3) and (4.4) also fail to ensure the conservation of angular momenta. This implies that, in order to obtain conserved angular momenta, one has to strengthen the asymptotic conditions. Observe that Eq.

(4.4) is equivalent to

$$\pounds_{\xi_U} g_{\mu\nu} = O(r^{-2}) \tag{4.27}$$

for every infinitesimal translation ξ_U in $(U,\overset{\circ}{g})$. Demanding that Eq. (4.27) should also hold for every infinitesimal rotation ξ_U in $(U,\overset{\circ}{g})$ one obtains conserved angular momenta.

9.4.4. Proposition. In addition to Eq. (4.3) assume that Eq. (4.27) holds for every Killing vector ξ_U of $\overset{\circ}{g}$. Then, for every infinitesimal rotation ξ_U, the right hand side of Eq. (4.25) converges in the limit as $r \longrightarrow \infty$ defining the angular momentum $J^{\xi}(\sigma_{ZX}|M)$ which coincides with the ADM angular momentum corresponding to ξ. If σ_{ZX} satisfies the field equations then $J^{\xi}(\sigma_{ZX}|M)$ is independent of the Cauchy surface M.

Proof. The asymptotic conditions (4.3) and (4.4) ensure the convergence of the second and the third integral on the right hand side of Eq. (4.25). The assumption that ξ_U satisfies Eq. (4.27) implies that the second integral tends to zero as $r \longrightarrow \infty$. Moreover, if ξ_U is tangent to ∂M_r for sufficiently large r, then the third integral also vanishes. Thus, the angular momentum $J_{\xi}(\sigma_{ZX}|M)$ corresponding to ξ_U is equal to the limit of the ADM expression given by the first integral on the right hand side of Eq. (4.25).

Using Stokes Theorem we obtain

$$\int_{\partial M_r} \pi^n_k \xi^k d_2x_n = \int_{M_r} \pi^n_{k;n}\xi^k d_3x + \int_{M_r} \pi^{nk}\xi_{k;n} d_3x \ , \tag{4.28}$$

where the semicolon denotes the covariant differentiation with respect to the physical metric $g_M = g_{ij}dx^i dx^j$ and the indices are raised in terms of its inverse g_M^{ij}. By direct computation one can verify that the constraint equation $J = 0$, where the supermomentum J is given by Eq. (2.19), is equivalent to $\pi^n_{k;n} = 0$, so that the first integral on the right hand side of Eq. (4.28) vanishes. The second integral is convergent because the asymptotic assumptions we have adopted imply that the integrand is of order $O(r^{-4})$ as $r \longrightarrow \infty$ while the volume of M_r is of order $O(r^3)$. Thus, the angular momentum $J^{\xi}(\sigma_{ZX}|M)$ corresponding to ξ is convergent and it coincides with the ADM angular momentum.

In order to verify the conservation of the angular momentum $J_{\xi}(\sigma_{ZX}|M)$ observe that if ξ_U is tangent to K_r, then the first integral on the right hand side of Eq. (4.26) vanishes. Eq. (4.27) ensures that the second integral vanishes in the limit as $r \longrightarrow \infty$. Thus, if σ_{ZX} satisfies the field equations, $J^{\xi}(\sigma_{ZX}|M)$ is independent of the Cauchy surface M, which completes the proof. ■

Though the assumptions made in this proposition are very natural, they are not satisfied by many interesting solutions of Einstein field equations. There are several approaches to the problem of angular momenta in general relativity, which are based on weaker assumptions[2].

From physical considerations one expects the energy of a relativistic system to be positive. The positivity of energy is quite apparent in many classical field theories in a given space–time manifold, for example see Eq. (7.2.28) for the energy in Yang–Mills theory. The energy of a gravitational field in asymptotically flat space–times is given by the limit, as $r \longrightarrow \infty$, of the right hand side of Eq. (4.23), and its positivity was a long standing conjecture in general relativity. At present there are several proofs of positivity of gravitational energy.[3]

---

[2]See, for example, Ashtekar and Hansen [1973], Regge and Teitelboim [1974], Geroch [1977], Ashtekar and Magnon [1984], and the references quoted there.

[3]Deser and Teitelboim [1977], Grisaru [1977], Schoen and Yau, [1981], Witten [1981].

As in the case of general relativity on manifolds with boundary, the field equations can be put in the Hamiltonian form with the Hamiltonian given by the momentum J^{ξ} corresponding to the direction of evolution described by ξ_X. In the case of asymptotically flat space–times the dynamical variables corresponding to a Cauchy surface M are defined on the interior of M only. Since the topology of smooth maps with open domains is rather complicated, one usually works with appropriate Sobolev space of dynamical variables.[4] However, the time propagation of Cauchy data on a Cauchy surface M may lead to singularities. In fact, under physically reasonable assumptions about the distribution of matter singularities are bound to develop in finite time.[5]

[4]See Fischer and Marsden [1972] and [1978].

[5]See Hawking and Ellis [1973], and the references quoted there.

References

ADM [1962], see Arnowitt, Deser and Misner [1962]

ARNOWITT, R., DESER, S. AND MISNER, C. [1962], "The Dynamics of General Relativity" in *Gravitation, an Introduction to Current Research*, pp. 227–265, L. Witten (ed.), John Wiley, New York.

ASHTEKAR, A. and HANSEN, R.O. [1973], "A Unified Treatment of Null and Spatial Infinity in General Relativity. I. Universal Structure, Asymptotic Symmetries and Conserved Quantities at Spatial Infinity", *J. Math. Phys.*, 19, 1542–1556.

ASHTEKAR, A. and MAGNON, A. [1984], "From j^0 to the 3+1 Description of Spatial Infinity", *J. Math. Phys.*, 25, 2682–2690.

BINZ, E. and ŚNIATYCKI, J. [1986], "Conservation Laws in Spacetimes with Boundary", *Class. Quantum Grav.*, 3, 1191–1197.

DESER, S. and TEITELBOIM, C. [1977], "Supergravity has Positive Energy", *Phys. Rev. Lett.*, 39, 249–252.

DIRAC, P.A.M. [1958], "The Theory of Gravitation in Hamiltonian Form", *Proc. Royal Soc.* (*London*), A246, 333–346.

EINSTEIN, A. [1915], "Die Feldgleichungen der Gravitation", *Sitzunsber. preuss. Akad. Wiss.*, 48, 844–847.

FISCHER, A.E. and MARSDEN, J.E. [1972], "The Einstein's Equations of Evolution. A Geometric Approach", *J. Math. Phys.*, 13, 546–568.

FISCHER, A.E. and MARSDEN, J.E. [1978], "Topics in Dynamics of General Relativity" in *Proceedings of Enrico Fermi Summer School of the Italian Physical Society*, Varenna, J. Ehlers (ed.)

FISCHER, A.E. and MARSDEN, J.E. [1979], "The Initial Value Problem and the Dynamical Formulation of General Relativity" in *General Relativity, an Einstein Centenary Survey*, pp. 138–211, S. W. Hawking and W. Israel (eds.), Cambridge University Press.

GEROCH, R. [1977], "Asymptotic Structure of Space–Time" in *Asymptotic Structure of Space–Time*, pp. 1–106, F. Esposito and L. Witten (eds.), Plenum Press, New York.

GRISARU, M.T. [1977], "Positivity of the Energy in the Einstein Theory", *Phys. Lett.*, 73B, 207–208.

HAWKING, S.W. and ELLIS, G.F.R. [1973], *The Large Scale Structure of Space–Time*, Cambridge University Press.

HILBERT, D. [1915], "Die Grundlagen der Physik", *Göttingen Nachschr.*, Mat. Phys. Kl., 395–407.

PALATINI, A. [1919], "Deduzione invariantiva delle equazioni gravitazionali dal prinzipio di Hamilton", *Rend. Circ. Mat. Palermo*, 43, 203–212.

REGGE, T. and TEITELBOIM, C. [1974], "Role of the Surface Integrals in the Hamiltonian Formulation of General Relativity, *Ann. Phys.*, 88, 286–318.

SCHOEN, R. and YAU, S.T. [1981], "Proof of the Positive Mass Theorem", *Comm. Math. Phys.*, 79, 231–360.

ŚNIATYCKI, J. [1970], "On the Canonical Formulation of General Relativity", *Proceedings of "Journees Relativistes 1970"*, pp. 127–135, Faculté des Sciences, Caen.

ŚNIATYCKI, J. [1988], "Conservation Laws in Asymptotically Flat Space–Times Revisited", *Rep. Math. Phys.*, to appear.

SZCZYRBA, W. [1976], "A Symplectic Structure in the Set of Einstein Metrics", *Comm. Math. Phys.*, 51, 163–182.

WITTEN, E. [1981], "A New Proof of the Positive Energy Theorem", *Comm. Math. Phys.*, 80, 381–402.

INDEX

GLOSSARY OF FREQUENTLY USED SYMBOLS

A	Yang–Mills potential
$\mathbb{A}$	atlas bundle (dynamics)
$\mathscr{A}$	atlas bundle (field theory)
Aut(P)	group of automorphisms of a principal bundle P
aut(P)	Lie algebra of infinitesimal automorphisms of P
B	boundary conditions
$\mathbb{C}$	constraint set (dynamics)
$\underline{\mathbb{C}}$	complex numbers
$\mathscr{C}$	constraint set (field theory)
Γ	torsion free connection
γ	Cauchy data
D	covariant differential
d	exterior differential
Δ	Laplace–Beltrami operator
δ	Dirichlet data
∂	boundary operator
$\nabla[g]$	the Levi–Civita connection of a metric g
exp	exponential map
F	field strength (Yang–Mills theory)
$\mathbb{F}$	fibre derivative
G	Lie group
$\mathfrak{g}$	Lie algebra of G
$\mathscr{G}$	group of symmetries
$\mathscr{g}$	Lie algebra of infinitesimal symmetries

g	metric
g_X	metric on a manifold X
g_X	action in X of an element g of a Lie group
$\mathcal{H}$	group of localizable symmetries
h	Lie algebra of localizable symmetries
$hor\, TP$	horizontal distribution on P
$hor\, \xi$	horizontal component of ξ
i_B	inclusion map $B \longrightarrow P$
J_P^g	momentum map corresponding to a Hamitlonian action of G in P
J^1Y	the first jet bundle of Y
$j^1\sigma_{YX}$	the first jet extension of σ_{YX}
L	Lagrangian
$\pounds$	Lie derivative
Λ	Lagrangian form
M	typical Cauchy surface
$\mathcal{O}$	reduced Poisson algebra
P	primary constraint manifold
$\mathbb{P}$	primary constraint bundle (dynamics)
$\mathcal{P}$	primary constraint bundle (field theory)
Π	form characterizing the boundary conditions
$Poiss(\mathcal{P})$	Poisson algebra of $\mathcal{P}$
π_{XY}	projection map from Y to X
Q	configuration space
$\mathbb{Q}$	configuration bundle (dynamics)
$\mathcal{Q}$	bundle of Dirichlet data

R	reduced phase space
$\mathbb{R}$	primary reduced phase bundle (dynamics)
$\underline{\mathbb{R}}$	real numbers
$\mathscr{R}$	primary reduced phase bundle (field theory)
R[g]	scalar curvature of the Levi–Civita connection of a metric g
$Ric[\Gamma]$	Ricci tensor of a connection Γ
$\mathscr{S}$	space of physical states
σ_{YX}	section of a bundle Y over X
T	absolute time of Newtonian dynamics
$\mathscr{T}$	space of Cauchy surfaces
tr	trace of linear maps
tr_g	trace of symmetric forms defined in terms of a metric g
τ	Cauchy surface
TQ	tangent bundle space of Q
T^*Q	cotangent bundle space of Q
$ver\ \xi$	vertical component of a vector ξ
$vol[g]$	volume form defined in terms of a metric g
X	base manifold / space–time
Ξ	Cartan form
ξ	element of a Lie algebra
ξ_X	vector field on X corresponding to the action of ξ
Y	fibre bundle over X
η_M	unit normal vector field of a submanifold M
Z	first jet bundle of Y
$\mathbb{Z}$	velocity bundle

$\mathscr{Z}$	space of Cauchy data
Θ	Hamilton form
θ_{T^*Q}	the canonical 1–form of T^*Q
Ω	DeDonder form
ω	symplectic form
ω_1	canonical form of the first jet bundle
$\wedge$	exterior product
$\lrcorner$	left interior product

A CATALOG OF SELECTED

DOVER BOOKS

IN SCIENCE AND MATHEMATICS

Mathematics

FUNCTIONAL ANALYSIS (Second Corrected Edition), George Bachman and Lawrence Narici. Excellent treatment of subject geared toward students with background in linear algebra, advanced calculus, physics and engineering. Text covers introduction to inner-product spaces, normed, metric spaces, and topological spaces; complete orthonormal sets, the Hahn-Banach Theorem and its consequences, and many other related subjects. 1966 ed. 544pp. 6⅛ x 9¼. 0-486-40251-7

ASYMPTOTIC EXPANSIONS OF INTEGRALS, Norman Bleistein & Richard A. Handelsman. Best introduction to important field with applications in a variety of scientific disciplines. New preface. Problems. Diagrams. Tables. Bibliography. Index. 448pp. 5⅜ x 8½. 0-486-65082-0

VECTOR AND TENSOR ANALYSIS WITH APPLICATIONS, A. I. Borisenko and I. E. Tarapov. Concise introduction. Worked-out problems, solutions, exercises. 257pp. 5⅝ x 8¼. 0-486-63833-2

AN INTRODUCTION TO ORDINARY DIFFERENTIAL EQUATIONS, Earl A. Coddington. A thorough and systematic first course in elementary differential equations for undergraduates in mathematics and science, with many exercises and problems (with answers). Index. 304pp. 5⅜ x 8½. 0-486-65942-9

FOURIER SERIES AND ORTHOGONAL FUNCTIONS, Harry F. Davis. An incisive text combining theory and practical example to introduce Fourier series, orthogonal functions and applications of the Fourier method to boundary-value problems. 570 exercises. Answers and notes. 416pp. 5⅜ x 8½. 0-486-65973-9

COMPUTABILITY AND UNSOLVABILITY, Martin Davis. Classic graduate-level introduction to theory of computability, usually referred to as theory of recurrent functions. New preface and appendix. 288pp. 5⅜ x 8½. 0-486-61471-9

ASYMPTOTIC METHODS IN ANALYSIS, N. G. de Bruijn. An inexpensive, comprehensive guide to asymptotic methods–the pioneering work that teaches by explaining worked examples in detail. Index. 224pp. 5⅜ x 8½ 0-486-64221-6

APPLIED COMPLEX VARIABLES, John W. Dettman. Step-by-step coverage of fundamentals of analytic function theory–plus lucid exposition of five important applications: Potential Theory; Ordinary Differential Equations; Fourier Transforms; Laplace Transforms; Asymptotic Expansions. 66 figures. Exercises at chapter ends. 512pp. 5⅜ x 8½. 0-486-64670-X

INTRODUCTION TO LINEAR ALGEBRA AND DIFFERENTIAL EQUATIONS, John W. Dettman. Excellent text covers complex numbers, determinants, orthonormal bases, Laplace transforms, much more. Exercises with solutions. Undergraduate level. 416pp. 5⅜ x 8½. 0-486-65191-6

RIEMANN'S ZETA FUNCTION, H. M. Edwards. Superb, high-level study of landmark 1859 publication entitled "On the Number of Primes Less Than a Given Magnitude" traces developments in mathematical theory that it inspired. xiv+315pp. 5⅜ x 8½. 0-486-41740-9

Math–Decision Theory, Statistics, Probability

ELEMENTARY DECISION THEORY, Herman Chernoff and Lincoln E. Moses. Clear introduction to statistics and statistical theory covers data processing, probability and random variables, testing hypotheses, much more. Exercises. 364pp. 5⅜ x 8½. 0-486-65218-1

STATISTICS MANUAL, Edwin L. Crow et al. Comprehensive, practical collection of classical and modern methods prepared by U.S. Naval Ordnance Test Station. Stress on use. Basics of statistics assumed. 288pp. 5⅜ x 8½. 0-486-60599-X

SOME THEORY OF SAMPLING, William Edwards Deming. Analysis of the problems, theory and design of sampling techniques for social scientists, industrial managers and others who find statistics important at work. 61 tables. 90 figures. xvii +602pp. 5⅜ x 8½. 0-486-64684-X

LINEAR PROGRAMMING AND ECONOMIC ANALYSIS, Robert Dorfman, Paul A. Samuelson and Robert M. Solow. First comprehensive treatment of linear programming in standard economic analysis. Game theory, modern welfare economics, Leontief input-output, more. 525pp. 5⅜ x 8½. 0-486-65491-5

PROBABILITY: AN INTRODUCTION, Samuel Goldberg. Excellent basic text covers set theory, probability theory for finite sample spaces, binomial theorem, much more. 360 problems. Bibliographies. 322pp. 5⅜ x 8½. 0-486-65252-1

GAMES AND DECISIONS: INTRODUCTION AND CRITICAL SURVEY, R. Duncan Luce and Howard Raiffa. Superb nontechnical introduction to game theory, primarily applied to social sciences. Utility theory, zero-sum games, n-person games, decision-making, much more. Bibliography. 509pp. 5⅜ x 8½. 0-486-65943-7

INTRODUCTION TO THE THEORY OF GAMES, J. C. C. McKinsey. This comprehensive overview of the mathematical theory of games illustrates applications to situations involving conflicts of interest, including economic, social, political, and military contexts. Appropriate for advanced undergraduate and graduate courses; advanced calculus a prerequisite. 1952 ed. x+372pp. 5⅜ x 8½. 0-486-42811-7

FIFTY CHALLENGING PROBLEMS IN PROBABILITY WITH SOLUTIONS, Frederick Mosteller. Remarkable puzzlers, graded in difficulty, illustrate elementary and advanced aspects of probability. Detailed solutions. 88pp. 5⅜ x 8½. 65355-2

PROBABILITY THEORY: A CONCISE COURSE, Y. A. Rozanov. Highly readable, self-contained introduction covers combination of events, dependent events, Bernoulli trials, etc. 148pp. 5⅜ x 8¼. 0-486-63544-9

STATISTICAL METHOD FROM THE VIEWPOINT OF QUALITY CONTROL, Walter A. Shewhart. Important text explains regulation of variables, uses of statistical control to achieve quality control in industry, agriculture, other areas. 192pp. 5⅜ x 8½. 0-486-65232-7

Math–Geometry and Topology

ELEMENTARY CONCEPTS OF TOPOLOGY, Paul Alexandroff. Elegant, intuitive approach to topology from set-theoretic topology to Betti groups; how concepts of topology are useful in math and physics. 25 figures. 57pp. 5⅜ x 8½. 0-486-60747-X

COMBINATORIAL TOPOLOGY, P. S. Alexandrov. Clearly written, well-organized, three-part text begins by dealing with certain classic problems without using the formal techniques of homology theory and advances to the central concept, the Betti groups. Numerous detailed examples. 654pp. 5⅜ x 8½. 0-486-40179-0

EXPERIMENTS IN TOPOLOGY, Stephen Barr. Classic, lively explanation of one of the byways of mathematics. Klein bottles, Moebius strips, projective planes, map coloring, problem of the Koenigsberg bridges, much more, described with clarity and wit. 43 figures. 210pp. 5⅜ x 8½. 0-486-25933-1

THE GEOMETRY OF RENÉ DESCARTES, René Descartes. The great work founded analytical geometry. Original French text, Descartes's own diagrams, together with definitive Smith-Latham translation. 244pp. 5⅜ x 8½. 0-486-60068-8

EUCLIDEAN GEOMETRY AND TRANSFORMATIONS, Clayton W. Dodge. This introduction to Euclidean geometry emphasizes transformations, particularly isometries and similarities. Suitable for undergraduate courses, it includes numerous examples, many with detailed answers. 1972 ed. viii+296pp. 6⅛ x 9¼. 0-486-43476-1

PRACTICAL CONIC SECTIONS: THE GEOMETRIC PROPERTIES OF ELLIPSES, PARABOLAS AND HYPERBOLAS, J. W. Downs. This text shows how to create ellipses, parabolas, and hyperbolas. It also presents historical background on their ancient origins and describes the reflective properties and roles of curves in design applications. 1993 ed. 98 figures. xii+100pp. 6½ x 9¼. 0-486-42876-1

THE THIRTEEN BOOKS OF EUCLID'S ELEMENTS, translated with introduction and commentary by Sir Thomas L. Heath. Definitive edition. Textual and linguistic notes, mathematical analysis. 2,500 years of critical commentary. Unabridged. 1,414pp. 5⅜ x 8½. Three-vol. set.
Vol. I: 0-486-60088-2 Vol. II: 0-486-60089-0 Vol. III: 0-486-60090-4

SPACE AND GEOMETRY: IN THE LIGHT OF PHYSIOLOGICAL, PSYCHOLOGICAL AND PHYSICAL INQUIRY, Ernst Mach. Three essays by an eminent philosopher and scientist explore the nature, origin, and development of our concepts of space, with a distinctness and precision suitable for undergraduate students and other readers. 1906 ed. vi+148pp. 5⅜ x 8½. 0-486-43909-7

GEOMETRY OF COMPLEX NUMBERS, Hans Schwerdtfeger. Illuminating, widely praised book on analytic geometry of circles, the Moebius transformation, and two-dimensional non-Euclidean geometries. 200pp. 5⅝ x 8¼. 0-486-63830-8

DIFFERENTIAL GEOMETRY, Heinrich W. Guggenheimer. Local differential geometry as an application of advanced calculus and linear algebra. Curvature, transformation groups, surfaces, more. Exercises. 62 figures. 378pp. 5⅜ x 8½. 0-486-63433-7

History of Math

THE WORKS OF ARCHIMEDES, Archimedes (T. L. Heath, ed.). Topics include the famous problems of the ratio of the areas of a cylinder and an inscribed sphere; the measurement of a circle; the properties of conoids, spheroids, and spirals; and the quadrature of the parabola. Informative introduction. clxxxvi+326pp. 5⅜ x 8½. 0-486-42084-1

A SHORT ACCOUNT OF THE HISTORY OF MATHEMATICS, W. W. Rouse Ball. One of clearest, most authoritative surveys from the Egyptians and Phoenicians through 19th-century figures such as Grassman, Galois, Riemann. Fourth edition. 522pp. 5⅜ x 8½. 0-486-20630-0

THE HISTORY OF THE CALCULUS AND ITS CONCEPTUAL DEVELOPMENT, Carl B. Boyer. Origins in antiquity, medieval contributions, work of Newton, Leibniz, rigorous formulation. Treatment is verbal. 346pp. 5⅜ x 8½. 0-486-60509-4

THE HISTORICAL ROOTS OF ELEMENTARY MATHEMATICS, Lucas N. H. Bunt, Phillip S. Jones, and Jack D. Bedient. Fundamental underpinnings of modern arithmetic, algebra, geometry and number systems derived from ancient civilizations. 320pp. 5⅜ x 8½. 0-486-25563-8

A HISTORY OF MATHEMATICAL NOTATIONS, Florian Cajori. This classic study notes the first appearance of a mathematical symbol and its origin, the competition it encountered, its spread among writers in different countries, its rise to popularity, its eventual decline or ultimate survival. Original 1929 two-volume edition presented here in one volume. xxviii+820pp. 5⅜ x 8½. 0-486-67766-4

GAMES, GODS & GAMBLING: A HISTORY OF PROBABILITY AND STATISTICAL IDEAS, F. N. David. Episodes from the lives of Galileo, Fermat, Pascal, and others illustrate this fascinating account of the roots of mathematics. Features thought-provoking references to classics, archaeology, biography, poetry. 1962 edition. 304pp. 5⅜ x 8½. (Available in U.S. only.) 0-486-40023-9

OF MEN AND NUMBERS: THE STORY OF THE GREAT MATHEMATICIANS, Jane Muir. Fascinating accounts of the lives and accomplishments of history's greatest mathematical minds–Pythagoras, Descartes, Euler, Pascal, Cantor, many more. Anecdotal, illuminating. 30 diagrams. Bibliography. 256pp. 5⅜ x 8½. 0-486-28973-7

HISTORY OF MATHEMATICS, David E. Smith. Nontechnical survey from ancient Greece and Orient to late 19th century; evolution of arithmetic, geometry, trigonometry, calculating devices, algebra, the calculus. 362 illustrations. 1,355pp. 5⅜ x 8½. Two-vol. set. Vol. I: 0-486-20429-4 Vol. II: 0-486-20430-8

A CONCISE HISTORY OF MATHEMATICS, Dirk J. Struik. The best brief history of mathematics. Stresses origins and covers every major figure from ancient Near East to 19th century. 41 illustrations. 195pp. 5⅜ x 8½. 0-486-60255-9